BEAST-PEOPLE ONSCREEN AND IN YOUR BRAIN

Recent Titles in
Brain, Behavior, and Evolution

Nightmares: The Science and Solution of Those Frightening Visions during Sleep
Patrick McNamara

Dementia
Patrick McNamara, Editor

Analyzing Criminal Minds: Forensic Investigative Science for the 21st Century
Don Jacobs

Spirit Possession and Exorcism: History, Psychology, and Neurobiology
Patrick McNamara

Science and the World's Religions
Patrick McNamara and Wesley J. Wildman, Editors

BEAST-PEOPLE ONSCREEN AND IN YOUR BRAIN

The Evolution of Animal-Humans from Prehistoric Cave Art to Modern Movies

Mark Pizzato

Brain, Behavior, and Evolution
Patrick McNamara, Series Editor

PRAEGER™

An Imprint of ABC-CLIO, LLC
Santa Barbara, California • Denver, Colorado

Library of Congress Cataloging-in-Publication Data

Names: Pizzato, Mark, 1960– author.
Title: Beast-people onscreen and in your brain : the evolution of
animal-humans from prehistoric cave art to modern movies / Mark Pizzato.
Description: Santa Barbara, California : Praeger, [2016] | Series: Brain,
behavior, and evolution | Includes bibliographical references and index.
Identifiers: LCCN 2015037427 | ISBN 9781440844355 (alk. paper) |
ISBN 9781440844362 (ebk)
Subjects: LCSH: Psychoanalysis and evolution. | Human evolution. | Human
evolution—Psychological aspects. | Evolution in motion pictures.
Classification: LCC BF175.4.E86 P59 2016 | DDC 155.7—dc23
LC record available at http://lccn.loc.gov/2015037427

ISBN: 978-1-4408-4435-5
EISBN: 978-1-4408-4436-2

20 19 18 17 16 1 2 3 4 5

This book is also available on the World Wide Web as an eBook.
Visit www.abc-clio.com for details.

Praeger
An Imprint of ABC-CLIO, LLC

ABC-CLIO, LLC
130 Cremona Drive, P.O. Box 1911
Santa Barbara, California 93116-1911

This book is printed on acid-free paper ∞

Manufactured in the United States of America

To David Bashor, PhD,
neurobiologist, long-term mentor, and key friend,
who encouraged the start of this project,
wrote early drafts of the introduction,
and continued to advise me for several years,
but did not live to see the published book
(yet his spirit enlivens it)

Contents

Series Foreword

Beginning in the 1990s, behavioral scientists—that is, people who study mind, brain, and behavior—began to take the theory of evolution seriously. They began to borrow techniques developed by the evolutionary biologists and apply them to problems in mind, brain, and behavior. Now, of course, virtually all behavioral scientists up to that time had claimed to endorse evolutionary theory, but few used it to study the problems they were interested in. All that changed in the 1990s. Since that pivotal decade, breakthroughs in the behavioral and brain sciences have been constant, rapid, and unremitting. The purpose of the *Brain, Behavior, and Evolution* series of titles published by ABC-CLIO is to bring these new breakthroughs in the evolutionary and brain sciences and their applications to the sciences and humanities to the attention of the general public.

In the past decade, some of these scientific breakthroughs have come to inform the clinical and biomedical disciplines. That means that people suffering from all kinds of diseases and disorders, particularly brain and behavioral disorders, will benefit from these new therapies. That is exciting news indeed, and the general public needs to learn about these breakthrough findings and treatments. A whole new field called evolutionary medicine has begun to transform the way medicine is practiced and has led to new treatments and new approaches to diseases, like the dementias, sleep disorders, psychiatric diseases, and developmental disorders that seemed intractable to previous efforts. In addition, evolutionary and brain approaches to understanding of culture and cultural products in the creative arts have also been appearing at an accelerating rate every year. The books in the *Brain, Behavior, and Evolution* series seek to contribute to this new evolutionary approach to brain and behavior and to bring the insights emerging from the new evolutionary approaches to psychology, medicine, and the creative humanities to the general public.

The *Brain, Behavior, and Evolution* series was inspired by and brought to fruition with the help of Debora Carvalko at ABC-CLIO. The series editor, Dr. Patrick McNamara, is the director of the Evolutionary Neurobehavior Laboratory in the Department of Neurology at Boston University School of Medicine. He has devoted most of his scientific work to development of an evolutionary approach to problems of sleep medicine and to neurodegenerative diseases. Titles in the series will focus on applied and clinical implications of evolutionary approaches to the whole range of brain and behavioral disorders as well as the whole range of topics explored in the humanities. Contributions are solicited from leading figures in the fields of interest to the series. Each volume will cover the basics, define the terms, and analyze the full range of issues and findings relevant to the topic that is the focus of the volume. Each volume will demonstrate how the application of evolutionary modes of analysis leads to new insights on causes of disorder and functional breakdowns in brain and behavior relationships as well as the creative cultural products of human ingenuity. Each volume, furthermore, will be aimed at both popular and professional audiences and will be written in a style appropriate for the general reader, the local and university libraries, and graduate and undergraduate students. The publications that become part of this series will therefore bring the gold discovered by scientists using evolutionary and neuropsychologic methods to understand brain and behavior to the attention of the general public, and ultimately, it is hoped, to those families and individuals currently suffering from those most intractable of disorders—the brain and behavioral disorders. Evolutionary and brain approaches to the arts and humanities will also be part of this series and will undoubtedly illuminate the creative cultural products of artists/scholars in these interdisciplinary areas and even perhaps increase appreciation of the common artistic inheritance of humankind.

Acknowledgments

An earlier version of Chapter 2 appeared as "Cave Rituals and the Brain's Theatre," in *Theatre Symposium* 21 (2013), edited by Bert Wallace, and is republished with the permission of the University of Alabama Press. My thanks go to Bert and others at the conference he organized at UNC—Wilmington, for their helpful insights, and likewise to colleagues at various conferences over the last several years, especially the Neuro-Humanities Dialogue in Catania, Sicily, and the Mind and Life Summer Research Institute in New York state, both in 2015. My thanks go also to Richard Schechner for answering my questions about *rasas* while we were at the Performance Studies International conference in China in summer 2014 (and to William Sun for organizing that meeting). I thank Lisa Perdigao for our conversations about vampire films and the *Buffy* TV show. I thank the editors at Praeger, especially Debbie Carvalko. I thank friends at UNC–Charlotte from various disciplines, who discussed ideas in this book with me: David Bashor (Biology); Mirsad Hadzikadic, Bill Chu, Celine Latulipe, and Ted Carmichael (Computer Science); Jim Cook, Mary Michael, Al Maisto, and Mark Faust (Psychology); Paul Youngman (Culture Studies); Jim Tabor (Religious Studies); Bruce Arrigo (Criminal Justice); Donna Lanclos and Jon Marks (Anthropology); and Teresa Scheid (Sociology). I thank my students at UNC–Charlotte (especially Carley Foster), who engaged with my theories and taught me about new films and videogames. I thank the K–12 teachers who were fellows in my Charlotte Teachers Institute seminar in 2015 (and Scott Gartlan for setting that up), allowing me to extend ideas from this book and my previous one into their various teaching situations. Also, I thank the UNC–Charlotte administrators, Ken Lambla and James Vesce, who gave me time and financial support for this and further research. And I thank my sons, Luke and Peter Pizzato, who helped me in 2011 (especially with Luke's Spanish-speaking ability and Peter's artistic sensibility) to find on the small roads of Spain and France, to fully experience, and to reflect upon the prehistoric cave art that I write about in Chapter 2.

Introduction

"You animal." This could be a joke, a tease, an insult, or flattery. We know we are animals, yet we often define "human" as distinct from "animal."[1] How can we be both? We encounter animals around our homes, sometimes also inside if we have pets or need an "exterminator." We meet them on farms or in zoos. We see them on television, movie, and Internet screens. Such *others* become like us when "tame" or unlike us, we think, when "wild."

I remember as a child being afraid of the dark, especially the darkness of my bedroom window at night. Sometimes, while lying in my bed, I would see in that window a wolf's face, both canine and human, peering into my room from the outside. I would try not to look, because it scared me, but I felt compelled to see whether it was still there. Beast-people in horror and sci-fi movies can affect us that way, too, attracting us toward what we fear *out there*—while reflecting something even darker *inside* us. In my case, I am still trying to understand that wolf face in my bedroom window as a child. Where did it come from? But I also recall that seeing the animated Disney movie, *The Jungle Book* (1967, dir. Wolfgang Reitherman), with Mowgli raised by kind parental wolves, made the scary wolf in the glass go away.

There is a wildness in human beings, *greater than that of wild animals,* through our evolution beyond instinctual patterns of behavior. Often, we see such wildness in those who are different from us: foreigners, political opponents, religious extremists, and terrorists. Yet their differences may also attract us, to meet across the cultural contexts that shape, in various ways, the animal drives in our brains. With popular films and TV shows, we imagine human characters turning into animals, or animals into humans, as werewolves, vampires, and lab hybrids, or apes creating their own civilizations. Paradoxically, these figures are both subhuman and super-natural, showing our basest drives and yet immortal aspirations. Why have they been revived so often? What insights do they offer, through fantasy worlds onscreen, about the animals in our brain's "inner theater" and its continuing cultural evolution?

THE THEATRICAL ANIMAL

In the 2001 movie, *The Animal* (dir. Luke Greenfield), Marvin Mange (Rob Schneider) is an obvious "loser" in the human social world: unskilled as a police trainee working in the evidence room, unattractive to women, and chased by dogs when he goes jogging in his neighborhood. But after a severe car accident, a Frankenstein-like scientist restores Marvin to life using parts from various animals, giving him new abilities, which gain him respect from others. At the park, he leaps like a dog and catches a Frisbee in his teeth, impressing Rianna (Colleen Haskell), who runs an animal shelter. At the airport, he sniffs out a smuggler who has heroin hidden in his rectum. At a prestigious party, Marvin saves a child from drowning in a lake, using sea lion and dolphin skills. But he also hears about, and thinks he may be causing, the nighttime mutilations of animals in his community, being investigated by his fellow cops. When he and Rianna have sex, he asks her to tie him up, so he will not lose control and hurt her.

Eventually, a mob chases Marvin like an animal (or like Frankenstein's monster in the classic 1931 film). But Rianna finds a way to save him, through her own animal nature, with the help of their black friend, Miles, who challenges the mob's racism. This satire, which includes a parody of animal magnetism in TV body-building infomercials, touches on many aspects of the human-animal paradox explored in this book. Marvin is part dog, sea lion, dolphin, and lusty, voracious ape, akin to werewolves and vampires in other films. Yet he is also a moral person, even an ideal one in the romantic comedy world, when he meets the right mate.

Movies and TV shows with werewolves, vampires, and human-like apes have become increasingly popular in recent decades, often with younger monsters and slayers, organizing in packs and falling in love, or simians facing social crises, across the animal/human divide. How do these chimeric figures in romance, comedy, horror, and science-fiction relate to earlier folk and film versions? Or to the beginnings of human artistry tens of thousands of years ago in prehistoric cave theaters? How do they express the theater within our heads, and its levels of animal ancestry, shared through ritual, stage, and screen media today?

As humans, we seem more angelic than other animals, flying higher than birds with our imagination, art, and technology. Yet at times, humans become devilish, too, acting wilder than the most ferocious animals. Our ancestors evolved "higher orders" of reflective reason, but our brains are still fueled by deep emotions, based in survival and sexual instincts, with pleasure/pain and territorial drives. We imagine possible worlds, reflect on how we think and feel, communicate in abstract ways, perform for others, and reshape our current environment—much more than other animals. And yet, through this continuing evolution, beyond instinctual patterns toward highly variable symbolic realms, we feel a "lack of being," with a need for purpose and social meaning.[2] Our little lives become a stage, rounded with a sleep, which our screen media extend into supernatural dreams and nightmares.

Humans bear strong emotional bonds to loved ones and particular communities. Other mammals also show attachment to kin, even after death. Elephants, for example, touch the bones of a departed relative and sometimes "bury" the corpse with branches, leaves, or dirt (Wise 170). A Japanese macaque in captivity, like other monkeys and apes in the wild, was observed carrying her dead infant with her for months (Nakamichi 429–30). But we humans anticipate loss, fear rejection, measure our aging, and see the passing of others as reflecting our own "being towards death" (as the philosopher Martin Heidegger put it). Belief systems and group identities become necessary, whether religious or secular, to give our lives meaning. But even with them, the stress of modern life, especially through our mass and social media, may become overwhelming—triggering animal emotions of anxiety, depression, and rage, involving a fragile sense of Self in relation to the Other,[3] sometimes with deadly results, contrary to the body's survival instincts.

I am a teacher of theater and film, and a creative artist, but I have also invested many years in reading neuroscience research, while talking with colleagues from various fields of science, technology, and the humanities.[4] I have written previous books about the brain's "inner theater" in relation to ghosts onstage and onscreen, and about gods, angels, and devils that appear in such external theaters between our brains.[5] Now, with this book, I want to go on a journey with you into the deepest levels of our brain's theater—into what we share as human animals—while considering various cultural artifacts that express and transform such inner playfulness, from prehistoric cave art to recent films about beast-people.

I mean "theater" in the broadest sense, both inside the brain and between brains, involving various media in the performance of Self for Other by our highly reflective species. Other animals, such as fish and birds, perform elaborate ritual displays. Many mammals play as youngsters in preparation for competitive mating rites and coordinated hunting as adults. Elephants and non-human apes will make paintings if trained by humans with tools and rewards. But only humans play from childhood well into adulthood, creating elaborate, reflective, and symbolic artworks, often with elements of ritual and belief.

This book might have included more stageplays, as well as films, with animal-human hybrids. In Chapter 1, I briefly refer to Eugene Ionesco's *Rhinoceros* (1960), and in Chapter 4, while exploring vampire folklore, I mention the bestial tricks of the ancient Greek theater god, Dionysus, in Euripides play, *The Bacchae* (406 BCE). But many other Euro-American plays might have been examined: Aeschylus's *Oresteia* trilogy with its Furies (458 BCE), the creation plays in medieval mystery cycles involving angels with wings and Satan as a "worm" in the Garden of Eden, William Shakespeare's *A Midsummer Night's Dream* (1590s) and *The Tempest* (1611), Peter Shaffer's *Equus* (1973), Lee Breuer's *Animations* trilogy with Mabou Mines (1979), A.R. Gurney's *Sylvia* (1995), Edward Albee's *The Goat, or Who Is Sylvia?* (2002), and Mary Zimmerman's *Journey to the West* (2004) and *The White Snake* (2013), based on Chinese stories. Popular musicals might have been considered, too, such as the media crossover *Beauty and the Beast*, with a stage version based on the Disney animation and an earlier film by Jean Cocteau (1946). However, this book focuses on silver-screen examples, and their precedents, as digitalized extensions of our inner and outer, individual and shared theatricality, with reflections also from the beginnings of human Self and Other awareness in prehistoric cave art. I hope this encourages theater, performance studies, and film scholars to see connections across their disciplines.

As humans, we are theatrical in at least four ways. (1) In the theater of the mind, we stage memories, future actions, dreams, and reality perceptions, using some of the same brain circuitry for both Imaginary and Real performances, framed in Symbolic contexts.[6] (2) The way our brain stages consciousness also has theatrical elements, with percepts and concepts in the spotlight,[7] with working memory as the stage space, with deep goal and conceptual contexts as the director/technicians backstage, and with memory systems as the audience, so that the vast majority of brain activity involves unconscious[8] circuits competing and cooperating to form what is conscious at this moment. (3) Between us, we create special spaces for fictional happenings on stages and screens, involving performers, designers, writers/directors, technicians, and spectators, who explore the best and worst in past events, in possible futures, and in fantasy worlds. (4) We radically transform our environment and view "all the world" as a stage—with theaters of war, ritual, politics, sports, and everyday life, relating to a cosmic theater as well, according to various systems of belief and morality. Odd-numbered chapters of this book will focus on the first two aspects, as the "what" and "how" of our brain's inner theater. Even-numbered chapters consider examples of the third aspect in prehistoric cave art and various films. All of the chapters, especially the final one, involve the fourth aspect.

When I say "brain," I also mean connections to the rest of the body, including the "second brain" in the gut (Gershon), through the feedback loops of the nervous system and hormones in the blood. The journey in this book will focus on the brain's anatomy, not the entire nervous and endocrine systems. But it is good to keep in mind that our brain's inner theater performs its imaginings, dreams, reality perceptions, and external actions in relation to an internal world below the skull as well.

Neuroscientist Bruce Wexler argues that conflicts between belief systems, producing intercultural violence today, come from a fundamental biological imperative that evolved in the human animal: to match internal and external structures (230). People will "fight to control the opportunity to create external structures that fit with their internal structures, and to prevent others" from changing their cultural environment (230–31). Neuroscientist David Eagleman suggests that humans have more instinctual, unconscious systems than we rationally like to think (87–88). The "speed and energy efficiency" of such unconscious structures in our brains leads to "instinct blindness," says Eagleman (88), borrowing that phrase from Leda Cosmides and John Tooby. "[W]e are not able to see the instincts that are the very engines of our behavior."[9] This book will explore such remnant instinctual structures as animal legacies in the human brain, transformed in social and symbolic ways—through each person's multiple, cultural environments.

Biologists and neuroscientists often use a bottom-up understanding of body-brain-culture feedback networks. But I will explore such networks in both directions—through evolutionary biology, neuroscience, psychology, cinema, and performance studies—to develop an "inner theater" model, including film aspects, in relation to such outer media. Recent theorists of film spectatorship have developed somewhat related approaches. Torben Grodal stresses evolutionary psychology and cognitive linguistics in exploring the "embodied visions" of cinema, through a "PECMA flow" model (perception, emotion, cognition, and motor action). With some consideration of bodily and neural structures, Greg Smith describes a "mood-cue" model for the "invitation to feel" in cinema, as a way to understand various popular and art-film genres. Carl Plantinga applies cognitive science and rhetorical theory to movie viewing, as a "cognitive-perceptual" approach, with a brief consideration of "catharsis," not as a purging, but as a "working through" or "reconceptualization," involving a "spillover" from negative to positive emotions with melodrama onscreen (8, 177–85).[10]

Likewise, Ed Tan, while rejecting a "cinema inside the head" notion, as tainted by psychoanalytic film theory, still argues for "catharsis" as emotion regulation and tension reduction, with the viewer's immersion in the film as fiction and yet a more distanced appreciation of the film as "artefact" (19–20, 35–36). Cynthia Freeland uses Kant, cognitivism, and neuroscience to explore "the sublime" in film, with a shift "from painful to pleasurable cognition, from dread to elevation," through a meta-awareness of the movie "*as* a movie," involving "moral reflection" through different emotion systems in the brain (73–81). I will explore such issues through cognitive, affective, and social neuroscience, as well as evolutionary biology and primatology. I will also apply ancient and modern *theater theories* to certain movies, focusing on potential cathartic effects on spectators.

Catharsis is often used in a simplistic sense as the *purging* of emotions or "blowing off steam." If we go to the movies for escapist entertainment, we seek distraction from our daily concerns through the fictional threats and thrills onscreen, especially with the promise of good winning over evil and villains getting punished. As researchers have found, when most people are shown clips from Hollywood films that portray injustice, they are "unsatisfied" by alternative endings "in which the victim found growth and fulfillment by accepting the loss and forgiving the transgression" (Haidt, "Moral" 856). They are "most satisfied" with an ending in which "the perpetrator suffered, knew that the suffering was repayment for the transgression, suffered in a way that matched the initial transgression, and, if possible, suffered in a way that involved humiliation" (857). But this apparently moral satisfaction, with poetic justice in getting righteous vengeance, is *not* what I mean by tragicomic catharsis. It may exercise, yet not exorcise malevolent emotions and violent impulses.[11] Later in this book I will refer to this as "cathartic backfire."

Hollywood is dominated by a melodramatic action formula: an underdog hero goes on a vengeful journey, fighting courageously and winning, despite enormous odds, against a clear-cut villain.[12] Often this perpetuates stereotypes of good and evil, with *contagious* fear and rage consolidating such circuits in the viewer's brain, on the way to a happy ending onscreen. Rather than purging aggression, this may vindicate vengeful violence, by nations, authorities, or vigilantes, in the viewer's everyday life.[13] And yet, as I explore with many examples here and in my previous books, melodramatic rituals, plays, and films may have tragic or tragicomic edges that provoke ironic twists in the viewer's perceptions,[14] with a more *complex* catharsis, as the purifying or *clarifying of emotions*.[15] Various feelings might be involved, beyond Aristotelian sympathy and fear, as ancient Indian theories will help us to investigate in the next chapter. For now, consider how erotic feelings could be evoked by vampires or other beast-people onscreen, yet mixed with horror or humor, drawing you toward the screen fantasies and yet protecting you from monstrous dangers with thrills at a safe distance or with comic exaggeration.[16] With tragicomic twists, this might combine Artaudian cruelty and Brechtian distancing effects, like a lens that becomes a mirror at the stage or screen edge, offering insights beyond entertainment.[17]

Such an application of theater theory to movie viewing is rarely done in film studies.[18] I hope to bridge this disciplinary divide, not just because these arts are connected historically, but also because similar performances occur in spectators' brains with a "live theater" space, "movie theater," or a "home theater"—even as stage paradigms differ from screen technologies. Stage actors, developing variations in their prepared performances, are affected by audience reactions through a sharing of space and mortal time, always involving some degree of improvisation. Film, on the other hand, presents a pre-performed, framed, and edited spectacle with more control of the spectators' vision and hearing, but not of *personal associations*. Movie viewers' inner theaters make a film "live" through their mortal sharing of imagined spaces with the immortals onscreen. Many potential associations occur in each brain, consciously and unconsciously, even in the same brain with repeated viewings of one film. And yet, the "cues" (or triggers) for emotional and cognitive associations in a film *reflect* its cultural contexts when made or later viewed. They also *affect* the evolution of viewers' brains and of the collective culture around them—through contagious emotions that might be cathartically refined, in "cognitive reappraisal," at key points in the film.

To explore such complex interactions of inner and outer theaters, I will make references to psychoanalytic terms, not only from Freud, but also from the French psychiatrist and philosopher, Jacques Lacan. Their theories have greatly influenced film and culture studies in the last half century, especially through the work of semiotician Christian Metz, philosopher Slavoj Zizek, and various feminists such as Laura Mulvey, Mary Ann Doane, Anne Friedberg, E. Ann Kaplan, Linda Williams, Joan Copjec, and Kaja Silverman, plus "new Lacanians" like Robert Samuels and Todd McGowan. But again bridging disciplines, I will take the unusual step of relating Lacan's terms to cognitive neuroscience, which has entered the discourse on "psychocinematics" *sans* psychoanalysis (Shimamura).

Indeed, there has been a pitched battle in film studies since the mid-1990s (Bordwell and Carroll) and then also in theater studies, between scholars who embrace the "cognitive turn" and those who still use Lacanian theories, which are based in continental philosophy and therapeutic studies, rather than empirical science (McConachie 9–16). I offer correlations between these fields as adding to our understanding of film and theater through different viewpoints—and as a further "consilience" between the sciences and humanities (Edward Wilson, *Consilience*; Slingerland and Collard).[19] But if you, the current reader, find such psychoanalytic terms distasteful, rest assured that they are extensions of my argument, not prerequisites for it. Also, if you find the neuroscience details tedious along this journey, especially in the odd-numbered chapters, feel free to skip toward larger issues at the end of each section or to the even-numbered chapters, which give illustrations of the animals in our brains through prehistoric and popular cultures. By combining different fields of research to investigate the brain's feedback loops in multiple directions, this book introduces a new model for exploring the inner and outer theaters that we create, including the performances inside and around you now, as you choose how to read this.

SUPER-NATURAL EVIL AND GOOD

In this book, I will use the term "super-natural" (meaning natural but also beyond nature) to encapsulate the paradox of our brain's remnant animal instincts, yet superhuman projections into technological and divine worlds. We inherit specific natural drives, but they develop in a hyper-reflective mental apparatus, shaped by various social influences and cultural frameworks, which extend today into virtual, mass-social-media realms. "Human nature" has changed radically in our animal to human evolution, and then super-naturally through numerous cultures around the globe. Yet it is based in a common brain structure, with distinctive theatrical functions revealed by neuroscience "mappings" and reflected in various film figures featured in this book.

What are the drives inside your brain and body that tie you to Mother Nature, yet push you, as a human being, beyond the natural world, making you feel alienated at times or envious of simpler animals that know what to do—that fit within a certain environment? How do the animal drives of your inner theater interact with those in your family members, friends, and rivals? And why do your fellow humans become inhumane at times, wilder than wild animals, in stories of the past and our current news media?

The horrors of human history, from ancient massacres to twentieth-century genocides, show a fierce animal nature continuing in our ancestors' evolution, even as "civilized" human beings. This may occur when resources become scarce and ideologies define certain others as inferior, yet threatening. Such a "shaky group self-concept" paradoxically combines "a belief in the superiority of one's group with an underlying sense of vulnerability and weakness" (Staub 297). It thus might involve projections of animal characteristics on other groups—to define one's own as cleansed of such bestiality. In Nazi Germany, for example, and in Rwanda more recently, a particular ethnic group was viewed as "rats" or "cockroaches," rationalizing the killing of certain men, women, and children as a subhuman, dangerously invasive species. Such vicious stereotyping is reflected, in critical ways, by some of the vampire, werewolf, and ape films considered here. Chapter 3 will also consider the role of the neurotransmitter oxytocin in group bonding and rivalries.

In the United States since 9/11, there have been calls to "hunt down the terrorists," as if they were prey animals. Innocent Muslims and Sikhs have been associated with them. Black people, Native Americans, Asians, "illegal aliens" from Latin America, and other immigrant groups have long been scapegoated and given subhuman characteristics as inferior, yet threatening to dominant white Americans. Sometimes, non-whites also scapegoat others, within their group and outside it, as less than kin. What are the sources, in our animal-to-human nature, of such fearful projections—and of the little evils that emerge in our impulses, habits, and "addictions," especially when disavowed and projected onto others? How do our animal drives become even wilder, through the violence of extreme political groups and religious zealots?[20] On the positive side, how might specific cultural morals, or a universal system of ethics and "human rights," continue to evolve from the animal heritage in and between human brains—as reflected in movies from various decades, affecting viewers then and now?

For humans, ego and reputation can become even more important than physical survival or genetic reproduction. As highly social animals, we may sacrifice our bodies to gain admiration and die without creating offspring, in order to live immortally through memorials, artworks, or other legacies. In our species, Life has found a new mechanism for perpetuating and changing itself: from genetic protein codes, epigenetic chemical markers, and social learning networks to cultural symbols and technologies that are uniquely human (Deacon; Jablonka and Lamb). We extend our brain theaters from physical to social to virtual environments, with increasing powers of transformation (Baumeister), as reflected by the cinematic monsters considered in this book. With each of us human-animal monsters, before birth and throughout life, these layers of repetition and change continue to develop—as the cells in our bodies interrelate with our brain's neural circuitry and the learning environment of particular cultures around us, which are also changing. Thus, we are shaped by and yet play a role in the "co-evolution" of nature and culture. We not only discover the reality of being here, but also remake it, through the current conflicts and connections of groups around us.

Imagine, just for a moment, the worst kind of evil: the torturing or killing of an adult, a child, or a large group of people, by a psychopath, a parent, or some hate group. To understand such evil, one might imagine a demonic madness possessing the villains, influencing their thoughts, feelings, and actions. Yet Mother Nature offers many examples of instinctual cruelty involving a logical purpose. Fights between individuals, especially males, occur in the social hierarchies of primate troops and sometimes lead to cruel results (such as faces bitten off or testicles torn away). Such behavior may have evolved because powerful, competitive traits in leaders eventually benefit the entire troop's survival.

These cruelties, along with other examples of genetic competition involving sexual selection and survival of the fittest, might suggest to some people a "natural law" justifying abortion or euthanasia. But I am *not* arguing for Social Darwinist competition as the only basis for morality or immorality. A greater awareness of both *competition and cooperation* in the animal heritage of human brains will help us to understand our still evolving moral values and those of others—regarding the ways that survival, reproduction, and territorial drives play out.

Each brain has about 100 billion neurons and 100 trillion connections. Its specific inner and inter-cranial networks are changing continually, especially with current social and artistic media. Popular movies with animal-human hybrids, expressing bestial drives and yet superhuman powers, reflect our hopes and fears as we continue to evolve culturally, with extended kinship and compassion, yet persistent dangers of selfishness and tribalism. Depending on how they are made and how we watch them (and explore them here), horror, comic, sci-fi, and romantic films with super-natural twists of plot and character may shift our identifications toward a greater cathartic awareness, especially of good and evil complexities, improving the evolution of our brain and social networks—between conscious and unconscious modes.

DRAGON TALES AND BRAIN MODELS

Several decades ago, in *The Dragons of Eden*, astronomer Carl Sagan popularized the notion of a "reptilian brain" in the human head—from neurologist Paul MacLean's triune brain theory. This was a way to understand the devilish "dragons" that humans become, as if tempted by the serpent in Eden, through our hopes for (and mythic origins in) paradise. Today, the notion of having a reptilian brainstem, beneath the old mammalian "limbic" system, as redefined by MacLean, and the new mammalian cortex, is often repeated yet not fully understood by the general public. The book you are now reading offers a more complete model of our animal, human, and super-natural brain, which structures the theater of everyday life and our species' future.

Another oversimplification about the human brain in recent decades has been the "left and right" distinction. Research in the 1950s and 1960s explored how certain areas of the neocortex function differently on the left and right sides, especially when the connecting fibers of the corpus callosum are cut to prevent the spread of epileptic seizures (Gazzaniga). Popular reports of such initial discoveries propelled a simplistic notion of "right-brained" creative, emotional types versus "left-brained" orderly, rational people. But a half century of further research now shows a much more complex asymmetry with brain networks on each side having distinctive functions that collaborate, yet sometimes conflict, through circuits that excite or inhibit one another. For example, specialized language areas are found in nearly all right-handers and most left-handers, involving familiar, literal, and categorical meanings for language production and comprehension in the left hemisphere.[21] But metaphor, context, prosody, and poetry are processed in the right hemisphere, especially when comprehending irony or jokes, which may compete with literal understandings (McGilchrist 51, 59, 70–71).

The left hemisphere's narrower focus with a bias toward ego expectations means it may confabulate—even while trying to be rational and objective. People with right-hemisphere strokes lose its contextual, holistic awareness and also have paralysis on the left side of the body (because each side of the sensorimotor cortex gets input from and controls actions by the opposite side). With the right hemisphere damaged and the left more active, they will deny their left-side paralysis or even claim that the paralyzed arm *belongs to someone else* (Ramachandran and Blakeslee 132–41). But this does not happen with left-hemisphere damage, because the right has a global awareness of the new context. According to neurologist V.S. Ramachandran, who has done experiments with such denial patients, the "left hemisphere is a conformist, largely indifferent to discrepancies, whereas the right hemisphere is the opposite: highly sensitive to perturbation" (141).

The distinctive functions of our left and right hemispheres, shared by most people, also relate to child development (with growth spurts on each side), to cultural depictions of masculine and feminine (with mechanistic and nurturing functions, left and right) or of good and evil (shown in the Latin term for the left side, *sinister*, relating to the right cortex, as opposed to the *dexter*ous right hand and left cortex). But the basis for such asymmetry goes far back in our evolutionary ancestry, aligned with predatory skills in the left hemisphere and prey or mating awareness in the right. Laterality of left and right brain areas has been found not only in mammals but also in fish, as well as in birds, reptiles, and amphibians—involving different types of vocalizations as well (Hopkins 250). The animal heritage and cultural manifestations of human brain asymmetry will be explored in this book through the right and left cortical characteristics of vampire/werewolf figures and their "slayers," and then through the primal and civilized elements in lab-monster and ape-planet movies.

New research in neuroscience offers many refinements to the triune and left/right brain models in how we think of ourselves, with angels and devils (or dragons), super apes and human beasts, emerging from our brain's potential for good and evil, individually and collectively. Cognitive psychologist Bernard Baars gives a theater model for the emergence of conscious concepts and perceptions inside the mind, as if on a stage with other, unconscious agents competing and cooperating for such presentations. This is based on Nobel Laureate Gerald Edelman's theory of "neuronal group selection" or neural Darwinism. It also relates to neurologist Antonio Damasio's description of dispositional "puppeteers" connected to "image spaces" in the brain. The chapters ahead will use such terms to revise the triune, animals-in-the-brain model, combining it with the theater-of-the-mind model, through new research on the left and right hemispheres, as well as other areas, to consider how the evolving, functional networks inherited by our brains are shaped by and affect family, social, and cross-cultural performances.

The MacLean-Sagan, triune brain offers an oversimplified parallel between instinct/emotion/reason and reptile/mammal/human.[22] Yet such animal metaphors provide further insights when refined through evolutionary evidence and current neuroscience. They also point at something more than metaphor. When we act "on instinct," which ancestral drives are we channeling through our brain circuits? When we have an "emotional outburst," or move together in subtle, unconscious ways through "emotional contagion," which animals are we akin to? How is human empathy related to that of other mammals and primates?

As humans, we evolved beyond fitting into the natural environment—with numerous cultures around the globe changing the fitness criteria for survival and reproduction, while also transforming the landscape around them. Yet we still have the basic brain structure and bodily equipment that evolved in the animal natures of our hunter-gatherer ancestors, hundreds of thousands of years ago, despite new habits and playgrounds, multiple status roles, mass-media identifications, and shifting ideological territories. What unconscious animal agents now drive human reason and cultural behavior, offstage or in the brain's inner theater, behind our conscious awareness and will to act?

Current reptiles do not lack the limbic and cortical (forebrain) structures that are further evolved in human brains, as the idea of our subcortical "reptilian-brain" suggests. Indeed, we share hindbrain, midbrain,[23] and forebrain structures not only with mammals and reptiles, but also with birds and fish (with all in the vertebrate phylum *Chordata*).[24] So it might be argued, as paleobiologist Neil Shubin says, that we have an "inner fish." In this book, I will sometimes use the term "fish/reptilian" or vertebrate brain, regarding the midbrain and hindbrain of the human brainstem (and cerebellum), which reflects our early evolution from fish, prior to a mammal-like reptile ancestor. This relates to reflex instincts, internal body-monitoring systems, and primary emotions (seeking, lust, fear, and rage), which we share, to some degree, with fish and reptiles (Davis and Panksepp 1948; Huber et al.). But the forebrain structures in humans have evolved, through our mammalian, primate, and ape (or hominoid) lineage, to be distinct from fish and reptiles, with *primary and social emotion* networks between the cortex and upper brainstem (including the thalamus), generally known as the "limbic system,"[25] and with an intricately folded, six-layered (rather than two-layered) neocortex along the inner surface of the skull (Killackey 1247). We have specialized left and right areas of the forebrain's neocortex that are different from other animals, too.

Parts of our brain (especially the brainstem and its ties to sensory and motor areas in the cortex) show our kinship with fish, regarding their schooling or "swarm" behavior. These areas of the human brain, connected with emotional and social networks (in the limbic system and frontal cortex) relate also to mammals living in herds, such as elephants, gazelles, wildebeest, cattle, and horses. Some parts (with limbic connections between the brainstem, thalamus, and cortex, especially through the amygdala in the temporal lobes) are expressed in aggressive behaviors, individually or collectively, and thus correspond to the ancient notion, made famous by Enlightenment philosopher Thomas Hobbes, that "man is a wolf to man" (*homo homini lupus*). There are also brain networks that reveal our "inner ape," as psychologist Frans de Waal puts it, involving our primate heritage of tool use, deception, politics, violence, and sexual play, especially in relation to chimpanzee and bonobo behaviors.

Today's models of "human nature," using animal comparisons in neuroanatomy and behavior, are not just a matter of academic debate. In our mass-media, virtual-reality, and planet-transforming age, with clashes of religious, communist, and capitalist cultures, with nuclear threats, continued slavery, repeated genocides, and long-term pollution effects, the *chimera* of our species is becoming godlike in power (though not as much in wisdom), in good and evil ways. How do the dragons of Eden interact today, in and between our brains, with their fish, reptile, herd animal, wolf, and ape aspects?

This book draws on what is currently known about the human brain, regarding its animal ancestry and neural circuitry, to explore prehistoric cave art and modern beast-people films, showing the framework of our bio-cultural evolution. (In my previous book, I explored many of the stages within that frame, in the history of Western theater leading to movie and television images of gods, angels, and devils.) The book you are now reading investigates the "inner theater" of our reality perceptions, fantasies, memories, and dreams in relation to the outer theaters of art, ritual, everyday actions, and cultural events—involving, for some, a cosmic theater as well. We will start by considering the angelic and demonic forces of Self and Other, in human nature and culture, regarding our brain's inner theater with its social networks of emotion and thought contagion between brains.[26] We will then look at Baars's theory of a "theater of consciousness," involving our evolutionary heritage and our daily performances, from childhood to maturity, with others as mirrors. We will also consider Damasio's explanation of the feeling of consciousness in the "theater of the mind," involving physical markers for emotions in the "theater of the body," as the basis for rational decisions and how "self" then "comes to mind."[27] This will give us a framework for exploring the human sense of Self as a *theatrical chimera*, involving the characteristics of our animal ancestors, from our brainstem, limbic, and neocortical systems to the many interconnections between us, reflected by beast-people onscreen. How do our inner neural networks become a collective theater through social actions and our current virtual worlds? How do we simulate the other's mind within our heads, unconsciously mimicking actions and expressions, as we interact?

The first chapter of this book develops a theatrical (and cinematic) model of the human brain, involving its vertebrate and mammalian structures. It thus develops a new theory of "catharsis," as the clarifying of emotions in neural networks within and between spectators of the stage or screen. This is based on Aristotle's notion of the spectator's sympathy and fear, on the ancient Indian tradition of *rasas* (emotional "flavors" in theater), and on the modern theories of Antonin Artaud (theater of cruelty) and Bertolt Brecht (epic theater with alienation effects for social change), along with recent neuroscience research. Chapters 3 and 5 explore further neuro-theatrics: emotional contagion, mirror neurons, inner simulation, outward mimicry, perspective taking, collective drives, and bodily identifications between our brains, regarding our mammalian and primate legacies—refining that initial model. These findings will also be applied, in Chapters 2, 4, and 6, to prehistoric cave paintings at the dawn of our species' animal-to-human awareness and then to recent films with vampires, werewolves, humanoid apes, and other fantasy hybrids.

Specifically, Chapter 3 explores the "hall of mirrors" between our mammalian brains, involving bodily projections, motor actions, and primal emotions that we share, mostly unconsciously. This builds upon an initial exploration, in Chapters 1 and 2, of the brain's "inner theater" and of the dawning of humankind's super-natural awareness, projected onto cave walls deep within the earth (shown also in a recent, "3D" documentary film by Werner Herzog, *Cave of Forgotten Dreams*). Chapter 4 investigates vampire and werewolf figures, from folklore to screen, as erotic, shape-shifting foreigners and teens (or "immortal children") with voracious attachment disorders. Such characters are super-natural, yet sometimes trainable, like the real-life wolves that evolved alongside various human cultures and became "man's best friend." Chapter 5 explores distinctive twists in our "ape ego," toward devilish cruelty or angelic empathy, with different left and right cortical functions in the brain's theater—along with possibilities of body-image transformations, even across racial stereotypes. This relates to the various *Planet of the Apes* movies and to hybrid species films (such as *The Island of Dr. Moreau*, *Cat People*, *The Fly*, and *Splice*), considered in Chapter 6 as visions of the animal-human-god potential in each of us. Chapter 7 compares the various types of beast-people from horror, comedy, and sci-fi films, summarizing the book's brain-theater model, while exploring further issues of morality, play, and catharsis, including implications for the classroom.

Together, these chapters consider one overall question. With our animal ancestry and yet super-natural aspirations, reflected in prehistoric cave art and today's mass-media monsters, what gives direction to our inner and outer theaters, especially the bio-cultural evolution of human cruelty and compassion? This is a key question, for we each play a role in that continuing evolution with choices of what and how we watch (or how we discuss and teach certain films), increasing our collective awareness of the beast-people onscreen and in our brains.

Chapter 1

Vertebrate and Mammalian Stages in Mind: A Neuro-Theatrical Model

This chapter combines insights from neuroscience, psychoanalysis, primatology, and evolutionary biology to show the basic elements of the brain's inner theater, from its animal heritage to its devilish temptations and angelic aspirations. How does a sense of "Self" develop within the brain, in relation to significant others and "the Other," the social and (at least for some) spiritual network in which the Self exists? How does this involve empathy and morality, regarding good or evil actions, and interpretations of the other's motives? How is the inner theater of Self and Other, with conscious and unconscious agents, shaped by early childhood experiences—and our species' evolution—toward potentially "cathartic" moments in art and everyday life?

EMOTIONAL CONTAGION, EMPATHY CIRCUITS, AND BRAIN PRUNING

As highly social, intelligent, and reflective animals, we make theatrical events to stage potential actions, relationships, and intentions. Not just in "theater" as an art form, but also in many other cultural situations—from childhood play to sports to adult rituals and ceremonies—humans share their inner mental worlds by staging fictional contests and symbolic performances. Such outer theaters relate to the "inner theater" (and cinematic aspects) of each person's brain. Imaginings, dreams, fantasies, and reality perceptions are staged for the "mind's eye"—with a sense of Self both acting and watching, along with the social Other. This also involves an inner voice of Self, or "internal monolog," being staged at each moment, with and for the Other, like a first- or third-person "voice-over" in film.[1]

Yet the Self is drawn from many potential selves in relation to others in different contexts. Thus, various aspects of Self compete internally to shape one's identity, supported by different impressions of how others might view such characters in certain situations, with confirmations and challenges coming through outward performances. One's sense of Self as angelic, bestial, or devilish, and of others as good or evil, as friendly or threatening, depends on an inner theater—before, during, and after acting in the world.

Like but beyond other animals, such as elephants and various primates, which fondle a relative's bones or carry an infant's corpse,[2] humans have developed an extensive awareness of death, as well as of self and other. Mourning for lost loved ones, along with the growing awareness of one's own aging and mortality, has led humans, in various cultural ways, to posit a cosmic theater where the dead Self and others continue to exist—as souls or reincarnations, here and elsewhere. Yet, the angels and devils imagined in such a cosmic theater and depicted by artists (with bird or bat wings, on a human or serpent figure) show different super-natural extensions of our vertebrate-reptilian-mammalian or avian relatives, even prior to a scientific theory of evolution.

We usually experience an integrated sense of Self in the body—sometimes considered to be the "soul." But according to current neuroscience, that Self is built upon the animal heritage of a proto self (assessing internal bodily states) and a core self (interacting with objects), in the homeostatic brainstem and emotional limbic system, plus an autobiographical self in the neocortex, especially in our right-brain imaging and left-brain language areas (Damasio, *Self*; Decety and Chaminade).[3] Indeed, there are various aspects of self in the body-brain nervous system and its social extensions: a body image involving brain circuits in the parietal and frontal lobes, a passionate self with limbic system emotions, a visceral self of "gut reactions" involving the insular cortex (a deep area folded between the temporal, frontal, and parietal lobes), an executive self in the frontal lobes, a mnemonic self of memories via the hippocampus (in the temporal lobes), a vigilant self in the limbic system, thalamus, and brainstem, and a conceptual, social self, unified to some degree and involving many of these areas (Ramachandran and Blakeslee 247–54).

In the brainstem, which evolved in our vertebrate ancestors, a primal sense of Self (or ego) emerges in us, mostly below conscious awareness, through the monitoring of bodily systems and the coordination of movements, also involving the cerebellum at the base of the skull (Damasio, *Self*). Through *emotional contagion* in the mammalian core of our brain and mirror neurons in the cortex (the brain's outer layers),[4] a social "manifold" develops, involving "shared circuitry" with other people, through motions and emotions (Gallese, "Roots"; Hurley and Chater 12–13). In extreme cases, some people have "mirror touch synesthesia," fully feeling the experiences of others, with greater empathy and less grey matter in the temporal parietal junction (TPJ), an area that distinguishes self from others (Banissy and Ward). Thus, the *inhibiting* of emotional contagion is needed to form distinct ideas of one's Self, especially for competitive situations, which male primates specialize in more than females (Baron-Cohen 34, 81, 136; de Waal, *Age* 214–18).

Such "self–other representations" involve the right cortex differently from the left and require inhibitory circuits in the prefrontal (front part of the frontal) cortex to distinguish self from other (Decety and Chaminade; Decety and Somerville). This can also be observed in the empathic behavior of monkeys and apes: from the "state-matching" of shared moods to "concern for others" in consolation acts, to specifically "targeted helping," which involves understanding the other's perspective as like and unlike one's own (de Waal, *Age* 209). In humans, a further degree of other and self (or group) awareness, enabling or inhibiting compassionate feelings and acts, occurs through the right and left association areas of the cortex, in its parietal and frontal lobes, and in the prefrontal lobes.

Contagious *group emotions* may also involve reactions "feeding back" between individuals, spreading through particular models and media images (Staub 390). This is sometimes called a "herd instinct" (by Nietzsche and Freud, for example), reflecting our mammalian ancestry, as in Eugene Ionesco's play, *Rhinoceros* (1960), which shows humans turning into herd-inclined rhinos. Likewise, with human mobs and riots, destructive group behaviors may develop, involving "both contagion and noncontagion periods" (Staub 390).

And yet, destructive twists of empathic understanding can also be found with *individuals*. A soldier using rape to torture an enemy onlooker may understand (through cognitive empathy), but not feel (with affective empathy) the pain of the objectified victim and her husband, except as the other's pain turns into the victimizer's pleasure (de Waal, *Age* 211–12). Thus, an angelic aspect of the human brain, the ability to fly out of the body, imaginatively, and into the other's viewpoint, becomes devilish—as in the fall of the Lucifer and his fellow angels in Christian depictions.

Many people today believe in a dualist sense of mind and brain, or of a *soul* inhabiting the body and yet extending beyond it, in this life and after it. However, most neuroscientists describe the conscious *mind* as emerging from and fully dependent on the living body—although it might someday be "transferable directly into computers" (Graziano 222).[5]

Either way, with an independent soul or a dependent mind, the brain's mechanism for experiencing Self and Other is crucial to our *current* life, identity, and society.

In certain situations, such as "road rage," any of us might experience a devilish part of the Self, triggered by physical dangers or ego threats. In more extreme cases, a raging madness might appear like devil possession—requiring exorcism or antipsychotic medication. Yet we can also cultivate the better angels of our nature in ourselves, our children, and others around us. An "empathy circuit" has been found in certain areas of the human brain (Baron-Cohen). Deficits in that circuitry, especially due to childhood traumas shaping neural networks, can produce narcissism, borderline personality disorder, or psychopathic evil. But the empathy circuit can also be developed in the other direction, toward a more compassionate morality, through therapy, artworks, and religious or secular meditation techniques.

Raised in a certain subculture by family and friends, while influenced by the larger culture through schools and mass media, children have their animal inheritance of primal brain circuits "pruned" by those environments (Solms and Turnbull 146–48). Nurturing acts, controlling punishments, tempting pleasures, and painful traumas experienced by the child, at the hands of bigger people, form the basis for beliefs in super-natural figures, as well as reality perceptions throughout life, often unconsciously. Such specific cultural prunings of inner-theater networks shape our later interactions in the theaters of everyday life and our mass social media, involving personal desires, identifications, associations, projections, and after-effects. But how do the primal drives, emotions, memories, and ideas in each of us—about self, other, and the super-natural (or super-animal)—reach a level of *conscious awareness*, from the brain's inner theater to the social networks and cosmic reflections of today's popular media, perpetuating stereotypes or provoking cathartic change?

STAGING AND SCREENING CONSCIOUSNESS THROUGH UNCONSCIOUS AGENTS

There is no single Self, as playwright, actor, or spectator, in a central brain area, creating or watching the internal and external show (Damasio, *Descartes'*). Even the conscious, autobiographical self is continually revised, as a series of "multiple narrative drafts," composed by an "audience," driven to ask questions and form tentative answers (Dennett 14, 113, 418). There may be a shifting, functional cluster of neural networks that gives us a feeling of Self, in relation to various emotional signals from the body (Edelman and Tononi). Yet the idea and image of Self, with a will to act, is often illusory—a role played for others, in changing situations, and a mask morphing in the mirror. Experiments by Benjamin Libet in the 1980s showed that the movement of a finger involves the firing of unconscious brain circuits four-tenths of a second *prior* to the "conscious wanting" of movement (Wegner 52–55). The biological drives and cultural desires of Self are mostly unconscious, prior to choices being rationalized by the conscious mind and interpreted by social relations.

The physical foundations of the Self are also unconscious, through the body's senses. For example, you do not see directly what your visual system perceives. There is a blind spot in vision where each eyeball connects with its optic nerve, which is automatically filled in by your brain. Each eye has clear vision only at its center, in the fovea, and twitches in continuous saccades to complete the visual field. The eyes' lenses turn the visual field upside-down, with the brain's visual system also breaking it apart into components, to then be reconstructed through memories, expectations, and fantasies (Ramachandran and Blakeslee).

Visual signals go from your eyes to the thalamus, at the core of your brain, and then to various primary visual areas at the back of your brain, in the occipital lobe. Area "V1" is sensitive to the contrast of light and dark, "V2" to color and orientation, "V3" to shape, "V4" to color, and "V5" to motion (Baars 65–66). From there, circuits extend to over 40 modules (or network hubs) of "higher" visual processing in the frontal cortex, temporal lobes, and related areas, analyzing objects, locations, faces, and much more, in relation to expectations and memories—as if the visual "actors," as neural signals, were performing for that "audience," and thus competing for consciousness (66).

Cognitive psychologist Bernard Baars uses this "theater" model to explain his Global Workspace Theory of consciousness, which he relates to Gerald Edelman's theory of Neural Darwinism or "neuronal group selection."[6] As Baars argues, "All unified models of cognition today suggest some sort of unconscious audience, including unconscious memory archives and automatic routines that are triggered . . . [by] working memory" (46).[7] This does not mean that a proscenium, thrust, or arena stage exists in your brain, like you might experience in an external theater. But consciousness is structured with a limited performance space in the neural circuitry of your mind,[8] as an attentional "spotlight," while supported by fringe consciousness at its edges and working memory as its "stage," which may hold only about seven random numbers or unrelated verbal items, and fewer mental images (41–44).

The brain's cognitive "audience" involves many specialized unconscious networks, like a vast society: automatic mechanisms for routine tasks; implicit memories for attitudes, skills, and social interactions; semantic (word) networks representing the world; declarative memories of beliefs and facts; and autobiographical memories for self-identity (Baars 43–46). This audience of memory systems, like a "legislature," builds coalitions for and against certain players to be on the stage of consciousness, as if applauding or booing them (46–47). It shapes short-term impulses into long-term patterns of external behavior, through "working hierarchies" of conscious significance—as to which ideas and perceptions are valued on the mind's stage.

More than 90% of brain activity is unconscious, with billions of neurons having trillions of connections (Ramachandran and Blakeslee 152; Solms and Turnbull 84). So there are numerous competing agents within your brain, behind your conscious, imaginary, and symbolic identity. Baars describes Deep Goal and Conceptual Contexts, with an unconscious "self" as agent and knower, plus executive controls (in the frontal lobes), as being like the "director" and "operators" of your brain's theater, along with its current intentions and expectations, behind the scenes in its staging of consciousness (42–45, 144–46). Certain "players" vie for presence within the limits of the conscious stage (42). These include outer senses (seeing, hearing, feeling, tasting, smelling, warmth, vibration), inner senses (visual imagery, inner speech, dreams, imagined feelings), and ideas (imaginable ideas, verbal ideas, intuitions). Thus, the imagery we see with our eyes closed, while awake or dreaming, appears as if in an inner (movie) theater of the mind—through the brain's *staging and screening* of consciousness, with perceptions both presented and blocked.

Indeed, the same brain circuits are used when perceiving reality, dreams, and willful imaginings, but with signals flowing more from top to bottom (from association areas to primary sensory areas) when there are no external stimuli (Baars 74; Kosslyn, "Einstein's" 263; Solms and Turnbull 210–11).[9] The inner theater of imagination also has an oval viewing portal, similar to what we see with our eyes open, about 45 degrees high and 120 degrees wide (Baars 73–74). Your mind's inner eye, *more like a movie camera* than your external vision, can zoom in and out when viewing imagined or remembered objects. It can also rotate such objects within the mind's theater, or pan across a scene (Kosslyn, *Ghosts* 128–29). Other *filmic* aspects are part of the brain's inner theater, too, such as "flashbacks" (quick memory associations), cuts between angles, variable framing with close-ups, temporal and spatial "chunks" of perceived scenes (Zacks 206–27), and superimpositions of fantasy and reality, such as animal-human figures.

Even with our basic perception of reality, there are theatrical and cinematic elements in the brain. Along with filling in the eyes' blind spots (as mentioned earlier), the brain backdates and rearranges our experiences, "like an editor cutting a film," because of a half second time-lag between perception and consciousness (Carter 26). In this sense, the brain has multiple stages moving toward the staging and screening of consciousness, "switching rapidly back and forth several times per second" (Baars 73). Cooperating and competing networks of perceptions, images, and ideas produce what you experience in each conscious moment. What appears on the main stage (or screen), through the *editing cuts* between substages, is based on "biological values," involving the cerebellum and hippocampus, for example, in movement coordination, emotion modulation, and the making of new memories (Damasio, *Self* 72, 74). But we also "fill in" our memories and future imaginings, through the "belief-transmission game" of contagious cultural values (Gilbert 78–81, 91, 216–22).

There can be interference or augmentation between the brain's substages. For example, when we watch a movie with the video and audio out of synch, this dissonance can jar our brain's acceptance or "suspension of disbelief" about the screen's fiction. Likewise, there can be conflicts between neural networks in their various sensory perceptions, imagined and remembered expectations, and verbal frameworks for conscious experience. Or they may conjoin to create dominant percepts and concepts, repressing others in the shifting stream of consciousness—as when we tap a foot while hearing music, or sing or dance with other people. Jeff Smith makes a similar point about *movie music*, with reference to cognitive experiments by Annabel Cohen, showing the emotional associations of a "polarizing" or "congruent" soundtrack, conflicting with or supporting abstract animated films (160). The tempo, pitch, and melody of the soundtrack (in a major or minor key) affect the viewer's rating of the films as "calm/agitated, sad/cheerful, villainous/heroic, or serious/humorous." This relates to our mammalian evolution of a unique auditory system with middle-ear bones that enable greater social and emotional communication, especially between mother and offspring (Carter et al. 172).

As with theater and filmmaking, rehearsals and editing are involved in the brain's staging/screening of percepts, concepts, and behaviors. Even with ordinary dialogue, the flow of what is heard must be cut by the brain into decipherable words. And then, turns must be taken in speaking, with ideas held back at times, in working memory, and tied to what others have just said, involving multisensory cues such as gestural mirroring and facial expressions—key theatrical elements in our "social intelligence" (Goleman, *Social*).[10] Indeed, the human evolution of less facial hair and more facial muscles than other apes shows the value of emotional signaling with the face, as well as other body parts, to our ancestors' mating and survival success (Greenspan and Shanker 144).

Within the brain's theater and throughout the external theaters of everyday life, networks of interactive agents collaborate and compete, augmenting or blocking what's conscious, individually and collectively, with unconscious "stagehands" signaling between brains, too. Consistency is usually enforced, with a sense of unity in one mind or between minds in a group. But there are, at the same time, a great variety of percepts, associations, and ideas—in the individual or social unconscious. For example, each of us has a "how" pathway in the brain, for unconscious vision and primal orienting behavior, as well as a conscious "what" pathway (Ramachandran and Blakeslee 74–79).[11] Some people, with neural damage to the "what" pathway, are consciously blind, yet can see unconsciously, even when putting a letter into a narrow mail slot. Likewise, there are numerous sensory stimuli that each of our brains gets at any given moment, but only a few, if any, reach the level of consciousness. And there are many networks of inner memories, associations, fantasies, and ideas that inhibit or excite one another—to produce the *players* onstage that we are conscious of.

People with Charles Bonnet syndrome, including millions worldwide with black spots in vision (scotomas), due to migraines, glaucoma, cataracts, diabetic retinopathy, or macular degeneration, have brains that automatically fill in such spots with fantastic or hyperreal figures (Ramachandran and Blakeslee 87, 104–12). Likewise, amputees with "phantom limb" pain have brains that project sensations onto a missing limb, from memory networks and personal associations, while lacking the return signals from that absent part of the body image (21–58). In this way, according to neurologist V.S. Ramachandran, all of us may be *hallucinating*, to some degree, all the time, but also getting sensory stimuli from our external contexts that inhibit the internally generated imagery (58–62)—until we sleep and dream, or reach another state of altered consciousness through sensory deprivation or other means.

Our social networks filter or "screen" what is experienced, and then accept it or reinterpret it as meaningful, in each brain's staging/screening of consciousness, especially through the mirror-neuron systems in and between brains. Even when we are alone, we carry that social context, or various superegos, within us. But we also have inflated, yet insecure egos that fight to maintain consistency and unity, with or against the social mirror. And we have id energies that drive us to conform or rebel. With an emotional id and a rationalizing ego at the heart of consciousness, especially through right and left cortical networks, various brain actors vie for its spotlight, as desires and fantasies of ego identity become pressured by social, superego demands.

ANIMAL PUPPETEERS FOR ID, EGO, AND SUPEREGO

Baars uses cognitive experiments to demonstrate the theatrical staging of consciousness, which involves unconscious contexts, goals, expectations, beliefs, intentions, and memory systems directing the percepts and concepts that vie like "players" for the "spotlight." This relates, too, in my view, to the Freudian notion of "ego," as conscious (and unconscious) Self, pressured by the "superego," as absorbed social contexts, and "id," as primal, survival, and sexual goals. The ego, or at least its conscious tip, would be a collection of the lead actors in the spotlight: the dominant ideas and sensations of the present moment, making up the Self, in relation to the audience of memories (as unconscious ego traces) and the director and operators of deep goals and conceptual contexts (as id and superego). "I am" what I think and feel now, regarding who I was and what I want in my environment—with my ego desires pressured by id needs and superego demands from the internalized Other.

According to Baars, conscious processes have a great range of contents, relating often to one another and to unconscious contexts, yet with a high priority on internal consistency and with limited processing capacity, while working in a serial fashion over time (182). Unconscious processes each have a limited range of contents, are relatively isolated and context free, and yet are much more diverse, operating in parallel with much greater capacities. Thus, in Freudian and cognitive terms, the conscious tip of the "ego," with its priority on consistency and serial unity, yet a limited capacity, resists (and is sometimes disrupted by) the isolated, diverse, and powerful anomalies that pop up from alternative networks, as they contest the current Self's ideas and perceptions.

Baars refers to Freud's influence on popular thought, art, and literature of the past century, with his exploration of "unconscious processes," but notes that Freud's claims "could not be tested in a persuasive way" (16). And yet, in the two decades since Baars wrote this, many neuroscience experiments have confirmed, corrected, and refined Freud's initial theories about unconscious processes—especially through the Centre for Neuro-Psychoanalysis, based in London, and its academic journal.[12]

Neurologist Antonio Damasio matches Freud's final, 1938 view of the *unconscious* with his own today, as the unrevealed aspects of the human mind, known only through a "narrow window of consciousness" (*Self* 177).

For Damasio, the self is a "broker" of that conscious experience, at the surface of the vast unconscious mind: "an internal and imperfectly constructed informer rather than an external, reliable observer." Ramachandran goes further, relating his scientific observations and experiments to various Freudian *defense mechanisms*: denial, confabulation, repression, reaction formation, rationalization, humor, and projection, especially with the left cortex inhibiting the right (Ramachandran and Blakeslee 131–56; McGilchrist). Likewise, neuroscientists Mark Solms and Oliver Turnbull relate many psychoanalytic theories to neuroscience research, including Freud's notion of the "ego," as partly conscious but mostly unconscious, with rational, reality-constrained, executive functions inhibiting the animalistic drive energies of the id (286–87). They argue that each cortical hemisphere involves different aspects of ego functioning—with the bilateral prefrontal lobes also bearing the inhibitory capacity of the Freudian ego against id drives and basic emotions, which they locate in "sub-cortical structures" (104, 271–72).[13]

Solms thus relates the ventromedial prefrontal cortex (VMPFC) to the Freudian superego, as a "stimulus barrier" against emotional drives from the subcortical and limbic id, protecting the ego "from the incessant demands of instinctual life" (Kaplan-Solms and Solms 275–76).[14] Other researchers, without using Freud's terms, point to the VMPFC, with its ties to the emotional limbic system (MacLean's "paleo-mammalian brain"), as a key area for linking moral cognition and moral affect (Roskies 192, 198). Damage to this area prevents the formation of new moral motives, although long-term moral knowledge is retained. The VMPFC is needed for developing "a moral sense," with the left VMPFC most active "in people who are compassionate and caring" (Changeux et al. x–xi).[15] Yet, damage to the right VMPFC, increasing the left's dominance, produces self-protective, broadly based denials, even of obvious things, as an ego defense mechanism (Ramachandran and Blakeslee 142–43). Researchers have also found that the VMPFC is crucial for emotional associations, self-representation, reward systems, and goals, including *beliefs* about both religious statements and ordinary facts (Harris et al.). With VMPFC damage, there are problems in moral motivation and social decision-making, including "social exchange" issues: impulsiveness, abnormal emotions, and faux pas (Damasio et al.; Kennett and Fine 178, 185; Prinz 266–67). This suggests that the *animal legacies* of the human brain involve subcortical drives and limbic emotions that are *inhibited* by the VMPFC, as it mediates between moral cognition and affect, to shape moral motivation and compassion.

Baars also suggests such animal legacies when he adapts MacLean's notion of the tripartite brain for his theatrical model of consciousness.

He focuses on the human brainstem's reticular formation, in the midbrain, as the "Reptilian Brain" (common to all vertebrates), which is needed for wakefulness (31–32). He describes the intralaminar nuclei, including the thalamus, at the top of the brainstem, as the "Mammalian Brain," needed for maintaining consciousness. (In MacLean's model, this would also include the limbic system, surrounding the top of the brainstem.) Baars views the sensory areas of the neocortex (in the top rear of your head) as the "Primate Brain," shaping mental contents, and our greatly expanded frontal areas (in the forehead), as the "Human Frontal Cortex," used for long-term planning, abstract thought, and voluntary executive functions (32–33). By combining Baars's model with Solms and Turnbull's neuro-psychoanalytic theory, plus the research mentioned earlier, we can locate the Freudian "superego" in the VMPFC, which will be explored in Chapter 5 as an inner "stage manager" or cinematographer. It is the key hub of a neural network where advanced, distinctively human areas of the frontal cortex meet the subcortical structures of the vertebrate (fish/reptilian) and mammalian brains. Here, consciousness is staged in a particular way, involving moral cognition and compassion, as crucial elements in the social dimension of ego consciousness, which also involves thalamocortical feedback loops that mediate sensory signals from the primate brain.

Baars argues that aspects of "self" are deeply rooted in the brain, involving mostly unconscious intentions and expectations, as contextual frameworks behind the scenes of our self-awareness (145). Likewise, Damasio describes the "protoself," as *unconscious* existential subjectivity, "the state of the organism," emerging from the human brainstem (*Feeling* 174). He locates *consciousness* in the brain's "composite 'performance space,'" which involves the "image space of the early cortical regions and upper brain stem" (*Self* 241).[16] He maps these early cortical regions in somatosensory, occipital, and temporal areas (152). As with Baars's director and operators of deep goal and conceptual contexts, and audience of long-term memories, Damasio finds that the brain's conscious *performance* space is "continuously engineered by interactions with the dispositional space that spontaneously organizes images as a function of ongoing perception and past memories." Damasio says that his account "compliments" Baars's Global Workspace model and he locates the "puppeteers" of the brain's dispositional space "in the association cortices of the frontal, temporal, and parietal sectors" (189–90). From here, the "puppet masters organize the show" in the brain's image spaces, involving other areas of cortex and the upper brainstem. To recall an experience, or perceive a new one in relation to memory categories, the dispositional puppeteers reconstruct sensory and ideational "maps" in the image networks (140–41). Thus, the elements of an experience are not stored and accessed directly, but are *reconstructed* as "explicit representations" on the mind's stage—each time the fragments of an event are re-membered.

Damasio mentions that the human fronto-insular, anterior cingulate, and dorsolateral prefrontal cortical areas have Von Economo neurons, also known as spindle cells or intuition neurons (signaling rapidly across long distances in the brain)—like the large brains of other, highly social mammals: dolphins, whales, elephants, and various apes (*Self* 242; Gazzaniga, *Who's* 39–40). Bottlenose dolphins, killer whales, elephants, and apes also recognize themselves in a mirror, to varying degrees, demonstrating self-awareness akin to humans (de Waal, *Our* 193–94; Rochat 87).[17] Of course, we show a much higher degree of self-reflection by creating images, words, and theatrical artworks. We also have Von Economo or "intuitive social decision making" neurons in much larger numbers, about 30 times as many as the average nonhuman ape (Gazzaniga, *Who's* 39–40). And yet, Damasio defines the sensation of "self" as more basic, speculating that even reptiles and birds may have conscious minds, from proto to core self, and that various mammals may have an autobiographical self: wolves and dogs, cats, elephants, and marine mammals, as well as apes (*Self* 26).

The performance space of the human brain's proto, core, and autobiographical selves, from its fish/reptilian (vertebrate) brainstem to mammalian thalamus and limbic system to advanced neocortical networks, produces a conscious "ego," with wide-ranging access to unconscious traces, yet a limited, serial capacity for staging ideas and perceptions. It struggles to maintain an internal consistency, in relation to its outer, social networks, and its internal "id" instincts and emotions, with a "superego" function in the VMPFC as a stimulus filter, involving various defense mechanisms. Thus, the lower-level vertebrates in our evolutionary heritage, with their successful erotic and life/death drives (of reproduction and survival) have given us the legacy of remnant instincts that percolate through our brainstem and thalamus—toward the mammalian limbic system.

A MAMMAL IN YOUR TANK: MOTIONS AND EMOTIONS

The mammalian drives for nurturing young and for social networking, in relation to reproduction and survival, become expressed as the "primary emotions" of joy, surprise, disgust, sadness, fear, and rage (Damasio, *Feeling* 50). To these Damasio adds sympathy, pride, indignation, contempt, embarrassment, shame, envy, guilt, and admiration, as "secondary emotions," or "social emotions," involving higher levels of social interaction *in humans*, along with "background emotions" or moods, on a scale from wellbeing to malaise, and calm to tense (51; *Looking* 45). Emotions "use the body as their theater," through the internal milieu of visceral, vestibular, and musculoskeletal systems, and then emerge, across the threshold of consciousness, as feelings in the theater of the mind (*Feeling* 51; *Descartes'* 28)—as with percepts and concepts competing for the spotlight in Baars's model.

Sensory association and higher-order cortices appraise any *external*, "emotionally competent stimulus" (Damasio, "Neurobiological" 51). The amygdala, VMPFC, and anterior insula also become involved in appraisal—and then trigger an emotional state. The basal forebrain, hypothalamus, brainstem nuclei, and anterior cingulate execute the emotional state in the body proper, through the central nervous system. Higher-level, social emotions, perceived as feelings, evoke innate and learned "scripts" (54). Such feelings and scripts involve intuition, reasoning, action tendencies, and the learning of stimulus-outcome links for future use. Regarding moral decisions, intuition often comes first and then is rationalized as "moral reasoning" (53). There can also be feedback to the body's somatic markers of emotion, from the mind's imaginary and conceptual theater of "as if" feelings, as a top-down pathway, according to Damasio. I would add that these are practiced by actors, onstage or onscreen, and spectators identifying with them as stars or characters, through alignment and allegiance (according to recent film theories).[18]

Psychologist Jonathan Haidt categorizes four "families" of moral emotions, which relate to Damasio's list and his use, along with Baars, of theater metaphors. Haidt lists the "other-condemning" family: contempt, anger, and disgust (with further children, such as indignation and loathing)—which we might view as *playing between* the body and mind, as primary and secondary ("Moral" 855). These emotions are instinctually based with survival triggers, such as goal blockage (anger) or food danger (disgust), and yet culturally refined with aspects of reciprocal altruism, including honor or ostracism (855–58). Haidt also lists the "self-conscious" family: shame, embarrassment, and guilt—which we can see as *moral feelings* playing in the mind's theater, through the brain's staging/screening of Self, regarding bodily emotions of survival, mating, and territory (involving proprioception, property, and propriety). Haidt lists, too, the "other-suffering" family of compassion and the "other-praising" family of gratitude and elevation—which are crucial, along with the primary emotions of anger and disgust, to the "as if" feelings evoked by stage and screen performances, especially with tragicomic *rasa*-catharsis, as explored later in this book. (Haidt's "moral foundations" theory will also be considered in Chapter 7.)

According to Damasio, the mind's feelings relate, in a top-down and bottom-up way, to social, primary, and background emotions in the body, to deeper drives and motivations, to pain and pleasure reactions in the dopamine neurotransmitter system, and to the deepest homeostatic (body-state monitoring) mechanisms: immune responses, basic reflexes, and metabolic regulation (*Looking* 37–45). Feelings "map" bodily states, especially through the right somatosensory cortex and insula (112–17). Some of this involves mirroring the emotional expressions of others. For example, facial expressions are mirrored automatically, with signals to mimic them sent to the viewer's facial muscles, at a level below conscious awareness (117). "False" body states can also be produced by as-if or filtered feelings, through the prefrontal lobes, as well as "actual" body states—with the brain's body-sensing (somatosensory) areas becoming "a sort of theater" (117–18). Thus, the internal homeostasis of bodily functions extends outward, too, in another feedback loop, through pain/pleasure mechanisms, instincts, emotions, and feelings expressed theatrically by the body (by the face, gesture, and vocal tone), interacting to fit with and against others, as a "sociocultural homeostasis" (*Self* 27).

Figure 1.1 combines MacLean's evolutionary triune brain, Baars's remodeling of that with the "staging" of consciousness (into four animal-human areas), Solms and Turnbull's Freudian terms, and Damasio's notion of unconscious emotions of the body playing as feelings in the more conscious mind, involving dispositional "puppeteers" and sensory "image spaces," which combine into a composite performance space. Some of the theater terms in Figure 1.1 will be explained further in Chapter 5, with other details given in Tables 5.1, 5.2, and 7.2. This also involves sociocultural homeostasis, as a cognitive, emotional, and bodily "hall of mirrors" between brains, explored in Chapter 3.

Many mammals, including apes, monkeys, wolves, and bats (plus birds), "demonstrate action programs that can be assimilated to those of human social emotions" (Damasio, "Neurobiology" 50). Neurologist Joseph LeDoux has explored specific emotional systems in the brain, especially the circuitry of fear, as a primal mammalian emotion that affects various social emotions in humans. He cites "basic emotions theorists" who argue that there are universal emotions across human cultures, which are also shared by "lower animals" (*Emotional* 114). But he also cites "social constructionists" who stress the different permutations of social emotions across human cultures (115). He mentions the work of Paul Ekman, which distinguishes certain facial expressions that are common worldwide from the bodily "emblems" (gestures with verbal meanings) and "illustrators" (gestures punctuating speech) that differ between cultures (LeDoux, *Emotional* 117).

Figure 1.1

human left-cortical *critic-scripter* & right-cortical *mime-improviser*, plus prefrontal & frontal areas for long-term planning, abstract thoughts, & executive controls (including LPFC *character*, DLPFC *editor*, rVLPFC *operator*, MPFC *actor*, DMPFC *director*, & VMPFC *stage manager*/"superego")

---

dispositional "puppeteers" in frontal, temporal, & parietal association cortices (with conceptual & deep goal contexts & memory traces)

---

mammalian "limbic system" *audience & stagehands* with social, basic, & background emotions in areas around the thalamus, above the midbrain & brainstem, with ties to the PFC (limbic system includes temporal lobes with amygdala as hub of the fear network, involving fast unconscious ties to thalamus & slower conscious ties through the sensory cortex to PFC)

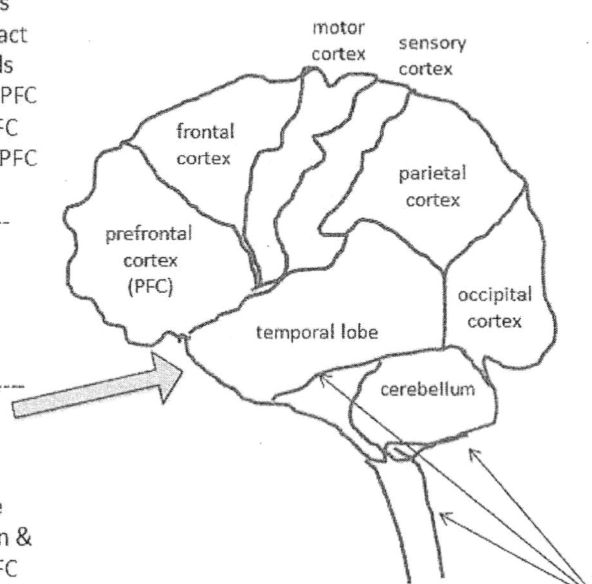

advanced **primate** sensorimotor, parietal, & visual (occipital) areas, from tree climbing & coordinated hunting & gathering behaviors

E

--- G

"image spaces" for the performance of percepts & concepts in early (sensory, occipital, & temporal) cortices with ties to the upper brainstem, in *the staging of consciousness*

O

---

mammalian limbic & subcortical vertebrate areas as Freudian "id"

fish/reptilian (vertebrate) brainstem area *stagehands*: midbrain & hindbrain (cerebellum, medulla, & pons) for instinctual drives of life & sex, pleasure & pain, internal body monitoring, & basic emotions

Levels, Stages, and Feedback Loops of the Animal Legacies in the Human Brain

But even "universal facial expressions" are regulated by social learning.[19] They may be interrupted, diminished, amplified, or masked—through "display rules" as automatic behaviors or conscious choices (117–18).

LeDoux finds that fear and defense behaviors are similar across mammalian species, even similar to those of birds and reptiles, arguing against MacLean that early evolved brainstem (hindbrain and midbrain) mechanisms are key to our basic emotions, not just limbic and cortical structures (*Emotional* 97, 102–3, 171). LeDoux has explored two distinct fear circuits in the human (and mammalian) brain: a fast, unconscious "low road" from the stimulus pathways through the sensory thalamus directly to the amygdala with its signals for emotional responses and a slower "high road" through the thalamus to the sensory and prefrontal cortices, with conscious interpretation and potential control of feelings, and then to the amygdala and its signaling (164).

For example, if I am walking my dog and suddenly hear another dog bark at us, my body reacts with a startle response, which my dog might feel through the leash, increasing her reaction as well. Fear, for her, also activates an aggression circuit and she barks. But my initial shudder, with a reflexive pulling in of my arms and spinning around to see where the other dog is, turns into a reassuring tone ("It's OK") to my dog—when I see that the big vicious beast is behind a fence and cannot harm us. The first reaction went through the fast circuit, in my brain and my dog's, even before I was conscious of the bark. The second went along the slower route, through my sensory association cortex and memory systems, evaluating the danger, through images and words (in the right and left hemispheres): bark, dog, big, but behind fence, not able to jump over.

The slower route, which LeDoux calls the "high road," evolved in mammals also (and perhaps, in different ways, in birds and reptiles). But only in humans are distinctive *left-cortical* language areas and more complex *frontal-lobe* interpretations and controls involved. Thus, the primary emotion of fear, in the thalamus-amygdala circuit, engages human social emotions through the cortex, about 200 milliseconds later.[20] Regarding the Baars/Damasio theatrical model, a perception of fear reaches my consciousness and has a spot-lit presence onstage, driven by somatic markers that have already occurred unconsciously: the muscle reflexes, adrenalin rush, increased heart rate, and gasp of my shudder response.

Such signals compete with my other thoughts and feelings while walking the dog. The higher, cortical route may also inhibit my body's emotional reaction (to flee from or fight with the strange dog) like the leash and soothing tone that I use as connections to my dog and her mammalian brain. This *inner* feedback loop, between lower and higher amygdala circuits—tied to the *social* feedback loop between my brain, my dog's, and our environment—involves deep goal and conceptual contexts, changing expectations, persistent beliefs, and long-term memories. The dispositional puppeteers (in Damasio's terms) coordinate how I perceive, conceptualize, feel about, and consciously react to the auditory and visual stimuli, as images in my brain play in its limited space of conscious awareness. Regarding the dog-barking example, I might feel alienated as I walk on that street, think about why I am walking my dog there, recall whether I have been barked at there before, and yet persist in believing I have the right to be there—as I continue walking further and then turn around, to pass by the hostile dog again (involving other people's potential perceptions of my race, class, and gender as well).

This example also shows how territorial drives are involved, as social and emotional expressions of biological, life and sex instincts, in humans and our fellow mammals, even when they are tamed as pets. I feel fear, and a bit of rage, then joy at survival—in relation to the more complex social emotions and territorial priorities of my cultural awareness—as I consciously realize that I am on a public street, yet near a private backyard where a big dog is barking at me from behind a fence. If a pretty lady is in that yard, too, sunning her body in a bikini, then my sexual instincts and further social emotions might be involved as well. The pleasurable lure of that perception, in my consciousness, might upstage any images of territorial danger, evoking feelings and thoughts about social fitness at that point, in the changing cultural environment, regarding my primal, core, and autobiographical selves. If her boyfriend comes to the fence, not just her dog, and we engage in a vocal or physical rivalry (like ungulate males, ramming heads in a rutting display), then the inner competition of ideas and senses in our brains will be externalized, as a conflict between us, over the rights of space and gazing. We will show our mammalian instincts toward her—as object of desire and yet spectator of us—in nature's theater of survival fitness and sexual selection.

As highly social animals, our human ancestors evolved memory systems with triggers for unconscious fears. These affect how we interact in various social contexts, through the feedback loop of our internal theater and external performances. But emotions also create *unconscious* (implicit) memories, through the "amygdala system," a different circuit than conscious (declarative or explicit) memory, which involves the limbic hippocampus and related cortical areas (LeDoux, *Emotional* 202). Traumatic events, coursing through the "low road" of fear via the amygdala, construct very stable networks that can be triggered later in life, as shown by animal studies of "conditioned response"—which may increase in potency due to repeated activation, like an "incubation of fear" (203). Such *amygdala-system memories*, involving deeper mammalian networks, affect the body as well, causing muscles to tense up, hormones to be released, and blood pressure and heart rate to change (202), as an "amygdala hijack" (Goleman, *The Brain* 30). While drawing on instinctual, automatic reactions, conditioned responses for fear, even in primates, are learned through family and social relationships, especially in the theater of infancy. For example, infant monkeys raised in a laboratory are not afraid of snakes—unless the mother is present when the snake appears (LeDoux, *Emotional* 237). But they learn more quickly to fear snakes, as ancient survival threats, than other objects.

According to Oehman's "preparedness theory," humans today may have a tendency to fear things that threatened our mammalian ancestors, such as snakes and insects, although with genetic variations causing some humans to be "super-prepared" for specific phobias (Oehman and Mineka; LeDoux, *Emotional* 237). Research has found that phobias for such primordial threats are stronger, or more resistant to "extinction" (desensitization), than a conditioned fear of flowers or modern dangers, such as guns and knives (238). Snakes and insects also prove to be stronger fear stimuli when masked from conscious awareness.

In relation to our animal heritage, LeDoux finds experimental evidence for the Freudian theory that some unconscious memories may never be accessible to conscious awareness (*Emotional* 239–46). The experience of stress triggers the release of hormones (adrenal steroids) that enable the mammal's body to react quickly to threats. Mild stressors, with adrenalin effects, actually enhance memory, in a "flashbulb" effect (243). But intense stress causes the shriveling of dendrites (neural connections) in the hippocampus, impairing conscious memory functions—due to a repeated feedback loop, in which the amygdala increases hormonal output (of the pituitary and adrenal glands) while the hippocampus decreases it. With prolonged stress, the hippocampus falters and damage can be permanent. This is shown with studies of stress in mice, rats, and even tree shrews (a mammal species related to early primate evolution), when subordinates are exposed to a dominant male in the social hierarchy. With monkeys in the wild, autopsies of subordinate males reveal stomach ulcers from stress and degeneration of the hippocampus. Human survivors of trauma, such as abused children and war veterans with PTSD, have a "shrunken" hippocampus and memory deficits (242). Excess stress in humans can cause depression, which is linked to poor memory, too. Thus, traumas, prolonged stress, and depression can cause physical damage to the hippocampal system, impairing the brain's conscious staging of emotional memories.[21]

Emotions are powerful, with or without ties to conscious memories. The networks of emotion from amygdala to cortex, producing feelings on the conscious stage, are much stronger than circuits in the other direction that inhibit emotions.[22] LeDoux suggests that psychoanalysis "may involve the control of the amygdala by explicit knowledge through the temporal lobe memory system and other cortical areas involved in conscious awareness" (*Emotional* 265). But the stronger emotive *players* in the brain's theater, competing for stage space against weaker, inhibitory controls in the temporal-lobe *audience*, show "why it is so easy for emotional information to invade our conscious thoughts." And yet, the weaker, cortex-to-amygdala connections are still "far greater in primates than in other mammals" (303). This suggests, according to LeDoux, that we are *evolving* toward a greater "control" of emotions by advanced frontal lobes, or toward a better "balance" between frontal and limbic systems, with "a more harmonious integration of reason and passion." This also relates, in my view, to theatrical notions of "catharsis" as the *clarifying* of emotional identifications between the audience and artwork.

Such clarification, as a fronto-limbic balance, is especially needed because our powerful mammalian amygdala (in the temporal lobes) projects to sensory areas from which it does not get input, increasing the inner imagery and effects of emotional experiences (LeDoux, *Emotional* 284). Extreme examples might be seen in temporal-lobe epilepsy, with a kindling of perceptual circuits making everything rich with meaning (related to orgasms and mystical visions), and in Charles Bonnet syndrome, when blind spots are filled in with hyper-real figures (Ramachandran and Blakeslee). Unconscious, emotional traces in the amygdala, projecting to current sensory areas, involve fears that extend far back into the past, even to the womb. For the amygdala reaches a high degree of maturity in the eighth month of gestation, allowing it "to associate a fear response to a stimulus *prior* to birth" (Cozolino, *Human* 56–57).

The amygdala, as limbic *stagehand*, also affects working memory, the *stage platform* for consciousness (Baars), with "strong connections to the anterior cingulate," a partner in the frontal lobe's "executive circuitry" (LeDoux, *Emotional* 285), which might be called the mind's *proscenium frame*. With various ties, behind the scenes of consciousness, to the sensory cortex, hippocampus, and prefrontal cortex, emotions from the amygdala affect ongoing perception, mental imagery, attention, short-term memory, working memory, long-term memory, and higher-order thought processes (287). Mediating pleasure and pain drives, from the earlier evolved brainstem, the amygdala has ties to the orbital lobes in the prefrontal cortex just above your eyes (i.e., the VMPFC *stage manager*), which "may be especially involved in working memories about rewards and punishments" (285). Specifically, ties between the amygdala and orbital medial prefrontal cortex (OMPFC) have evolved in humans beyond direct danger signals, to "evaluate the reward or punishment value of highly complex social interactions" (Cozolino, *Human* 319). Such neural connections, revised by our species, exemplify why we strive for meta-meanings, through politics, religion, and art. With the gift and curse of our brain's theater, with its hyper-awareness of self, Other, and mortality, we need a purpose to life—based on, yet evolving beyond the pleasures of survival and reproduction.

PLEASURE, RITUAL, AND NEURO-AESTHETIC LAWS

In humans, the basic vertebrate (fish/reptilian) brainstem instincts—toward pleasure and away from pain—take many permutations. Pleasure or its anticipation, in the dopaminergic system (with specific circuits excited by the neurotransmitter dopamine), may even unite with pain pathways, through primary and social emotions. Thus, painful punishment becomes a pleasurable reward, as with various masochistic rituals in covert places or the theaters of everyday life. With other animals, pleasure fuels healthy survival and reproductive behaviors, while pain helps with avoiding danger and learning hierarchical contexts. Animals aim for a balance between pleasure seeking and pain avoidance (Leknes and Tracey 318). But humans have evolved to be prematurely born, in comparison with other apes, due to our big brains and the narrow birth canal in females that upright walking necessitates. Humans are normally born at less than half the proportional gestation period of other apes. We should be in the womb for 18–24 months, to reach their level of physical maturity at birth (Cozolino, *Human* 21; Shlain 8). In our species, the theaters of family, friends, and mass media powerfully shape the infant's developing brain outside the womb, beyond prewired instinctual behaviors for pleasure seeking and pain avoidance.[23] In its first few years, the human infant has billions more neurons than later in life (Solms and Turnbull 146–47). Its brain is gradually "pruned" through cultural interactions, with nonuse causing neuronal death and activation causing growth. This rewires the brain's circuitry through feedback from others, in the lifelong staging of desires, fantasies, and repeated behaviors—in hedonistic or masochistic directions that far outweigh basic survival needs.

Such repetition, pruning the brain circuits, occurs early in life through interactions with the child's primary caretaker—in the neuro-aesthetic "ritualization" of baby talk (Brown and Dissanayake 46–48). With humans and other animals, ritualized behaviors alter specific actions, in a new context, to make them simplified, repeated, exaggerated, and elaborated, sometimes by manipulating expectations (47). Thus, *rituals* take ordinary actions and give them new meanings—as in dance, poetry, music, visual art, and theater. But primal ritualized behaviors also arise "naturally" between a mother and child, according to culturally evolved patterns. The "vocalizations, facial expressions, and head/body movements of adult caretakers" trigger an infant's instinctual responses and vice versa. The mother communicates feelings by her "smiling, open eyes, eyebrow flash, head bob, head nod, soft undulant vocalization, touching, patting, [and] kissing" (48). This initially involves the limbic system and right cortex, more than left, in "attunement" between mother and child, which shapes the infant's ego through the mirroring of conscious and unconscious desires (Cozolino, *Human* 191–96). It also involves "instrumental parenting" with adults challenging the child to gradually develop new skills, while acting like an *external frontal cortex*—with memories and future goals for the child (Wexler 108).

But such artistic rites in child to adult development, eventually involving left-cortex language areas and today's virtual technologies, create an alienation from nature, as the cultural sense of self develops through the Other's desires and demands. Often, the ritual art of family life, as an extended womb pruning neural circuitry, is not enough to make up for the loss of instinctual rules for existence, which "wild" animals still have. Although our basic needs are securely met, the animal in us wants more—as *pleasure and purpose*—through consumer passions, media fashions, political movements, religious ideals, and artistic transcendence.

Influencing the new field of "neuro-aesthetics," Ramachandran has defined nine trans-cultural "laws" of art, based on the animal heritage of our neural structures: grouping, peak shift, contrast, isolation, peekaboo (perceptual problem solving), abhorrence of coincidences, orderliness, symmetry, and metaphor (200). These neuro-aesthetic laws are defined in relation to visual art, yet they also suggest aspects of perception and meaning in other art forms, including theater and film. Ramachandran argues that *grouping*, as meaningful in historical paintings, developed from a much earlier evolved advantage for predator and prey animals, perceiving one another despite camouflage and "cluttered scenes" (203). Fragments of an enemy, barely visible through tree leaves, but grouped into a threatening form by the brain of our ancestors, gave them an evolutionary advantage, passed on to us. Indeed, the firing of neurons in monkey brains become "synchronized" when fragments are grouped into a whole object, producing a collective effect on "the emotional core of the brain" in humans as well—to give attention, identify the form as positive or negative, and then take action (205).

Likewise, caricatures are appealing to us today because they capture "the essence" of a significant face, a function that Ramachandran relates to the Indian notion of *rasa* (from the ancient theater manual, *Natya-Shastra*), evoking a particular emotional flavor and spirit in the viewer (Ramachandran 206). Exaggerations of form and movement, as with an Indian sculpture of a goddess or nymph, "hyperactivate" mirror neurons in the superior temporal sulcus, moving the viewer with beauty and awe (208). This *peak shift* effect (Ramachandran's second law) can be found with seagull chicks, too, which peck instinctively at the red spot on the mother's yellow beak (to make her regurgitate food), but will peck even more at several red lines on a yellow stick as an abstract "superbeak" (210–11). Ramachandran argues that modernist abstract art taps into "the figural primitives of our perceptual grammar" (212). Yet, personal quirks of aesthetic preference may vary greatly. Ramachandran agrees with Freud that they are based, especially when involving sexual attractions, on "templates" that develop from family experiences and later "accidental encounters" (213).

Attention to *contrast* (Ramachandran's third law) is part of the human brain's universal grammar for art because it helps in an animal's survival to perceive edges and boundaries, or figures against a background (219–20). Artistic *isolation* (fourth law) also helps the brain, especially the right parietal lobe, to focus on meaningful, emotional items in the perceptual landscape (221–23)—for the spotlight of the brain's inner theater, in Baars's terms. The aesthetic joy of the *peekaboo effect* (fifth law) stimulates the brain's feedback networks between higher and lower visual areas in various stages of perceptual "problem solving." This involves signals to "limbic reward structures"—with art becoming a "foreplay for the grand climax of object recognition"—when the elements in view produce cognitive and emotional meaning (228–29). Ramachandran suggests how aesthetic beauty may become different from, yet still be related to erotic pleasure, as the peekaboo effect of collective meaning becomes a more "multidimensional experience" with art, rather than just a primal reward for individual survival and sexual reproduction (230–31).

According to Ramachandran's sixth law, *abhorrence of coincidences*, aesthetic interpretation "feels good" when alignments in the artwork make sense, or do not seem too coincidental if depicting a natural scene (231–33). *Orderliness* (his seventh law), through repetition and rhythm, also becomes pleasing through predictability (233–34). Indeed, the firing of neurons at a certain rhythm called a "gamma" range (about 40 Hz), in the brain's theater of remembered, imagined, and perceptual representations, becomes crucial to the "binding" of neural networks—especially in relation to rhythmic actions and social networks: "rituals, chanting, dancing, drumming, and so on" (Fost 37–39). This feels valuable and meaningful through serotonin modulation of the dopamine reward system, with ties to transcendent experiences in art, religion, or temporal lobe epilepsy (40–48, 102–4; Ramachandran and Blakeslee 179–88).

And yet, art balances at times between regularity and chaos (Ramachandran 234). It moves from "familiarity" to a novel "uncertainty and ambiguity," and then to the "resolution" of that tension (Fitch et al. 66). Fauvist paintings, for example, present a "conflict to resolve" between color and object, against conventional expectations, activating certain brain areas (Kirk 320–22). Surrealist paintings do that with object and context (321–23). In music, studies show a brain response, even without conscious awareness, to "mismatch negativity," when a repeated note is suddenly replaced by a different note, or by "sounds deviating from a majority of tones in their timbre," or by notes in a melody played out of tune or out of key (Tervaniemi 222–23). Thus, the conscious experience of artistic pleasure involves perceptual analyses, implicit memory integration, and cognitive mastering, "the moment at which interpretation or meaning is imposed on the artwork" (Vartanian 266–67)—or when new meanings are found in it and through it, rather than imposed upon it.[24] I would call this "cognitive remastering" (akin to cognitive reappraisal): a change in human-animal, fronto-limbic networks, with a *catharsis* of primal and social emotions, as meaning *emerges* from the artwork, in the visual and acoustic experience of movies as well.

Symmetry in art (Ramachandran's eighth universal law) may relate to evolved values of beauty, signaling health, because facial asymmetry can be a sign of parasitic infection, which reduces fertility in a potential mate (235). Even babies, a week after birth, look longer at attractive faces (Chatterjee 304). Symmetrical beauty may also be involved in collective rituals, from baby talk to public performances, along with exaggeration, repetition, and embellishment, to signify and reinforce social cohesion. It has been found that symmetry "enhances processing fluency," helping the viewer to makes sense of the stimulus, through reward circuits involving the orbital frontal cortex (Chenier and Winkielman 281, 285). Yet the brain's preference for symmetry applies to objects more than large-scale scenes. This may be due to different visual networks: the "what" pathway through the temporal lobes for discerning objects and features within objects, which also detects symmetry, versus the "how" pathway through the parietal lobes for relations between objects and surroundings (Ramachandran 236).

For his ninth law, *metaphor*, Ramachandran notices how "multiple metaphors amplify each other" in classical Indian sculpture—perhaps bridging the "translation barrier" between the left brain's language logic and the right's more dreamlike intuition, for a "communion" between them (237, 243). This relates also to different tendencies of male and female brains, and of Western and Eastern cultures. Left and right brain functions, through frontal and parietal ties to the limbic system, may help to explain certain aspects of "catharsis," with complex artworks purifying the viewer's sympathy and fear, as suggested by Aristotle about Greek tragedies, over two thousand years ago.

RASA-CATHARSIS

Catharsis has developed a bad reputation in psychology because it is often understood as simply "venting" emotions in the hope of purging them, which may increase aggressive circuitry in the brain, through what I call "cathartic backfire." This is a risk in film, TV, and videogame viewing/playing, as many studies have shown—with the habitual learning of violence, from screen to life, along with an addictive desensitization toward it onscreen, increasing the viewer's appetite (Zacks 113–24). Especially through the melodramatic formula of a sympathetic hero avenging a victim (usually female) by battling an evil villain, stereotypes are confirmed and vigilante violence encouraged. Sometimes, copycat violence is triggered by such a film, even if it has an ironic anti-hero, through many factors in a spectator's brain—such as John Hinckley's shooting of President Reagan and others, inspired by *Taxi Driver* and an obsession with actress Jodie Foster, who played the avenged female victim in that film.[25]

And yet, the Aristotelian sense of tragic catharsis, as clarifying or purifying emotions, not simply venting or purging, might be experienced at the tragicomic edges of melodramatic movies, regarding the ancient Indian notion of *rasa* (refined emotional flavor), which Ramachandran mentions. Primary and social emotions may be refined in the spectator not just through the peak-shift effect, but also with metaphoric twists of left-cortical logic and right-cortical intuition, involving Ramachandran's other laws as well. For example, symmetry may trigger erotic desire or nurturing care systems in the viewer. Yet asymmetric, bestial elements of the same character, like vampire fangs in a beautiful face, or a human body transforming into a wolf's (with peekaboo and isolation effects also), might evoke fear and disgust. Ironic twists in the brain's emotional systems may challenge assumptions of orderliness, involving binary stereotypes, through the scene and story.

Certain "motion pictures" will be explored in detail in the chapters ahead—in order to discover such tragicomic edges in beast-people depictions and their potential *rasa* effects. (Spoiler alert: endings will be given away.) One might explore each film deeply, almost frame by frame, to consider how visual compositions, rhythms, and editing cuts affect the biological universals of spectators' inner theaters, according to Ramachandran's laws. And his theories will be applied to prehistoric cave paintings in the next chapter. But because this book offers a comparison of many films in several genres, it will then focus mostly on narrative and dramatic elements, character imagery, and tragicomic twists in melodramatic frameworks—especially regarding how violent emotions, with the potential for cathartic backfire, might turn toward *rasa*-catharsis. In many of the films, a specific mix of emotional flavors will be investigated, as a *rasa* recipe, eliciting inner-theater networks. Examples will show ironic twists of *conflicting emotions*, which might evoke a shift in spectators' brains: from right-cortical Artaudian passions to left-cortical Brechtian critiques (and back)—for more attentiveness to one's own emotional impulses, as a *rasa* refinement of cognitive reappraisal and remastering.

Although this neuro-theatrical theory of ironic twists, with passionate empathy and yet critical distance, involves the modern views of Antonin Artaud and Bertolt Brecht, it is also suggested by the ancient Indian texts, the *Natya-Shastra* (by Bharata Muni, dated between the 400s BCE and 200s CE) and the later commentary, *Abhinavabharati* (by Abhinavagupta, c. 1000). According to Keijo Virtanen, *rasa* involves a "generalizing process of emotions, in which a spectator can at the same time participate in enduring feelings . . . and still preserve the necessary distance for understanding" (64). This includes three stages: first a fluid "melting," then "enlargement and folding," and then "expansion" of consciousness. Virtanen thus explains the affective to cognitive development of a *rasa* experience, relevant to my theory of catharsis as emotional purification and mindfulness, through left and right, prefrontal and limbic areas: "the 'tasting' . . . [as] the perception of one's own awareness" (67).[26] Likewise, Ramendra Kumar Sen relates both Aristotelian catharsis (with a balance of pity and fear) and traditional *rasa* theory to the "cure by opposites," as a dominant technique in ancient Greek and Indian medicine (92–97),[27] akin to my neuro-theatrical theory of ironic emotional twists, with film illustrations described in the chapters ahead.

But this also requires a spectator in "the right frame of mind," who is "*budha*, 'wise,' . . . with all senses undistracted and capable of critical observation" (Virtanen 67). The *rasas* involve unconscious archetypes (*vasanas*), which are "universal," like Ramachandran's neuro-aesthetic laws, as "potential functions of the mind"—evoked through the causes, effects, and transient emotions shown onstage (65–66). Savoring the *rasa* includes both *resonance*, "in the same way that the strings of a guitar will respond, untouched, with similar vibrations to those produced on another guitar," *and aesthetic distance* (with painful emotions becoming pleasurable) through depersonalization and detachment, toward "Pure Consciousness" (Maillard 35–37; Virtanen 64, 70). *Rasa* is also defined as a skillful blend onstage of a particular emotion's determinants (immediate causes in real life), consequents (direct effects), and transitory states (fleeting or secondary manifestations)—like the flavors in a cooked dish (Maillard 33–34). This paradoxical progress of *rasa*, through sensual pleasure to its resonant source and yet also toward transcendence of it,[28] sounds more like Artaud's theater of cruelty, with actor-audience communion, than Brecht's epic theater (influenced by European Enlightenment ideals), with his goal of social change through historical parallels in the plot, involving an actor's gestures onstage to provoke ironic questions about character and context. Yet I would include both of these modern views in a neuro-theatrical theory of *rasa* catharsis, with potential, personal and social effects today. The degree of Artaudian or Brechtian results in the audience would depend on the plot, character, and emotional twists onscreen (or onstage) *and* the spectators' various inner-theater attitudes and associations—toward a particular form of shared awareness with changes occurring then in the performances of everyday life.

The eight emotions (*bhavas*) of ancient Indian theater, with their related *rasas*, include Aristotle's two cathartic feelings, tragic sympathy (or sorrow) and fear, but add six more, variously translated as love/happiness/delight, humor/mirth, anger/wrath, heroic vigor/courage, disgust, and wonder/awe.[29] According to the *Abhinavabharati*, the goal of Sanskrit drama is to display a certain dominant *bhava* (emotion or mental state), as well as secondary ones, through specific facial expressions, hand and body gestures, and the story (Goodwin 177–82). This evokes the corresponding *rasa* in spectators[30] and moves them toward union with *Brahman* (the Oversoul), through *tanmayibhavana* (universal communion). Influenced by Buddhism, with calmness as the goal of Enlightenment, beyond sensual attachments, the *Abhinavabharati* names a ninth *rasa*, "peace" (*shanta*), as the source and destination of the eight (Byrski 176–77, 186).[31]

Plays were defined in ancient Greece as a tragedy, comedy, or farce (satyr play), but in India through the eight *bhavas* and their corresponding *rasas*—with peaceful Enlightenment, beyond emotional ego pleasures, as the cathartic goal.[32] Specific facial expressions, poses, and gestures were used in ancient India to convey each *bhava* and evoke its *rasa* in the viewers (along with other ideas through a coded sign language), a tradition that continues with various dance-dramas today, such as *Kutiyattam*, *Bharata-natyam*, and *Kathakali*, which have also influenced Bollywood movie musicals. A recent experiment with American and Indian college students showed that recognition of such coded facial expressions for the *bhavas* may be transhistorical and cross-cultural (Hejmadi et al.).

Current neuroscience focuses on a similar array of six primal emotions: sadness, fear, anger, happiness, surprise, and disgust, according to Damasio's list mentioned earlier (comparable to Panksepp's list considered later). Damasio's further list of social emotions—sympathy, pride, indignation, shame, envy, guilt, and admiration—also relates to the *bhavas* of courage and awe, as well as to Aristotle's example of Oedipus evoking admiration and yet sympathetic shame, through the tragic flaw of hubris in Sophocles' play. (Recall that awe is involved, too, in Ramachandran's peak shift effect, with the hyperactivation of mirror neurons in the upper temporal lobes.) Like the ancient Greek practice of performing three tragedies and then a satirical satyr play, all in one day, Indian Sanskrit dramas, even if focusing on sorrow, fear, or disgust, were always striving "to end on a note of fulfillment and happiness" (Bhat 32–33).[33] But the same might be said of most movies today, which may have tragicomic twists but rarely end as tragedies, whether from Hollywood, Bollywood, or elsewhere. That final note of fulfillment and happiness, along with a refinement of emotional tastes throughout the show, creates a potential *catharsis* of spectators' inner theaters, altering brainstem-limbic-parietal-frontal, right and left cortical, animal-human networks, including personal and cultural memories, within and between brains.

ANIMAL SCRIPTS OF PANIC, FEAR, RAGE, AND SEEKING—WITH DREAMLIKE MEDIA DEMANDS

Most humans around the globe, especially in the "First World," have developed technologies of survival that make the animal necessities of food, water, and safety easy to obtain. Ironically, however, our higher-order consciousness, language, and technology makes us more aware of death's inevitability, more alienated and vulnerable to depression, while stressed by the demands and distractions of modern life and mass-media marketing. The human ego extends our survival and reproduction drives far beyond biological concerns. Our premature birth, even as normal humans, along with our self-awareness, means that our ego is often at risk—as it performs for others. The inflated, insecure, human ego draws on pleasure/pain mechanisms and emotions, from the proto to core to autobiographical self, in the brainstem, limbic, and cortical systems.

The amygdala's fear circuits, fast and slow, unconscious and conscious, may be engaged even when our bodies are not in danger—but our egos are. Social fitness becomes more crucial than environmental adaptation, because culture is our dominant environment. We need to belong and have purpose in our lives, not just physically survive. Reproducing parts of our identities, even without offspring, we influence others and leave symbolic yet material traces of our egos with them, through monuments, artworks, and daily interactions with their neural networks, beyond our body's mortality.[34] Thus, the current cultural evolution of consumer desires, artistic challenges, and virtual media powers builds upon the neural scripts of our vertebrate and mammalian, as well as later human ancestors. But what climax do they seek, and what cathartic effects do our *animal scripts* produce, in and between our brain theaters?

As members of a complex social group, young mammals play, in a "rough and tumble" way, practicing competitive and collaborative skills (Panksepp, *Affective* 284–86). Humans go much farther with childhood and adult sports, theater, cinema, television, videogames, and other mass media, fashioning their fear and rage drives in socially acceptable ways. Even our dreams, with a mechanism and perhaps some contents inherited from our mammalian ancestors, may be a threat perception and avoidance "rehearsal" system for relating primal survival networks in the brain to the current cultural environment (Revonsuo). This helps to explain the prevalence of wild animals in children's dreams today (or the wolf I saw, as a child, in my bedroom window while awake), as inherited threat figures in the brain's imaginary theater, even if they are not encountered in everyday life, except in zoos or nature shows onscreen.[35] Such dream figures eventually shift toward modern threats as the child matures. Research also shows that dreams consolidate new skills and emotional experiences from recent daytime events, integrating them with long-term memories, although in a world of much greater imaginative flexibility, where "the normal waking rules don't apply" (Linden 210–20). This encourages creativity in waking life, especially in the realms of art and entertainment, yet that also hinges upon basic survival and reproduction scripts.

Today, we have multiple, proliferating, live and screened, fictional and "news" media. Each of these *dreamlike theaters* stages a particular collective consciousness—with extensive social networks, yet a limited spotlight of attention, like the staging/screening of consciousness in our brains. Through these theaters, including online social media, we present idealized and demonized ego-images of ourselves and others, as beautiful or grotesque, proper or perverse characters, in pleasurable or threatening scenes. The "superego" that we absorb through such media may demand enjoyment, even more than moral obedience, in perverse twists of pleasure and pain, to perpetuate endless desires and marketing lures (Zizek, *Enjoy* and *For*). "Be patriotic and go shopping," Americans were told by President Bush after 9/11. But the pressure of "shop till you drop," like the War on Terror, never seems to end—and increases during certain seasons as a super-natural rite.

Behavioral economists have demonstrated that people will bid up the price of a product, such as a coffee mug, higher than they initially offered for it, once they have been given it for free and owned it for an hour. This increases its value to them, as a gift and possession, along with the desire of others for it. People will even bid $28 for a $20 bill, in an auction of it, competing with others' desires. And experimental subjects will overbid on a stock in a collective computer game—even when told that its value will automatically drop, at each step in time, toward the game's end. Thus, our inflated, insecure egos, drawing on primal emotions of fear and rage, through competitive feelings of identity and greed, create "bubbles" in the marketplace, far beyond survival needs (or sexual reproduction instincts). Yet the collapse of such a bubble causes real survival problems, as well as ego crises and economic depressions.

"Panic" in the stock market, like the infectious threat of terrorism or an individual's symptomatic anxiety, draws on a particular circuitry in the brain that we have inherited from our mammalian and earlier ancestors. Experiments with rat pups have shown that, when separated from their mother for a few minutes in the first two weeks of life, their vocalizing increases until she returns, triggering her nurturing behavior of licking, grooming, and carrying them (Kandel, "Biology" 514). Such consistent maternal care in infancy lowers the stress response (measured by the production of glucocorticoid hormones) in the pup's brain, throughout its life, reducing fearfulness and stress-related disease. Panic in the pup, or "separation-distress," with instinctual calls for mom, increases the likelihood of the mother's better "attachment behavior" and the pup's future survival, even with fearful situations. Likewise, the panic system in our brains serves an important purpose, as mammals that are helpless at birth and very dependent on nurturing throughout infancy.

Newborn monkeys that are isolated from their mother for the first six to twelve months of life act "like severely disturbed or autistic children," even when returned to the company of other monkeys, though the same procedure with older monkeys has little effect (Kandel, "Biology" 513). And as humans, we may have a higher susceptibility to panic than other primates, given our normal prematurity at birth, especially those of us who have suffered the absence of maternal care at critical points in development. The death of a parent, for example, places human children "at higher short-term risk of major depression, anxiety disorder, PTSD, conduct problems, and impaired academic and social competence," plus long-term risk of major depression and anxiety in adulthood, although quality care giving after the loss may reduce that risk (Luecken et al. 310). Attachment trauma in childhood, "from abuse, neglect, or profound misattunement between parent and child," may also produce "borderline personality disorder" later in life, with terror triggered by real or imagined abandonment, "similar to what any young primate experiences when physically abandoned by its mother" (Cozolino, *Human* 256–57). On the other hand, a study showed that infants in a neonatal ward, born more prematurely than normal, who were touched and stroked by caregivers (like the rat pups being groomed and licked) for just 15 minutes, three times a day, grew 50% faster, were more active and alert, and matured faster in behaviors than those who got the same medical care but were not touched (Sapolsky 95–96).

Prolonged separation of the rat mother and pup (three to six hours per day for two weeks) causes her to ignore the pup, increasing the pup's glucocorticoid response to stress throughout its life. During such separation, after calling to the mother for a period of time, the pup will then withdraw into a physically inert state, which appears like human depression (Solms and Turnbull 131).[36] Instinctually, this may help it survive, by hiding from predators, until the mother finds it. The panic or "separation-distress" system in human brains operates in a similar way (129–30). It is centered in the anterior cingulate (at the top of the limbic system) with ties to other areas involved in maternal and sexual behavior in mammals. Thus, it is "intimately connected with *social bonding*," as well as parenting. When the human infant is separated from its mother, it first expresses anxiety and anger, shown by clinging, searching, and crying, like the rat pup. It then shifts to sadness and despair, instinctively "conserving energy and withdrawing from danger" (Kandel, "Biology" 514).

Individual anxiety in adult humans, or collective panic with the stock market or terrorism, draws on the same mammalian panic system, but adds the "bubble" effect of a particular ego and culture. Deep survival goals direct the amygdala's quick, unconscious fear responses, amplifying phobic objects. This produces images and ideas on the cortical stage of consciousness (the higher fear route) through offstage contexts: long-term beliefs and prejudices, immediate expectations, and personal memory traces—anticipating further threats in the media's *collective dream theater* toward the bubble's eventual collapse.[37]

Jaak Panksepp has researched the neural circuits of primary emotions in rats and other mammals, to find the corresponding circuits in the human brain for its panic/grief, fear, rage, and seeking systems, including lust, care, and play subsystems.[38] The panic/grief circuits are concentrated in the anterior cingulate,[39] fear and rage circuits in the lateral and medial temporal lobes, and seeking in the ventromedial frontal lobes (*Affective* 316). Yet fear and rage are tied to our fish/reptilian heritage and seeking relates us to invertebrates (Davis and Panksepp 1948; Huber et al.). Indeed, all of these networks converge on emotional and self-representation zones in the human brainstem: the periaqueductal gray (PAG) area of the midbrain. Thus, Panksepp offers neural correlates for the *rasas* of sorrow, fear, and rage, as well as the play impulse of theater and movie making. He also helps to distinguish the lust and care networks of the brain, in relation to primal reproductive drives, which are combined in the *rasa* of romantic love.

Solms and Turnbull relate Panksepp's four "primitive instinctual mechanisms" (panic/grief, fear, rage, and seeking) to the Freudian "id,"

Table 1.1

Modern Social Neuroscience and Ancient Theatre Theories of Emotions

Jaak Panksepp's research on emotion networks in human brains, as related to animals:

- **seeking** (from invertebrate to fish and reptilian brains)
- **fear**, **rage**, and **lust** (in reptiles and mammals)
- **sadness/grief/panic**, **caring**, and **joyful play** (just in mammals as highly social emotions)

[These involve stimulus-memory-prediction patterns that become hyper-theatrical in us.]

Antonio Damasio's list (including some of Panksepp's terms, shown in bold):

- background emotions (moods): well-being to malaise, calm to stress
- primal or primary emotions: ***joy***, *surprise*, *disgust*, ***sadness***, ***fear***, and ***rage***
- social or secondary feelings: *sympathy, pride, indignation, shame, envy, guilt, & admiration*

Jonathan Haidt's "families" of moral emotions (with Panksepp's terms in bold, Damasio's in italics):

- the "other-condemning" family: contempt, anger (***rage***), and *disgust*
- the "self-conscious" family: *shame*, embarrassment, and *guilt*
- the "other-suffering" family: compassion (related to *sympathy* and **caring**)
- the "other-praising" family: gratitude and elevation (*admiration*)

[These are from Haidt, "Moral." But see also Table 7.3 in chapter 7.]

Aristotle's ancient Greek theory of <u>katharsis</u> as purifying certain emotions (through plot, character, ideas, language, sound effects, and spectacle):

- tragic *sympathy*/**caring** and ***fear*** in the audience, through the hero's **seeking** and conflicts

(*Continued*)

Table 1.1 (Continued)

eight ancient Indian bhavas (emotions), each with a specific rasa (a flavor evoked by art):

- tragic ***sadness***/*sympathy*/***caring*** (as in Aristotle's theory)
- ***fear*** (as in Aristotle's theory)
- anger (or ***rage***)
- erotic/romantic love or happiness (involving **lust** & **caring**)
- humor (**joyful play**)
- courage or heroic vigor (with obstacles to the hero's **seeking**, in Aristotle's model)
- awe (suggested by Aristotle with the audience's *admiration* for the tragic hero)
- *disgust* (indicated, too, by Aristotle's notion of the hero's tragic flaw or hamartia)

& the ninth rasa of peace (shanta), as the goal of mindful attentiveness to such feelings [union with Brahman (the Over-soul), through tanmayibhavana (universal communion), as influenced by Buddhism, with calmness as the goal of Enlightenment, beyond sensual attachments, according to the Abhinavabharati, regarding the earlier Natya-Shastra]

involving vertebrate to mammalian levels of self and emotions (137). They relate frontal-lobe mechanisms to the Freudian "ego," especially in the ventromedial and orbitofrontal areas, involving "voluntary programming, regulation, and verification of action," which distinguish humans from other animals, and adults from children (136–37). In their model, the ego of the frontal lobes is in a continual balancing act with the id of the limbic system and brainstem (especially the seeking, rage, fear, and panic circuits), mediated by the superego of the ventromedial prefrontal cortex (VMPFC). They associate the id with Damasio's notion of the "core self" and ego with the "autobiographical self," yet remind us that only a small part of the ego is conscious (97).[40] They say the "aim" of psychotherapy (and of artistic "catharsis," I would add) is to "*extend the functional sphere of influence of the prefrontal lobes*" (287–88). Such a material change in the brain, in precisely that area, has been demonstrated with functional-imaging studies (288). Studies also show that depressed patients, successfully treated with psychotherapy, have achieved "metabolic normalization in their frontal and temporal lobes" (Cozolino, *Human* 337). Patients with anxiety disorders, thus treated, "have reduced activation in cortical and subcortical areas." Therapeutic and theatrical *catharsis* might change not just our conscious ideas, but also how consciousness itself is staged by unconscious agents, including our mammalian, emotional networks—thus increasing the reflective awareness of certain feelings.

Following Damasio, Solms and Turnbull locate the "content" of consciousness in posterior cortical association areas (the advanced primate brain within us), monitoring the outside world, and the "state" of consciousness in "the ascending activating system of the brainstem" (our vertebrate, fish/reptilian inheritance), which monitors the body's interior (90). The evolution from animal to human consciousness, with the brain representing thoughts, not just inner and outer sensations, through complex, abstract, and internally generated images, means that our conscious state "is generated by a *virtual* body."[41] Human consciousness, with its mammalian heritage, "consists of feelings (evaluations) projected onto what is happening around us" (92). Such projections are increased by "repression," in the Freudian sense, which removes certain experiences from conscious memory access, but allows primary "id" processes, with "compulsive, stereotyped" emotional networks, to return with a force beyond our secondary, rational, ego controls (286–87), as LeDoux shows with fear circuits. Here again, the inner extension of prefrontal networks, through therapeutic awareness and artistic catharsis, may help humans to better integrate the brain's various animal and cultural mechanisms—across the staging of proto-self to core-self to autobiographical consciousness.

Such a catharsis (as the tragicomic clarifying of emotions) may provide an antidote for addictive ideals of inflated ego power in our mass media. These power ideals often rationalize compulsive stereotypes and justify the quick reactions of *fear* and *rage* circuits (or the *bhavas* of fear, anger, and heroic vigor) in film, television, videogame, and news-story melodramas of heroes versus villains. Extending subjective awareness about our animal emotions, by questioning particular cultural fantasies, could counter the ecstatic *seeking* of money and fame, by investors and self-promoters, from Wall Street to "reality TV," Facebook, Twitter, and YouTube. For these drives repeatedly produce economic and social-media bubbles, which become like Pandora's Boxes, spreading *panic* and formerly repressed feelings when the bubbles collapse.

Akin to the cultural seeking of money and fame, the biological "seeking system" that Panksepp maps, in rat and human brains, is also known as a "*foraging/expectancy* system," and more generally as an "incentive or appetitive motivational system," which involves "'wanting' as opposed to 'liking'" (*Affective* 145). It is associated with curiosity, interest, and expectations—or the anticipation of a reward, even before getting it (Solms and Turnbull 115). Reward anticipation is also related to *trust*, especially through activity in the caudate nucleus (in the striatum, near the top of the brainstem), as shown in a neuro-economic study, involving an investment game (Lone Frank 196–97). Also, women tend to have a longer activation of expectation-reward areas than men (198), perhaps through their evolutionary role in child raising, with *long-term* trust and delayed rewards in the *seeking* of a mate and security.

Like Freud's notion of "libido" and "libidinal drive," in the animal heritage of our brains, the seeking system involves survival and reproduction instincts: foraging for food and water, plus sexual interests (Solms and Turnbull 116–17). There are internal "need-detector mechanisms" in the hypothalamus (in the brainstem, just under the thalamus), sampling for hunger, thirst, and sexual needs (117). These, as well as perceptual and cognitive triggers, activate the seeking system and its bodily behaviors (118). But in humans, the drive for food, water, or sex is not fully regulated by instincts, or restricted to healthy limits, as in other animals.[42]

The brain's *seeking* system operates in either an "on" or "off" mode, as part of the mesocortical-mesolimbic dopamine system (Solms and Turnbull 116–19).[43] It relies on memory networks to stage specific desires and actions. Dopaminergic synapses perform like "gatekeepers," fueling desires from unconscious stages to the conscious or obsessive *spotlight* (Panksepp, *Affective* 144). When abundantly active, they make a person feel "as if he or she can do anything." Over-activation of the seeking system also produces psychotic delusions, which can be addictively triggered by cocaine and amphetamines. Antipsychotic medication thus lowers the dopamine level. But L-dopa raises it, in the treatment of Parkinson's disease, which can also produce psychotic symptoms. So dopamine at the right level is crucial for "people and animals to operate smoothly and efficiently in all of their day-to-day pursuits."

As experiments with rats show, dopamine levels increase in the brain with its correct anticipation of a pleasurable reward, even more than with the reward itself, but dopamine activity also increases with an unexpected reward (Politser 88–89). As learning progresses and the stimulus-reward association becomes reliable, there is a "fast-acting *subthalamic* loop connecting to dopamine neurons in the striatum" (in the basal ganglia around the top of the brainstem), producing the pleasurable "expectation component of the predicted reward" (94).[44] This triggers excitement, through connections to the hypothalamus (just under the thalamus at the top of the brainstem). But that is also inhibited by another circuit when the reward becomes more predictable (95). Thus, the *seeking* system has an expectation-reward component, with a "plastic" (highly flexible) learning mode, yet it also connects with a "hardwired" habit network, "engaged by repeated rewards" (124). With this network engaged, certain rites of seeking and consuming become driven by *habit*, even when involving pain or "repeated punishment," since "cells in the dorsal striatum that modulate action often resist adaptation." This may still serve an evolutionary purpose, since "habits can be precise, reliable, and economical choice mechanisms. Yet, they are prone to bias."

As in the neuro-aesthetic models considered above, a tension between familiarity and novelty is shown in this neuro-economic view, which may increase the brain's dopaminergic engagement in *learning*, or be countered by the neural rutting of *habit*. And yet, through artistic *catharsis*, the artist's and viewer's inner theaters might break the habitual networks of unconscious fantasies that produce repeated desires and actions. Such reevaluation and dopaminergic excitement, as adaptive cultural learning, also involves the brain's prefrontal lobes with memory networks in the orbital, ventromedial, frontomedial, and dorsolateral areas (Politser 96–97).

Likewise, in the neuro-economic "actor-critic" model, a *critic* (or stage manager) in the brain "updates the expectations of reward and reevaluates policies according to experience," which may relate to "learning signals encoded in dopamine neurons" (Politser 100). The *actor* in this system then "binds the dopamine-mediated predictions of the hedonic [pleasure] value of stimuli to the sequences of acts needed to obtain them." Again, the prefrontal cortex is involved with representing and perhaps changing current goals. Serotonin may also play a role in focusing this actor-critic, learning mechanism, with dopamine involved at the "wanting" level. Expectations and goals, from critic to actor, then connect with a "comparator" in the brain, which functions like "an 'audience reaction' to the 'actor's' play"—producing a sense of "liking" when the desired object is consumed or obtained (101). And yet, habitual actions may result when the brain's "automatic pilot" overrides the actor-critic-audience's appetite for novelty (102).

I would argue that today's marketing lures, with ego-survival and sex-drive displays, from magazine ads and billboards to TV commercials and Web interactions, may stimulate our dopamine levels and seeking systems to high levels.[45] This may have effects in our brain like cocaine and amphetamines, to some degree, which "produce a great burst of energy and increased interest in objects in the world" (Solms and Turnbull 206). Events in the world have a special significance and yet arouse suspicions (206–7). At higher doses, these drugs (or the internally produced dopamine levels) cause a "stimulant psychosis," with patients becoming paranoid and experiencing auditory hallucinations. As teenagers, we are especially susceptible to this, since our prefrontal networks do not mature until our mid-twenties (282; Cozolino, *Human* 69–72).

At all ages, we may become habitually addicted to the hallucinatory images and energies in our evermore virtual, paranoid-schizoid, and mass-mediated world. Our fear and seeking systems may need more and more stimulation for excitement to be felt—and we may suffer withdrawal pains if we turn away from the movie, TV, laptop, and handheld screens. Yet, that appears to be the collective environment for most of us in the "developed world." It has become normal for us to consume mass-media rewards, and sacrifice our survival energies to get them, for our dopamine fix, with other medications prescribed to modify the side effects. Our brains, for better or worse, are changing in this hyper-screen environment. But a re-evaluation is still possible, through traditional art forms and these newer media, involving our brain's panic, fear, rage, and seeking/lust systems (and perhaps other *bhavas*) toward a greater awareness of emotional triggers and habitual practices—with cathartic insights about the current cultural twists in our animal natures.

AGGRESSION AND ANXIETY REHEARSALS—WITH *RASA* RECIPES

We inherit a "cold" type of aggression from our predatory ancestors, aligned with the seeking system and its drive for food and sex (Solms and Turnbull 124). This is distinct from the "hot" aggression or rage system, but both may be evoked by the competitive marketplace of our consumer culture and geopolitical news media. These animal systems in our brains also relate to a third type of aggression (in the lust system), involving male dominance behavior, territoriality, and social emotions, which are significant, too, in our mass-mediated environment. In many mammal species, males are larger than females and they fight for dominance, especially for mating rights during breeding times, when testosterone levels are high. But with humans, unlike most animals, females have evolved a hidden ovulation—not displaying when they are in heat, ready to breed.[46] Humans compete for breeding possibilities beyond seasonal, instinctual rules, using cultural paradigms that sublimate such aggression, selection, and pleasure principles into many other fields. With fluctuating hormonal levels, humans compete for dominance, or to be desired objects, in numerous arenas and rites, such as sports, politics, popular entertainment, marketing, and even religion.

The *rage* system, involving hot aggression, is triggered when a goal is frustrated. This may be specific as in fighting for an object, such as food, or abstract as in fighting for an idea, like "freedom." Even a human baby will express rage, struggling for freedom without knowing the word, when its arms are held at its sides. When a goal of the seeking system is thwarted, "neural patterns of frustration" in the frontal cortex send downward signals to rage circuits, in the temporal lobes, amydgala, and brainstem (Panksepp, *Affective* 189). Our brains are thus prepared by evolution to project "the cause of anger" upon an external source, to blame others and fight with them, unless higher cognitive appraisals are made (189–90).

This is often expressed in stage and screen melodramas (especially action, horror, and sci-fi/fantasy movies with monsters and slayers) with a "villain" blocking the hero's goal and thus becoming the object of the hero's righteous rage. In horror films, there is also a shift from panic/grief, with paranoid uncertainty about danger, to focused fear and courage. The cold, rational aggressiveness of goal seeking turns into hot wrath, with vengeful battles for dominance and territory, evoking similar rage and pleasure-lust networks in spectators' inner theaters as they cheer the hero's violence. Projecting the cause of the hero's and spectator's fear/anger onto the villain, while a great formula for selling tickets, along with violent spectacle and an emotional soundtrack, may confirm stereotypes represented by that monstrous antagonist. This may encourage preemptive strikes or vengeful acts, by vigilantes and governments, against "terrorists" in real life—as threats like vampires, werewolves, or beast-people to the hero's allies and loved ones. But when heroes and villains are presented with a tragicomic mix of intimacy and distancing effects, shifting audience networks between cold, hot, and lustful aggression, or between paradoxical *bhavas* of fear, grief/compassion, and humor,[47] heroic vigor and disgust, or (especially with vampires) awe, love, and disgust, a more complex catharsis might result, toward the peaceful *rasa* of reflective emotional awareness—as examples in Chapters 4 and 6 may show. This potentially positive cathartic effect of violent movies, involving slayers and monsters, also comes with a contagious danger, however, of spectators extending melodramatic identifications and emotions toward real-life stressors as fear and rage triggers, drawing on memory networks of personal and cultural traumas.

Studies show that persistent stress activates rage in rats, with an increase of the neurotransmitters dopamine and norepinephrine (which triggers the hormone adrenaline), especially through a "trauma-sensitized" alarm system (Niehoff 126–33). The rage circuit extends, like fear, from the amygdala through the hypothalamus to the PAG (Solms and Turnbull 124–26).[48] Rage also involves hormone signals between the body and brain, through the HPA (hypothalamus-pituitary-adrenal) axis, particularly testosterone in male rodent, primate, or human status seeking, which can lead to dominance battles, and estrogen in both men and women (Niehoff 157–59, 171–73). The difference between rage and fear, triggering fight or flight instincts, is determined by interactions between the medial and lateral central parts of the amygdala (Solms and Turnbull 126).

Even at low levels, activation of the rage system can cause irritability, priming us for aggressive outbursts—with our mammalian body shifting from chronic scowling and muscle tension to grimacing with teeth bared, mouth growling, and fists (or claws) extended. This benefitted our animal ancestors with automatic action routines for surviving violent encounters. But the frequent frustrations of modern life, with multiple consumer desires and work anxieties, may prime us toward quick anger, like with the quick fear route, between sensory areas and the amygdala. This is shown, for example, in "road rage" or with holiday mobs in shopping malls. And yet, if we realize the signs of our body changing, with feelings rising in us, such as fear or anger, we can activate prefrontal networks inhibiting automatic responses—even when faced with the emotional contagion of someone else's rage: "Is my survival really threatened or just my ego?"

Neuropsychologist Louis Cozolino encourages us to see such choices in the long-range view of evolution, especially regarding two tendencies that we inherit as a species. We are primed by Mother Nature to project our flaws: "to attribute our own unconscious processes, feelings, and motives to those around us" (*Human* 341), thus battling inner demons by seeing them in others. Also, our brain's power for "abstraction, conceptualization, and imagining the future"—along with the stressors of our individualistic, competitive culture and advanced, multitasking technologies—increases our anxieties today, giving us chronic stress and irritability. Anxiety and stress are evolved mechanisms for survival, yet they have dangerous consequences in our brains and bodies when chronically activated.

Ironically, through the success of our species, we often seek such chronic experiences in daily life, or in dreams and the mass media, as if continuing to rehearse modes of bestial survival, through "omnivorous consumerism and thrill-seeking behavior" (Cozolino, *Human* 342).[49] And yet, we may now be on an "evolutionary frontier," as we change our neural wiring and social networks (341)—perhaps through the *rasa*-catharsis of mindful attentiveness, practiced in theater and film viewing. Such cultural and neural changes involve the primal emotions of panic, fear, rage, lust, and seeking, plus the social emotions of care, sadness, and joyful play. But these also connect with the left and right cortical hemispheres in different ways, relating to the evolved specialties of males and females, from our hunter-gatherer ancestry, for hundreds of thousands of years—and the primate, mammalian, and vertebrate structures of our inner theaters, which took millions of years to evolve.

STAGES OF (IN-)FLEXIBILITY: RIGHT AND LEFT, EAST AND WEST

There was a bottleneck in human evolution 70,000 years ago, possibly due to climate swings (after eruptions of the Toba volcano, in Sumatra, Indonesia), which could have reduced the human population to 10,000 adults, or as low as 1,000 breeding pairs, and selected for a more flexible brain to survive the extended periods of cold and drought (Ambrose; "Supervolcano"). Although there were many earlier migrations of hominins out of Africa, humans on other continents today are descended from a relatively small group that left around that time, 70,000–60,000 years ago. They spread across the world, even to the northern climates of Eurasia during the Ice Age. (Neanderthals evolved in northern Europe and Asia separately from humans, from a common African ancestor, and survived there until 30,000 years ago, yet also mixed genetically with our species.) Something changed in our ancestors' brains—enabling their survival through the climate bottleneck and their reproduction across the globe, into numerous cultures in vastly different environments.

That change probably involved a higher degree of *flexibility* in adapting to new environments, with expanded frontal lobe and left-brain networks, conducive to creativity, and yet also to selfish impulses, especially as humans evolved from hunter-gatherer groups toward increasingly larger, settled societies (Gazzaniga, *Who's* 31–36, 151, 157–58). Eventually, humans developed different cultural attitudes, regarding self-agency or collective harmony, as in the individualistic Euro-American or more communal Asian and African traditions (183). But already, 32,000–11,000 years ago, when humans created cave art in Europe, they had bigger brains on average than ours today,[50] with larger frontal lobes than those of Neanderthals. How did the connections in those large brains become specialized for greater adaptability—leading to higher levels of artistic self-awareness and technological transformations of the environment, radically escalating in the last hundred years?

Chimpanzees and orangutans, but usually not gorillas, show self-recognition in mirror tests (with a red dot on the face). According to Daniel Povinelli's arboreal environment theory, apes that continued to dwell in trees despite their large size evolved a "kinesthetic self-concept," through their dangerous interactions with arboreal branches, which gorillas lost or never developed because they spent most of their lives on the ground

(334). Our common ancestor with these apes, "a large-bodied, highly arboreal species," may have seen itself in a three-dimensional (3D) "skein of highly variable fragility, compliance, and space," while swinging and bouncing on tree branches, six million years ago.

Regarding a later stage of evolution, Merlin Donald theorizes that our hominin ancestor, *Homo erectus*, developed a "mimetic" culture, about two million years ago, involving a "kinematic" imagination, gestural and tonal communication, and playacting (273). Such self-awareness in humans today, with facial recognition and mimetic expression in the mirror, is centered in the right prefrontal lobe (Keenan 134–40). Prosody in human language is also a right-cortex function: "deliberately raising and lowering the voice, and producing imitations of emotional sounds" (Donald 261). Thus, our *Homo erectus* ancestors, as big apes that moved to the ground and walked erect, brought their self-aware, poetic, right-brain functions with them and developed an early form of theater—through mimetic interactions.

About a half million years ago, according to Donald, humans evolved a "mythic" culture with verbal language, which now involves distinct areas in the left hemisphere (Broca and Wernicke's areas).[51] Donald also finds the start of today's "theoretic" culture of technology and art 40,000 years ago—with complex tool use, art objects, and art on cave walls, externalizing human memory and retrieving cultural knowledge through symbolic devices (262). Thus, humans' right-brain self-awareness, as a visuo-spatial, kinesthetic, imaginary skill, extended into a left-brain, verbal, narrative, symbolic dimension of self/other recognition, communication, and theatricality, with devices that transformed the environment. This may have occurred through the hardships of climate change and migration, selecting for a much more mutable human consciousness—to interact collaboratively and creatively in new settings.

However, such successful adaptations in human brains and cultures came at a cost: the loss of instinctual patterns and purposes for life in a stable environment—or the "loss of being" (Lacan's *manque-à-être*). This produces our human "want to be," our endless desire for more, to fill our hollow, insecure egos, which evolved as kinesthetic, narrative, and technological extensions of Self for survival and reproduction. Prematurely born and then powerfully transformed by the external womb of a particular culture, our vertebrate drives, mammalian emotions, and ape egos became twisted in extraordinary ways, especially through left and right cortical interactions, within and between brains.

Humans are amazingly collaborative, living in populations far larger than other apes. But we are also highly territorial. Our collective egos draw neighborhood, class, ethnic, racial, religious, gender, and national borders—idealizing our allies (or exotic others) and stereotyping our enemies, to commune with or battle. Some of these categories encourage us to cross the lines of conflict and join hands, when we see others as different and yet our kin. But we often find the need to *demonize* enemies, as outside threats, in order to bind ourselves within the group more cooperatively—with mimetic and mythic identities inherited across many generations (Volkan). Or we make a *scapegoat* to focus reciprocal violence in the community on a certain sacrificial object, a person or group that can be vilified, punished, or exterminated (Girard).

Wild male chimpanzees use their collective skills not only to hunt and kill monkeys for food, but also to patrol the borders of their territory and kill, trample, tear the skin off, and drink the blood of a lone chimp, which might have been a former family member and playmate (Wrangham and Peterson 5–17). Yet, human groups have gone much further, with the mass murder of neighbors, many times in history, using right-brain kinesthetic and left-brain symbolic stagings of identity and desire. Even the best cultures, with peaks of "enlightenment" in ancient to modern Europe, Asia, Africa, and the Americas (or Australia and Oceania), bear tragic flaws and legacies, with trails of destructive errors, as they continue to revise their neural and social networks, in each new generation.

Psychological tests show that Westerners today, especially those of northern European heritage, are more individualistic, analytical, and linear thinking, whereas East Asians are more holistic, relational, and cyclical thinking (Nisbett). For example, Americans focus more on foreground objects in pictures and the shapes of sculptures (with arbitrary names), while Japanese attend more to background scenes and the sculpture's materiality, even at the age of two (81–95).[52] Americans interpret abstract moving figures on a video screen as having more internal causal factors like shape and weight; Asians see them affected more by context, such as gravity and friction (116–19). Americans believe more in fixed personality types and consistent trends, Koreans in situational influences and changing fortunes (119–20). In random association tests, Americans often infer a false rule, whereas Chinese did not (95–96). Yet Americans are better at judging the position of an object (a rod) without being influenced by its field (a box) than Chinese, who are more contextually oriented. Surveys also show that Asians feel less in control of their lives than Westerners, and instead of trying to control situations, they are more likely to adjust to them (97). Feeling "in control" of their lives is associated with health more by European Americans than by Asians or Asian Americans. And Asians express more well-being than Americans when they are part of a group that might provide control. Women are also found to be more holistic and group oriented than men in both Eastern and Western cultures, yet the differences are greater between cultures than between genders (99–100).

I would extend such findings to argue that Western cultures have shaped the human brain toward left-hemisphere, egoistic, analytical, cause-and-effect functions, stressing individual freedom and greed, especially with American "free market" ideals involving "less government." East Asian cultures have shaped brain networks toward right-hemisphere (or more bilateral), communal, relational, holistic processes, emphasizing group harmony, with individuals submitting to systematic organization.[53] These current inclinations, exemplified by Western capitalism and Eastern communism (or the different forms of capitalism in various Asian countries), relate to traditional belief systems, especially Greco-Christian and Daoist-Buddhist/Shinto-Zen (Nisbett). Of course, such traditions are mixing today, more and more, with increasing cross-cultural influences. Research shows that an individual Asian-American or Hong Kong resident (influenced by English culture) may be primed in either direction with Eastern or Western cues, such as a dragon or Mickey Mouse, calligraphy or a cowboy, or with personal memories—toward contextual or individualistic, self-motivated perceptions (118–19).

In the last century, Nazi Germany and Imperial Japan demonstrated that Westerners and Easterners are both susceptible to the dangers of emotional contagion and ideological rationalization (as right and left brain processes) in a herd mentality inherited from our mammalian ancestors, which may also produce predatory groups and leaders in a pack or troop system, like wolves and apes, but more wildly and systematically destructive. The recent American development of PACs (Political Action Committees) raising huge funds for campaigning politicians who are then beholden to donors and corporations may be another sign of a herd-pack-troop culture, even in the individualistic West. The contagion of fears and desires, in a herd of antelope (or school of fish), where each member influences others in rippling waves of mostly unconscious psycho-physical gestures, to eat grass, migrate for water, or flee from danger, may shift in humans from a "grassroots," bottom-up democracy toward a pack or troop hierarchy, with brutally competitive leaders (or investment bankers) operating beyond laws and ethics. This happens also when the business interests and "global security" of nations or religious groups involve violent tactics, with troops or suicide bombers, causing civilian sacrifices, which then call for more and more. Yet such collective, self-perpetuating stress, with the sacrificial demands of bio-cultural survival, draws on mechanisms in our prey and predator, herd and troop heritage that might be altered—if we become more aware of them.

While investigating our brain structures and nervous systems, as inherited from mammalian relatives, Robert Sapolsky found that "zebras don't get ulcers" but primates do. When alerted to danger as a herd, through emotional and physical contagion, zebras' stress systems are activated and they flee from predators. One of their weaker or younger members might be lost, sacrificed as a meal to the lions. Yet the surviving zebras, once in safety again, return to munching grass. Their stress system is easily deactivated. But baboons, which Sapolsky studied directly in the wild, forage for grass, tubers, and fruit only about four hours per day (257). They spend much of the rest of the day in stressful competition about the shifting social hierarchy of the troop. And they get ulcers like us.

One might think that the least stressed baboon would be the alpha male. But glucocorticoid hormone levels show that alphas are highly stressed, because baboon leadership changes every seven to eight months, or weekly at other levels of rank and mating rights (Cheney and Seyfarth 57–58, 116–17). The least stressed males are those who gain social control by initiating fights, who best assess when threats are real and predict the outcome, and who use others as outlets for frustration (Sapolsky 263). These baboons "give rather than get ulcers." But another, more peaceful type of male also has low stress damage, through patterns of "social affiliation." They spend the most time grooming females who are not in heat, and are groomed by them, while also playing with the young. With other primates, too, it has been found that both genetics and mothering style shape the infant's behavioral personality in later life (264). Offspring of mothers who are "less anxious and more permissive . . . grow up to be more confident and exploratory."

Sapolsky argues that humans are likewise "ecologically buffered and privileged enough to get stressed mainly over social and psychological matters" (257). But feelings of predictability or control can reduce stress. Indeed, a nursing home study showed that when visits by college students were predictable or controlled by the elderly, their health benefitted more than when visits were not (270). Yet, Sapolsky warns that feeling in control about a disastrous event might increase the stress afterward, causing *guilt* because "it's all my fault." In relation to Nisbett's research, this shows a significant danger to the Western, left-cortical bias of self-control.

CATHARTIC CONTROL OR CHALLENGE AND CHANGE?

The staging and screening of Self and Other in our inner (movie) theater involves various levels of our animal heritage: fish/reptilian brainstem areas, mammalian limbic system, primate parietal awareness, and distinctively human right- and left-cortical functions, especially in the executive control areas of the frontal lobes, as "director" and light/sound "operators." Such animal levels in our brains produce proto, core, and autobiographical selves. These involve dispositional "puppeteers" and other director/operators of deep goal and conceptual contexts, as well as the "audience" of long-term (automatic, semantic, declarative, and autobiographical) memory systems. They shape the competition of neural actors to reach the "spotlight" (and silver screen) of consciousness, as percepts and concepts, regarding the platform of working memory.

The backstage (or film production) elements of our inner theater include primal drives of survival, reproduction, and territoriality. These remnant animal instincts are radically altered by human cultural environments, especially with right- and then left-cortical attunement to caretakers at an early age, in home rituals as our first theatrical experiences. Our animal drives become expressed through pleasure/pain signals, primary emotions (seeking, fear, rage, panic/grief, and lust), background moods, social feelings, and collective cognitive norms.

Along with visual and auditory cues, smell may trigger collective fear between humans through "alarm pheromones," which activate the right amygdala (Radulescu and Mujica-Parodi).[54] Thus, emotional contagion, akin to that of our mammalian relatives, may contribute, through right- and left-cortical networks, to moral intuition and reason, regarding dangers and errors in the social environment. Emotional contagion with visual and auditory cues from the screen might also convey "as if" dangers and tragicomic flaws in a movie theater, with pheromone cues from other viewers as well. (Similar effects might have occurred in prehistoric cave theaters with rock images and resonant chambers.) The mirror neuron systems involved in such contagious signaling of unconscious backstage agents between brains also relate to the "peak shift" and other neuro-aesthetic effects in visual art and music (as key cinematic elements), plus a potential cathartic clarification and *rasa* refinement in prefrontal areas of conscious mindfulness, with tragicomic twists and various emotional ironies in particular films. But a *quicker* signaling may occur between our subcortical stagehands, through emotional contagion with screen images and sounds, even as we experience a *slower* cognitive empathy with more cortical awareness (Eisenberg and Eggum 73). This depends, too, on the "emotional content of pain," but not the pain itself, mirrored in our inner theaters when we watch scenes of someone suffering, through "deliberate acts of imagination" that increase mirror-neuron activity, drawing on the audience of memory systems resonating with the "context" onscreen (Watson and Greenberg 127–29).

The lure of personal control over stressful situations helps to explain the appeal of horror movies, action films, or farce onscreen and related shows onstage. Identifying with the hero in danger from a monster, villain, or banana peel arouses survival drives through primary and social emotions, in a dreamlike threat rehearsal, where pain becomes pleasurable. With the viewer's safe distance from the scene and subsequent sense of relief, in the victim's escape and the hero's victory, or in collective laughter, the theatrical threat also gives a dose of individual control and communal belonging. Like religious believers, who find relief from anxiety and mourning by imagining how God watches over them and decides what is best—theater, cinema, and television viewers, or videogame players, find hope in a godlike role, while watching characters akin to themselves struggle under stress. The right cortex and limbic system may catch contagious horror, rage, panic/distress, or other *bhavas*, but the left brain's belief and control networks can practice their circuitry of being in charge. This may also involve distinct fear circuits: the quicker "low road" through the amygdala with immediate bodily reactions, such as the startle effect, shifting toward the slower, more reflective "high road" through frontal and left cortical networks (LeDoux, *Emotional* 164). Thus, a further catharsis of complex, tragicomic *rasas* might challenge left-cortical binaries of good and evil, through right-cortical openness to new perspectives, altering VMPFC prejudices. But virtual stress also demands the real sacrifice of emotion, time, and money for its pleasure and practice rewards, confirming or challenging stereotypes—while drawing on survival and erotic drives in communal and individual ways.

As many Eastern, Western, and other traditions mix in a jet airline and Web-linked world, the animal legacies in our brain theaters continue to evolve, on earth and in the screen cosmos. But traditional ideals of the divine are still attractive to many. They are symptomatic of the current "clash of cultures" between "Matter First" secularists and "Mind First" believers (Richards 260–67)—as well as East and West, or "developed" and "developing" worlds, involving the Party, Buddha, Allah, Yahweh, God, and various other ideals of human transcendence. Figures of the divine mostly have human attributes today. Yet, they emerge from a worldwide tradition, which extends to prehistory, of seeing the animal, human, and divine as a continuum, not just a binary of man versus animal or god versus man. In the next chapter, we will consider how prehistoric cave art, tens of thousands of years ago, shows a critical point in our cultural evolution: a crucial shift in self and Other identity, from natural to supernatural and technological, through the animal within the human. This suggests, too, how a catharsis (and *rasa* refinement) of cultures might evolve today—not as a battle for survival or an ethnic cleansing of scapegoats, but an awareness, acceptance, and renegotiating of the animal drives in each of us, even perhaps an extended kinship with those who seem to threaten "our way of life."

Chapter 2

Prehistoric Caves as Emotion Picture Theaters

A primal form of theatrical and cinematic performances emerged in certain caves, 32,000 to 11,000 years ago, during the Ice Age or "Upper Paleolithic Period," as evidenced by paintings and etchings on the walls, musical instruments, and other artifacts.[1] One of the earliest, with some of the finest art, is shown in Werner Herzog's 3D documentary film, *Cave of Forgotten Dreams* (2010). It explores prehistoric art in the Chauvet (or Pont d'Arc) cave in France, which was discovered in 1994 and is off-limits to the public. This chapter considers Herzog's film and the Chauvet cave, plus details from lesser-known caves in France and Spain, which I visited in June 2011.[2] (A detailed reproduction of the Chauvet cave opened recently to the public and I visited it, along with Grottes d'Arcy, in July 2015.) Such caves offer evidence of a primal form of performance with theatrical and cinematic elements: the evolution of human subjectivity based in, yet moving beyond mammalian playfulness, through shared emotions, neural filters, and simulations of the Other—with pictures that *moved* in the flickering firelight within acoustic chambers. How do the cave spaces, along with images made in them, reveal the birth of a distinct sense of human character (like and unlike other animals), using the rock wall as mirror and map for personal and collective visions, as scenery for performance and as screen medium?

THEORIES OF CAVE THEATER

In his recent book, *Palaeoperformance*, and in earlier essays, Yann-Pierre Montelle explores the evidence for many aspects of theatricality in the prehistoric caves of France: not only the paintings, etchings, and carvings on the rock walls, but also the various sizes and shapes of caverns, the bone flutes, scrapers, and bullroarers found in them, and the resonant tones produced today by tapping stalactites of different lengths (*Palaeoperformance*, "Mimicry," and "Paleoperformance"). Prior to Montelle, Jean Clottes and David Lewis-Williams developed a theory, based on such evidence, plus neuroscience research and anthropological comparisons with recent African traditions, that prehistoric cave art shows the recording, display, and ritual use of hallucinatory trance experiences, involving animal-spirit guides.[3] Ice Age peoples may have produced cave art through visions that occurred spontaneously in the extreme darkness with its echoing sounds, evoked by the flickering fires of torches and animal fat lamps, painful treks deep into the earth, loss of oxygen and increase of other gases, and the rock's natural shapes and crevices.[4] Shamanic drugs might also have been involved. The mysterious geometric lines and dots in prehistoric cave art, along with realistic animal figures and hybrid animal-human creatures, correspond to the types and stages of hallucinatory visions that can be evoked in the laboratory *today* through sensory deprivation or flashing lights. Such hallucinations, now with objects from our cultural context, are experienced, too, by people with migraines and other disorders. Thus, cave art reveals at least two possibilities of prehistoric theater: (1) visionary experiences of geometric, abstract, and supernatural figures from within a human head projected onto or through the rock surface as an emotional *screen medium* and (2) shared performances in the cave using the rock art as scenic background, mythic illustrations, and *moving* characters in the firelight with its shadows.

The types of cave-art spaces also demonstrate various aspects of prehistoric theater. Some are easily accessible in huge chambers. Others are hundreds of meters inside the earth, requiring crouched walking or crawling through narrow passages. In more accessible chambers, *processions and large ritual gatherings* may have occurred. But for those with tight corridors deep into the darkness, with slippery surfaces, low ceilings, and sudden cliffs, and with bears, hyenas, and lions potentially present, *fearful, courageous, and painful ordeals* were involved for a few individuals—perhaps shamanic leaders putting initiates through transformative nightmares. Montelle relates this to male initiation rites practiced by various tribes elsewhere in the world historically ("Paleoperformance" 134–36). Combining Montelle's argument on initiation rites, Clottes and Lewis-Williams's theory of rock art as a super-natural membrane, and my own experiences underground (and with Herzog's film), I will give details from specific caves, showing the possible origins of human theatricality in distinctive forms of self/Other awareness.

A CAVE ONSCREEN

Herzog's voiceover, during his documentary film, calls Chauvet a "proto-cinema," for this cave shows some of the earliest and finest paintings yet discovered, with repeated figures that become *animated* in flickering light. Some are carbon-dated between 30,000 and 32,000 years old. Various animal figures have three-dimensional shading in the legs, bellies, and heads, sometimes turned toward the viewer, showing a *perspective* technique long before the Renaissance. There are many overlapping and echoing images, suggesting movement and perhaps scenes of predation (with lions stalking bison and rhinos) or of sparring (with rhinos knocking horns).[5] Yet, most of Chauvet's figures, as in the caves of France and Spain I visited, are floating on the rock surfaces—not tied to contextual scenes or obvious narratives. This emphasizes *contrast* (Ramachandran's third neuro-aesthetic law) and indicates very personal, dreamlike visions in the paintings, rather than general representations of reality outside the cave. But the figures may also contain a collective language of symbols and imagery that we can no longer read. They seem to involve Ramachandran's fourth neuro-aesthetic law, *isolation*, in stimulating our right parietal and emotional meaning networks, like Cro-Magnon brains tens of thousands of years ago.

The etchings are stunning, too, with detailed images of horses and other large animals cut into the rock, as white outlines against the darker surface layer, while using the natural curves of the walls, like the paintings. (The 3D technology of Herzog's film enhances the viewer's experience of this.) Such etchings suggest an *immediate* artistic impulse—perhaps outlining on the wall the animals and other super-natural figures seen in or through the rock, during shamanic trances. Later, some of the etchings were painted, with black or red outlines, and sometimes with more detailed shadings and interactions, using black charcoal or manganese rock and red or yellow ochre. "Crayon" sticks of such materials and colors have been found in some caves. Mineral pigments were also ground into powder, mixed with liquid, and painted with a brush or sprayed from the mouth, sometimes using bone tubes that have been found. This technology, along with the scaffolding needed to make paintings 5 meters high in caves such as Font-de-Gaume (15,000 years ago), required prior planning and materials collected in or brought into the cave. The carved sculptural reliefs in rock shelters like Abri du Cap Blanc (of horses 15,000 years ago) would have taken even more time and work to produce. Yet, the numerous overlapping engravings in many caves might have been created quickly—in direct response to personal visions—and then shared with others in that form or with further artistic elaborations, showing a spectrum from individual to collective theater experiences.[6]

Herzog's film involves interviews with scientists working at Chauvet, including a statement by Jean Clottes that Upper Paleolithic peoples were different from us in their perception of reality as super-natural, with "fluidity" between human, animal, and non-animal forms and with "permeability" between performance spaces, rock walls, and other worlds. And yet, this fluid and permeable context for prehistoric art, akin to our virtual experience of the Chauvet cave in a movie, or other caves in person, suggests a connection with humans tens of thousands of years ago, whose brains were anatomically "modern." Today, we enter the cinema chamber in near darkness and feel transported through the images on the wall. We add personal associations, memories, and fantasies to the scenes onscreen, while photos on a filmstrip, or digital dots in a video, repeated yet subtly changing, evoke the illusion of moving figures in our heads. With or without 3D technology and glasses, we see much more of a fluid and permeable world in the motion picture theater than is really there—with the images onscreen, the acoustics around us, and the lures of cinematic cuts, moving camera shots, and diegetic framings.

Seeing a photograph of cave art, or a film documentary of Chauvet, does not convey the full experience of being inside a cave with prehistoric theatrical imagery. Yet Herzog's film illustrates how humans have experienced theater for at least 30,000 years, extending the inner performance spaces of reality perceptions, memories, fantasies, and dreams toward shared artistic, ritual, and entertainment events. The multiple echoing images of horse and lion heads, or of a single repeated head, of rhino head(s) with horn(s) and partial body outline(s), show a *movie theater impulse* to present various moments together at once, collapsing and yet extending time and space. This also relates to Ramachandran's law of "orderliness" in the natural alignments and pleasing rhythms of multiple, emotional images.

Regarding the neuro-aesthetic laws of "symmetry" and "metaphor," as well as rhythmic orderliness, there are chambers in the Chauvet cave, as shown in Herzog's film, that look like ritualized spaces with a central totem for people to gather around. On a large rounded stalactite, hanging from the center of the rock ceiling, in the deepest cavern that has paintings, are the black outlines of a woman's legs (no feet) and vulva, akin to carved prehistoric figurines that have been found in Germany, such as the "Venus of Willendorf."[7] Yet there is also a dark, filled-in bison head, just above the vulva and legs, as if emerging from them, or as a bison-man with one leg (and perhaps a human hand) overlapping the woman's. A lion's head and body emerge from the other side of the woman's legs. Around this central stalactite painting are many images on the walls, including vulvas and repeated animal heads. All this indicates that Chauvet was not a prehistoric art gallery, but a theater for gendered, animal-human myths and rites.

Another chamber, which "resembles an amphitheatre," contains further evidence for ritual performances in Chauvet (Clottes, *Chauvet* 99). A bear skull placed on a flat rock, which has charcoal on it dated to 30,000 years ago and dozens of bear skulls around it, including one with black lines on it, suggests a staging of life/death, human/animal, predator/prey meanings.[8] Such symmetry and possible metaphors may relate to the many paintings and engravings throughout the cave, as a network of meanings: lions, bears, rhinos, horses, bison, ibexes, reindeer, and mammoths (including one with a thin chest and strangely rounded feet).

Perhaps these artworks were also inspired by the spectacular natural formations of the cave, its huge columns, cake-like layered stalagmite mounds, numerous hanging stalactite needles, and rock curtains glistening with calcite crystals. The Chauvet cave today, like others with prehistoric art in southern France and northern Spain, conveys the sense of a majestic cathedral, with high arches and spires drawing the eye upward, through womb-like openings. Its painted figures may reflect ancestral totem spirits, possessing performers who embodied them for others, as well as the minds of visual artists who left such images—sharing the animal legacies of their inner theaters with us, too, as their future Other.

CAVES IN SPAIN

The Tito Bustillo cave, in the Asturias region of northern Spain, has artwork created over a long stretch of time, from 20,000 to 12,000 years ago, with most of it made 15,000 to 14,000 years ago. There is a chamber, with a very narrow opening, that bears red triangular signs on its walls, perhaps with mythic, performative meanings, like the Chauvet stalactite. This room has been dubbed the "Chamber of Vulvas," although the mysterious signs look like hoof prints as well. Maybe they reflect a hunter's imagination, tracking the deer, bison, and horses that are represented in outline or fully shaded forms elsewhere in the cave. The evidence from bones left at campsites shows that the killing and eating of such large animals was a rarity during the Ice Age (Lewis-Williams, *Mind* 258). Yet, the theory of Henri Breuil, from half a century ago, that the cave art demonstrated an attempt to gain power over such animals, as hunting magic, might have validity here—especially if these were rare and prized game. Indeed, there are black outlines of bison and horses on a red background in Tito Bustillo, which is unusual for Paleolithic cave art.[9] The red wash, which was painted first, might represent blood or the living tissue of the animal's inner body, as experienced by hunters. Inside the legs of some of the figures, red paint also suggests an exploration of musculature, using the natural texture of the rock. Perhaps this involved shamanic performances of animal spirits as well, with the human actor wearing an animal skin and mask, as in many tribal cultures around the world.

The famous cave of Altamira, near Santillana del Mar, about a 90-minute drive east of Tito Bustillo, also has red and black shading within its black outlines of bison, but only a little red wash behind them in some areas. At least, that is my perception from photos[10] and a life-size reproduction of Altamira's main polychrome chamber, since the actual cave is now closed to the public, due to bacterial problems brought into it by twentieth-century visitors, which damaged the art (as in France's Lascaux cave). The paintings also use natural protrusions in the cave's ceiling, with very detailed, curled bison within these round shapes, as if sleeping there. Perhaps both active and sleeping (or dying) animals were envisioned through the rock membrane and performed in this cave.[11] Campsite evidence near the entrance of Altamira shows Ice Age peoples living there 22,000 to 13,000 years ago. The art is dated at 16,000 to 14,000 years ago.

At the smaller, El Buxu cave to the west, nearer to Tito Bustillo, there is also evidence of prehistoric habitation close to the cave's entrance, 20,000 to 18,000 years ago. Deeper into the cave, through a one-meter, crawl-space passageway (with the floor lowered now for modern visitors), there is a chamber with etched and black-outlined goats on a natural arch. Further inside, matt-like, cross-hatched forms have been etched over the outlines of deer, as if covering them or performing magic upon them. In that area there is also a red, sideways E sign. Further into the cave, the most detailed horses are drawn, plus more goats and a large fallow deer above them. These etchings are very difficult to make out—until the Spanish guide, like a modern shaman, outlines them with the shadow of a finger or a red laser pointer, etching the images into the viewer's brain as well.[12] This relates to Ramanchandran's neuro-aesthetic "peekaboo effect" with sudden object recognition evoking the *rasa* of "awe" through problem-solving reward networks in the limbic system and creating a distinctive memory pattern in the viewer's temporal lobe, as the *inner audience* for restaging such figures in the imagination.

Near El Buxu and Tito Bustillo, the El Pindal cave entrance has a wonderful view of the Atlantic, from a cliff overlooking the sea, though in prehistoric times the water level was much lower, so the view would have been different. Along the walls inside the cave, which has no evidence of habitation, there are significant red animal outlines and markings. Moving into the depths one finds: triangles (perhaps akin to the vulva or hoof marks in Tito Bustillo), bison and a horse that involve the natural concave surfaces, abstract outlines and repeated "claviforms" (key or club shapes like the letter "P"), many dots, a deer or goat figure with hash marks on the rock ledge over it, five vertical lines crossed by two thick horizontal ones, and the outline of a mammoth with a red spot outside it and another mammoth nearby with a similar spot inside it (both facing deeper into the cave).

Such animal outlines and abstractions relate to Ramachandran's law of a "peak shift effect," like the red lines on a yellow stick as a super beak for a seagull chick's pecking instinct. The red spots inside and outside the mammoth outlines may depict a super-natural soul or heart, as a visionary trigger for the *rasas* of "awe" and "heroic vigor." The one inside could also be a hunter's target for the primal emotion of *seeking* and the need to feed. But such an encounter outside the cave would have been a very rare event, since most of the hunting involved smaller game. Only three mammoth depictions have been found in the caves of northern Spain (but there are more in France, especially in the black outlines of Rouffignac). The artwork here is from 20,000 to 14,000 years ago.

These caves, along with others I describe later, show a long-term engagement with animal figures, floating on cave-wall surfaces, as paintings or etchings. Ice Age humans must have identified more with animals in their environment than we do today, since they were struggling to survive, like them, as predators and prey, in very harsh conditions. Performances with the cave art—as isolated experiences of visions in narrow caverns, such as El Buxu, or with more collective rituals in larger cave spaces, such as Tito Bustillo, Altamira, and El Pendal—probably involved both shadow play, as in Plato's parable of the cave,[13] and redefinitions of character, as in Lacan's mirror stage. But if they were seeking ideal truths beyond the cave's hallucinatory shadows and artistic images, prehistoric peoples were going *deep inside*, through the rock wall and its crevices, rather than outside to the sunlight, as in Plato's myth. They may have been exploring something ontologically prior to Plato's antitheatrical prejudice and his suspicions about art luring us away from truth. While inventing visual art on the cave walls, they were exploring *transformations of identity*: from shared mimetic emotions and movements, with animals of the environment outside the cave, to finding Self in distinctive human forms, through Other spirits, with desires like and unlike their own, captured in various *movie-theatrical* representations.

Mobile artworks have been found in many caves, such as El Pendo in Cantabria, which has evidence of Neanderthal and then human habitation, from 84,000 years ago until the Bronze Age. Occasional flooding from rainwater in the large mouth of that cave, and many fallen rocks, make excavations of the different layers of occupation difficult. But at the Paleolithic layer, hundreds of deer horns with various engravings were found, including a rare snake representation, also jaws eating a deer etched into bone from 20,000 years ago and a thin human female etched onto another bone (unlike the fat "Venus" sculptures found elsewhere). A Cro-Magnon child's skull was also found. Such artistic figurines, along with evidence of prehistoric human habitation, may relate to Victoria Nelson's argument about later puppets, robots, cyborgs, and virtual entities as "Neoplatonic daemons," which still bear an "uncanny aura . . . even in a secular context," thus showing us "the underground history of the soul" (31).

There are dozens of paintings deeper inside the El Pendo cave, from the same time period, beyond the reach of sunlight, along a 25-meter wall and another 9-meter wall. The higher paintings would have required scaffolding for the artists (as at Font-de-Gaume in France). But the red paintings later became covered with black bacteria, making them difficult to see, even after the walls were cleaned in modern times. There is a red deer with a small head, filled with fingertip dots of paint and yet empty at the center, perhaps suggesting pregnancy. There is also an incomplete horse without a head, a bull, and the disembodied head of a deer, plus two deer facing opposite ways outward, one with a long neck, the other with front legs up as if leaping. And there is another painting of a deer that uses a crack in the rock for its back. There is a possible snake form, too, and one deer head inside the outline of another deer, at its front quarters. Within this cave one can see partial remnants of a large gallery of prehistoric artworks, on the walls and in mobile objects left in the cave. They reflect the inner theaters of many prehistoric humans, mirroring various forms of the animal Other, and yet also the beast-person within, like figures from dreams.

More elaborate red paintings, with a possible anamorphic image, appear in the narrower passages of the Covalanes cave, which is also in Cantabria.[14] Its rock formations include fossils of marine animals, such as seahorses, strange calcite cubes, and swirled surfaces, produced over the millennia by ocean and rain water, rushing or seeping through the earth. But it has no evidence of habitation. Its artwork dates from 20,000 years ago and becomes more vividly experienced today, as with the Niaux cave in France, because the cave is unlit, except by visitors' flashlights. (Radioactive uranium in the calcite that formed naturally under and over some of the Covalanes artworks allows for objective, not just stylistic dating.) A rock panel shows a group of six female deer in red outlines, though they only appear gradually, through the guide's tracings: a deer with a long neck, painted around a protruding rock for her stomach (suggesting pregnancy), a deer outlined with fingertip dots like in El Pendo, the full outline of a deer with legs in perspective, another deer with head turned and body colored in, and two deer with necks overlapping while facing opposite ways, so that one is clearly behind the other.

This mix of figures in faint red outlines and dim light appears fragmentary, like etchings in other caves (especially the many overlapping lines in France's Combarelles cave). The fragments that form into wholes, but are cut up again by overlapping outlines, relate to Ramachandran's neuroaesthetic law of "grouping" as a survival practice for prey and predator identification outside the caves. But such fragments may also reflect the illusory formation of the Self (as ideal ego) in the Lacanian mirror stage, masking the contrary experience of an alienated, uncoordinated, libidinous "body in pieces." This occurs in infancy, as shown by the jubilation of the child in the mirror, but it also repeats throughout one's life, with numerous desires of the Other on stages, mirrors, and screens that offer visions of a whole Self. Thus, one can see mirror-stage reflections of a prehistoric internal theater or "intra-organic mirror" (Lacan, *Écrits* 78) in caves such as Covalanes,[15] with the emergence of a new form of human identity, tens of thousands of years ago, like yet beyond the perspectives of other animals.

On the opposite wall from that six-deer panel is an auroch (Ice Age bull) with the edge of the rock as its back. There are other deer further into the cave and then a horse with detailed mandible and mane. To the left and below it are a red colored-in triangle (or female sign) and three deer. On the opposite wall, under a rock curtain, there is another deer with long neck and lifted legs as if leaping. There are big deer on the inside of that curtain, plus a filled in triangle on top of a box (perhaps a phallic sign). And then, in a narrow passage, wide enough for just three people, another deer appears. But its full form with a white (calcite) eye is seen only at a distant angle, down the passageway, like the anamorphic illusions in later Western paintings, such as the two-dimensional skull in Hans Holbein's *The Ambassadors*, which becomes three-dimensional when viewed at an angle.

Lacan uses this skull as an example of how "consciousness, in its illusion of *seeing itself seeing itself*, finds its basis in the inside-out structure of the gaze," with an object reflecting the ghostly lack of being in the subject (*Four* 82, 86–89). Likewise, according to current neuroscience, human consciousness emerges, beyond biological determinism, through the recursive complexity of the brain's neural networks interacting with outer social networks (Gazzaniga)—in a fractal *hall of mirrors*. One can see this illusory nature of human consciousness in Covalanes, emerging in a distinctive way 20,000 years ago, with the gaze of the shamanic artist and spectator reflected in the eye of the deer as Other. The uncanny deer, in this sense, is a "phallic ghost," as Lacan says about the Holbein skull, signifying the lack of being and "want to be" in humankind's early evolution beyond the natural environment. Ironically, humans may have glimpsed their increasing power *over other animals* by envisioning animal spirits within their brains and projecting them as artistic representations, making cave walls into emotional, movie-theatrical mirrors.

Another Cantabrian cave, El Castillo, shows evidence of being inhabited as long as 120,000 years ago by Neanderthals and then, much later, by prehistoric humans. Originally, there was a tight crawl space for the entrance, but it is now deeply excavated into a wide opening, with an "Interpretation Center" nearby. Inside, there is a large chamber with horses in red outline on a flat cracked surface, plus partial outlines of deer and bison, floating on the rock canvas and pointing downward. There is also a black bison pointing down, with a small deer coming out of its back and a red *negative handprint* on its left, from 30,000 years ago. (This term refers to the Paleolithic practice of putting a hand against the rock and blowing or brushing paint around it, leaving its imprint as a negative space within the painting. A "positive handprint" involves putting paint on the hand and then placing it on the rock.) Lower down, in a wide chamber, there is also the outline of a red bison, with negative handprints below it, on its left. Further inside can be found the print of a right hand made with the back of the hand to the wall, two bison in black outlines on a rock ledge, a smaller chamber with an auroch outline, horizontal spots, another auroch, spots in a horn shape, and then, after 10 meters of blank wall, some diamond shapes with tails, like kites, plus further images deeper into the cave, not accessible to the public. The large chambers in El Castillo would allow many people to participate in ritual performances, though there is a steep grade within and between some of the areas. The 60 handprints throughout the cave may show some kind of formal sign system, or "I was here" gesture, or simply playfulness.

Language, identity, and play are key elements in the new form of human mind that evolved with cave art.[16] As mentioned in the previous chapter, Merlin Donald has mapped the stages of cognitive and cultural evolution in our hominin ancestors: from episodic animal consciousness toward mimetic communication with kinetic gestures and playfulness in *Homo erectus* two million years ago, then to a mythic culture with verbal language in *Homo sapiens* a half million years ago, and then to a theoretic culture of technology and art 40,000 years ago, as evidenced by the cave and mobile artworks (120, 255, 260–74).[17] These levels of consciousness, communication, and identity are also stages that we pass through today, from childhood to adulthood, forming *layers* of our internal and external theaters. Archeologist Stephen Mithen gives a similar timeline and adds that humans about 40,000 years ago evolved a new "cognitive fluidity" between various types of intelligence that had been isolated in the hominin brain before that: natural history intelligence (such as interpreting animal hoof prints), social intelligence (with intentional communication), and technical intelligence (producing artifacts from mental templates).[18] This led to new productions of symbolic meanings that mixed natural, social, and technical brain mappings of inner and outer worlds, with movie-theatrical performances in the caves, involving animal, human, and abstract figures.

The cognitive fluidity of Ice Age cave theater also involved *music*. Las Monedas cave, a short walk from El Castillo, has many geological spectacles and a small area of manmade art, with a smiling reindeer, horses, a fox, a bison, and various other shapes from 12,000 years ago. When I took a tour there, a guide illustrated the performative qualities of that cave by tapping several hollow stalactites of varying sizes with his knuckles, like a xylophone (or "lithophone," as he called it), showing the different notes they contained and the soft echoes of the cavern, as if in a small concert hall. He pointed to much larger organ-like columns, closer to the entrance of the cave, suggesting musical possibilities in various caverns. Also, the *echoing of the human voice*, in this cave and others, evokes musical and super-natural qualities, which may have influenced the placement and performance of Paleolithic art (Díaz-Andreu et al.).

Hornos de la Peña, which is also in Cantabria, offers a very different experience. One has to crouch or crawl with the guide into very small chambers (some with room for just a few people), which have mostly etchings, but also a few paintings. A horse and auroch are engraved in an area near the entrance, which was inhabited 28,000 years ago. Then there is a tight space to crawl through, into a somewhat larger chamber, which might hold a dozen people, where another horse is engraved on the wall, also "macaroni" finger fluting (produced by human fingers moving through the soft clay of the cave, in straight or wavy lines), a goat from 20,000 to 18,000 years ago, with its eye involving the rock's natural indentation, and a bison from 15,000 years ago. Deeper into the cave are a goat and more abstract lines from 30,000 to 28,000 years ago on one side of a small chamber, with larger horses and a deer head to the right, plus a possible serpent figure. Above the horses, there is an "anthromorph": a human figure with a bird head, raised hand, and tail, from 20,000 years ago. Nearby is a horse with cross-hatches over it. And in the low ceiling, there is a horse with a hole in its stomach or womb. The guide explained that *horno* means "oven," and is also a reference to "womb" in Spanish, as in the English phrase for a pregnant woman "with one in the oven."

The experience of entering this cave and its small chambers with difficult to discern, yet intriguing engravings, and then crawling (sometimes backwards) out of the cave, also makes it a womb-like space, or *chora*, in Plato's and Kristeva's (Lacanian) sense: an enclosed space of becoming for the thetic, Imaginary subject, via the art on the walls as performative, interpretive mirrors. The Self of each visitor today is changed, to some degree, by performing with the guide and others in a small group, squeezing their bodies into a space inside "Mother Earth" and perceiving the paintings and etchings—as lines and scratches in the rock, made tens of thousands of years ago, which gain Gestalt forms once again. Although we live in a very different world now, with mass-media screens, electric lights, climate-controlled homes, food in grocery stores, vehicles for speedy travel, and many other transformations of the environment, we can find a kinship with our prehistoric ancestors through such cave theaters. (The term "ancestors" here assumes that similar artistic impulses developed in prehistoric humans around the world, not just in Europe.) We can glimpse how their sense of Self was emerging, through reflections on animal and human nature, in the various reptilian/mammalian forms and a peculiar bird-man with a tail—in this cave as a *chora* "where the subject is both generated and negated" (Kristeva, *Revolution* 28). We can see evidence of the prehistoric mind, with its cognitive fluidity, creating imaginative and sometimes *hybrid* creatures, as characters for Self and Other, yet also seeking a "cognitive anchor" in the cave images or symbols, as in later religious art and ritual performances (Mithen, "Evolution" 48–50).

CAVES IN FRANCE

There are various caves in France (like Chauvet) with elaborate artistic expressions that I have written about elsewhere: Lascaux, Font-de-Gaume, Combarelles, Rouffignac, Cougnac, Pech Merle, and Niaux. But there are also smaller French caves with details worth considering.

The several caves of Grottes d'Arcy offer the closest location to Paris for viewing prehistoric art. Unfortunately, 80% of the art there was destroyed in 1976 by a foolish attempt to clean it. The paintings that remain, in the single cave open to the public, are located near the back, on the sides and ceiling of wide chambers, with floors that would have been lower 28,000 years ago when the art was made. This suggests collective ritual experiences, after individuals placed red or black outlines of mammoths and deer on the rock surfaces—like the medieval cathedrals of France with stained glass, sculptures, and artworks that raise viewers' eyes upward.

The Merveilles cave is located farther south, near a popular tourist site, the medieval cliff-side town of Rocamadour (worth a visit also). The cave was inhabited 20,000 years ago and has some interesting natural formations, along with paintings and etchings near the entrance, mostly in a single large chamber. They are difficult to see, even with the guide's help. There is a negative handprint over a deer that appears to be leaping through a calcite (white) barrier, although that may have formed in the thousands of years after the artwork was made. (The natural spectacles in various caves considered here, stalactites and stalagmites, columns and drapes, grew very gradually, taking thousands of years to form.) To the right of that handprint and deer is a stag with an oversized head, a horse facing the opposite way, a negative handprint that strangely has six fingers, a big horse, and perhaps a hyena.

Gargas, a cave in the Pyrenees,[19] has handprints, plus animal outlines and etchings, from around the start of the cave art era, 30,000 years ago. There is the negative print of a child's hand near the small living area by the original entrance, then many adult hands deeper into cave, especially on one wall. There is also a handprint at the opening to an inner chamber, with a red spot on the other side, perhaps signifying something about that inner space (which is off limits to visitors today). There are a total of 231 hands in the cave, with 124 clearly visible and just 10 of those having all their fingers (Curtis 196–97). Over half of the hands are missing all four fingers, with just a thumb showing. None have a missing thumb. This may signify the importance of an opposable thumb, or some other hand-sign language that we cannot read today. But the art throughout the cave seems to have started with a child's gesture, near a campsite home at the cave's entrance, perhaps making a playful handprint—before it became an adult performance rite. This may have involved hands disappearing into the rock wall, communing with the spirits there as they were painted over, leaving negative prints (Lewis-Williams, *Mind* 216–20).

It has been observed by primatologists that humans exhibit "neoteny": both physically and behaviorally we look like young apes (de Waal, *Our* 240–41). Adult humans, with round heads and flat faces, extend youthful play into many performance activities, from theater to sports to videogames, while also displaying the curiosity and inventiveness that led to vast transformations of culture and environment in prior generations.

Prehistoric artists may have had very serious reasons for painting their hands onto the cave wall, especially with missing fingers. Some paleontologists today speculate that the fingers were lost due to frostbite or disease, or were deliberately cut off in sacrificial rites. But it is more likely that the cave artists were displaying the significance of digits, as present or absent, by bending their fingers while making the handprint—to make it appear fingerless. Such playful fragmentation in art, involving potential metaphors or signs, shows a crucial element in the growing flexibility of our human ancestors as big-brained apes. For better and worse, humans found ways to survive and reproduce far beyond natural instincts, by *playing* with their inner drives and brain theaters, projecting them outward as spirits and cave images, and eventually changing the world. But then the problem became, as it still is now, how to *limit* human creativity and destructiveness, in performance spaces beyond the cave.

North of the main cave-art area around Lascaux, the Villars cave[20] has spectacular natural features, including needle-like stalactites and rock drapery, enhanced by the modern tour's sound and light show. Through a low and narrow passage, the guide takes visitors to half a dozen black horse outlines, including one that has changed to blue over the millennia, with a few millimeters of natural white calcite covering it. Close to the cave's prehistoric entrance, a wall painting shows a bison apparently charging at a thinly drawn man, almost a stick figure, whose hand extends over his head, perhaps in fear or to provoke the bison, or with magic power over it.[21]

The art here is recent, compared to Chauvet and Gargas. It is from about 18,000–17,000 years ago, almost as close to our time as it was to the artwork of those other caves. This approaches the end of the Paleolithic cave art period in Europe and start of the Neolithic agricultural revolution, 11,000 years ago in Mesopotamia, which led to the domestication of certain prey animals, whose images are in the caves: goats, aurochs, and horses. They were captured in ways beyond artistic representation and turned to other uses, as humans created increasingly large settlements and then civilizations, involving the storage of wealth with a warrior class and weaponry to defend it or take it away from others. As the Judeo-Christian scriptures put it, mankind gained "dominion" over the animals, through God's creative will. Yet some animals, such as charging bison, could still be a threat. The *cognitive anchors* of religious beliefs, with gods as super-parental figures (perhaps deriving from cave images), gave limits to the reckless playfulness of the human ape and conceptual constraints to our big-brained plasticity. Such beliefs, making up for the lack of instinctual patterns of perception and behavior in our species, are still needed by many today, even as science has given other cognitive anchors in the last half millennium, which have also led to great destructiveness, as well as creativity and good works, in theaters beyond the cave.

Pair-non-Pair, north of Bourdeaux, had a small prehistoric entry hole, which required crawling, but prevented bears from entering the cave. Inside, there is a man-made ring in the low stone ceiling, perhaps for an animal-skin curtain as a barrier against the cold. Such evidence of habitation at the entrance of this cave relates to the steady temperature inside, near 60 degrees Fahrenheit all year, as compared to the Ice Age extremes outside. There is a second chamber beyond the entryway here, with etchings (plus a hole above, giving light), and a third room with a natural spring and pool, a vent in the rock ceiling for a campfire, and more engravings. With the help of the guide, one can find a mammoth with a goat above it, a bison, and several horses. Two of the horses have long necks, one with an open mouth, looking back at the pool area. There is also a megaceros (giant elk) and a large horse over it, involving differences in the natural rock colors for shading. There are seashells in the rock used for the eyes of the horse and mammoth. This cave was inhabited over a long period, from 80,000 to 18,000 years ago, by Neanderthals and humans. The art is from 30,000 years ago, in the earliest period of cave theaters, like Chauvet.

Ice Age humans saw their own faces much less than we do today with our mirrors, photos, videos, and Facebook sites. Yet, with a pool of water in caves like Pair-non-Pair, prehistoric infants might have experienced the mirror stage in that horizontal mirror, as well as through the mother's touch, smell, cooing, and facial expressions. The further imaging of Self and Other, through perceptions and etchings of animal figures on the cave walls, may relate to the good fortune that adults felt in finding this cave, where they could live with a campfire, water, shelter, and a steady temperature. The engraving of the horse with a long neck and open mouth, looking back at the pool area, may speak to this—and to touch, sound, movement, and vision as aspects of performance in the small ritual spaces of this cave theater.

MIRROR STAGE CAVES

Why are there very few human images in prehistoric cave paintings (other than handprints)[22]—and none in the finely detailed, lifelike style of certain animals in Chauvet, Lascaux, Font-de-Gaume, Niaux, and Altamira? Findings from neuroscience about the brain's basic anatomical functions, which we share with Paleolithic peoples, along with the evidence of theatrical spaces and images in the caves, interpreted through psychoanalytic cultural theory and dream research, may provide an answer.

The performances of everyday life, even today, involve "emotional contagion" and mimicry between brains, often at a subconscious level, through intuition (Von Economo) and mirror neurons, like in a herd, pack, or flock of animals moving collectively, with individuals reacting to subtle cues from others.[23] Simulations in the brain mirror another person's actions, facial expressions, gestures, vocal tones, and phonemes, beyond (or beneath) verbal language. Our brain theaters play the character of the other we observe or interact with, sending signals to perform the other's actions and emotions, as "somatic markers" (in Damasio's phrase) of the other's thoughts and feelings.

Thus, our prehistoric ancestors may have felt a kinship with herd animals in their environment, such as horses, deer, elk, bison, aurochs, ibexes, mammoths, and rhinos, which they etched and painted inside certain caves. Yet these creatures are often depicted as individuals, floating on the rock wall and sometimes overlapping—as if with a degree of theatrical self-reflection, based in, but moving beyond a continuous herd mentality. The prehistoric sense of a Self was probably continuous with the natural environment and its impressively large herd animals. A distinctive image and symbolic order of "human being," with super-natural powers, may have emerged through identifications with such herd animals and their contagious mimicry, yet individualized movements.

Today, I (like other humans) have a sense of my own character that inhibits most of my mimicry at the overt level, unless I choose to copy someone's gestures. This is especially so in the dominant convention of Euro-American spectatorship. I watch the show with others moving around the stage, or onscreen, while I stay in my seat, participating vicariously. Even with my brain sending signals to mimic the movements and emotions onstage, I remain still—like most people while dreaming, when motor and emotion circuits are active, but the body's movements are inhibited. However, some people suffer from "echopraxia," due to damage in the "braking mechanism" of the frontal lobes; they compulsively mimic others or do typical things in certain contexts (Rizzolatti and Sinigaglia 151). This shows how mimicry is signaled in the rest of us, too, but blocked by our frontal lobes.

The human sense of Self in our prehistoric ancestors emerged from proto to core selves (which other animals have) to an autobiographical identity, through emotional contagion and mirror neurons—using inner simulations of the Other to form the conscious Self. As with infants today in the mirror stage and all of us later in life, the character of Self in our prehistoric relatives developed through interactions with the Other: from "the specular *I* . . . into the social *I*" (Lacan, *Écrits* 79). The Self, unifying various selves in the brain and nervous system as a character onstage in the inner theater, uses reflective images from the Other's apparent perceptions to form simulations of the Other's simulation of the Self, putting those into the developing plot of autobiographical memories, fantasies, and dreams.

The Other for Ice Age humans was very different from ours. It involved few contacts with other persons, and no mass-media images, just a small tribal group, as extended family, occasional meetings with other groups, and the natural environment with many fellow animals, including impressively large creatures, which became the dominant images in the caves. And yet, like our prehistoric ancestors, we have remnant animal drives within our nervous system, as a "fragmented body," moving in contradictory ways behind the mask of Self and its body image (or Gestalt). Lacan sees this exemplified in the human infant's normally premature birth, with an uncoordinated, libidinal body at the age of 6–18 months, finding a jubilant, yet illusory wholeness in its mirror image—and in its multisensory interactions with the (m)Other. The proto-self, monitoring interior body states, and the core self, interacting physically with the world, are not fully unified in their mappings. Their dispositions are divergent as well as convergent. But with the development of an autobiographical Self, in each human child and in our ancestors' evolution, a newly unified (although hollow) ego emerges, involving right-cortical intuitions and left-cortical language areas, like the holistic animal images and yet abstract, possibly symbolic figures on the cave walls.[24] Or like the herd of hands in some caves, with absent fingers as a sign language of distinctive personal marks.

Theatrical spaces and images in prehistoric caves show particular instances of mirror-stage, contagious interactions between early humans and their minds' *projections* onto the rock walls. These probably involved (1) their basic animal drives, (2) remembered animal forms from their environment, and (3) animal characters or fluid animal-human hybrids from their inner brain theaters, including mythic plots, as well as handprints and other signs, which we can no longer read. We cannot know what the bison/lion-headed man/woman in Chauvet meant to its creators, or the full plot of the bison charging the stick-man in Villars, or the *rasas* of the bird-headed man with a tail and raised hand in Hornos de la Peña. But we can see how the animal or animal-human Other depicted on the cave wall, or envisioned through the rock membrane, became a way for early humans to develop an image of Self as *part* of the natural world, yet evolving *beyond* it. Our ancestors, tens of thousands of years ago, were just starting to develop the illusions and realities of human ego power, over animals and nature. So they made far fewer images of the human form.

As mentioned in the first chapter, children today often dream of wild animals, while having little or no contact with them in reality. Dream researcher Antii Revonsuo theorizes that our dreams, like those of other mammals, may have evolved as "threat rehearsals" for survival in real life—and that we carry in our brain the prehistoric images that haunted our ancestors for thousands of years, yet helped them to survive and produce us (85–111). As we mature in our modern world, we absorb images from its threats and goals to rehearse in our dreams, masking our ancestors' interactions with "wild" animals as natural dangers, food sources, and spirit guides. But they may appear again, as urban invaders, tame livestock, lovable pets, or beast-people in our dreams and current emotion picture media.

The many *overlapping* images in cave paintings and etchings might show the prehistoric threats of humans to one another, as well as the depths of each person's visionary perceptions through the rock wall. Like graffiti artists today, cave performers may have competed to present their tribal or personal totem, overwriting those of others. And yet, the mixture of animal outlines also suggests the source of such ego and group aggression in each of us: the fragmented body behind the Self, which finds an illusory wholeness in the mirror or theatrical image, as display and gaze of the Other. The Self is thus a fragile character, alienated by the Other's misrecognitions and social demands. This produces, even in the mirror-stage infant, a defensive "armor" of Self, masking its own fragility and inherent paranoia, which sometimes results in aggression toward others, especially those figured as evil (Lacan, *Écrits* 78–79).

The cave theaters of Spain and France, along with Herzog's film of the Chauvet cave, show particular mappings of the "extended minds"[25] of our prehistoric ancestors. These relate to (1) the mirror stage that we each pass through as infants, reflecting an idealized Self and Other, yet also the animal scripts of primal emotions within us, (2) the media screens where we find ideal and fragmented ego images, sometimes as animal or animal-human monsters, and (3) the mappings that our inner brain theaters perform in everyday life, with backstage signaling, biased projections, and conscious communications between us, as explored in the next chapter. Prehistoric cave art shows the brains of our ancestors—like ours today—filtering their perceptions of reality through memories, goals, and fantasies, involving primal (animal-drive) mappings and higher-order dispositions, from the unconscious mind to the conscious staging of Self and Other, through *pictures* on rock-wall *screens*, moving in the firelight and evoking emotions in performers and watchers.

Subsequent religious traditions repressed the animist insights of prehistoric cave performers and "pagan" shamans, with sky gods and monotheisms. Hell, on the medieval European stage, was represented by a monstrous cave.[26] But if we dig back through the layers of such cultural ideals, past the ancient Greek "birth" of theater, imagining cave theatricality in prehistory, then a primal Thespis can be glimpsed in the many handprints, abstract shapes, animal forms, and animal-human hybrid figures of specific chambers and their rock-wall scenery. Such artworks, like the beast-people in many movies today, remind us that in becoming human, technological, and virtual, we are still animals—and sometimes "wild" beyond our dreams. Prehistoric performers' paintings and engravings reflect how we use each other, as animals and media figures, to map the fragments of Self into transcendent yet illusory characters, which have real effects on our continuing biological and cultural evolution.

Chapter 3

The Other Mammal in Me: A Hall of Mirrors between Brains

Look into the eyes of another person. Whether a stranger or a familiar friend, there is a mystery in how that *other* sees you at any given moment. He or she mirrors your extended identity, through an inner theater behind the eyes, which you cannot see. That mysterious darkness, involving many people in your life, reflects your own inner theater's imagining of the other's viewpoint, as like and unlike yours. Yet, the backstage areas of our inner theaters often signal one another, unconsciously, through facial expressions,[1] gestures, vocal tones, and the emotional contagion of mirror neuron networks, forming a *hall of mirrors* between our highly social, mammalian brains—with fractal reflections at the edges of our brain's staging and screening of consciousness.

There are many animal legacies in your brain's theater—as considered in previous chapters here—affecting the dreams, imaginings, and reality representations that you construct of Self and Other, objects and environments. In prehistoric times, our ancestors painted and etched animals (mostly herd mammals) on cave walls, deep within the earth, perhaps communing with such figures in trance, through the rock as a membrane to another world or a *projection screen* from the inner one. They probably shared such images, in personal shamanic ordeals or larger communal rites, in narrow or wide areas of the caves. In the cave theaters, they may have tried to gain a magic hunting power over large herbivores, which were rarely their successful prey, or traveled to a spirit realm with such figures for healing powers (according to Breuil's and Lewis-Williams's different theories).

Prehistoric humans gradually evolved an identity beyond other animals, through new technologies of survival, such as the hunter's spear thrower (about 15,000 years ago) and artist's paintings (much earlier), as ways of capturing and communing with the Ice Age creatures around them. By representing powerful animals from outside the cave, or animal-human hybrids, prehistoric peoples built super-natural identities within and between their brains—as a superior species of watchers and actors with cave walls as emotion-picture screens. Today, how are such animal-human theatrics still evolving within and between us, through the reflective framework of Self and Other? What can neuroscience show us about the morphing projections between our inner theaters, relating to the monstrous transformations in beast-people movies, which shock us and yet draw us into that cave-wall screen?

THE DEVELOPMENT OF SELF, OTHER, AND FAIRNESS

In the first 2 years of life, while responding to others in developing ways, humans start to recognize themselves in the mirror.[2] Most animals never do this, even dogs and cats that become our pets, with household mirrors around them. Only apes, dolphins, and elephants seem to pass the mirror test of self-awareness. But human babies, even at the age of 4 months, behave differently when seeing their own body movements in a video or a person imitating them (Rochat and Striano, "Who's"). Thus, a primal sense of theater begins with human infancy—in the mirror or other people's eyes.

At 6 months, infants start to attend to the goals of others (Decety and Chaminade 579–80). At 9 months, they follow the gaze of others toward "joint attention" with objects and events, showing an awareness of others as intentional agents through pointing, imitative learning, and social referencing about objects (Fernandez-Duque and Baird 82).[3] Full awareness of Self in a mirror, at about 18 months, coincides with the emergence of *social emotions*, such as shyness and pride, plus an awareness of others' emotions (Decety 145; Rochat 87). At that time, infants also develop "an essential aspect of the adult theory of mind, namely that people (and not things) act in purposeful, intentional ways," different from oneself (Meltzoff 35).

Prior to that, and beneath our current self-awareness, we may experience other people, especially primal caregivers, as a *continuation* of ourselves. The mother's breast, caressing hand, or smiling face may be sensed as a nurturing, womb-like environment, continuous with the baby's body—when she is present. When she is not and the baby's brain signals an inner problem, such as hunger, a cry erupts, calling for the restoration of that nurturing external womb. Gradually, the other person becomes recognized as different from oneself. Yet, less than an hour after birth, a baby's face will *instinctively mimic* an adult's facial expressions (Goldman 13). This begins a continued shaping of personality, moods, and neural networks by such mirroring interactions, eventually becoming a self-aware relationship with specific others—forming a family foundation for future social connections, with the Self like and unlike the Other.

As in the early childhood of each of us, *a continuum of self-other relations* evolved in our primate ancestors, prior to self-awareness in some species. Baboons, who live in large social hierarchies like humans, perceive others' intentions and emotions, but not others' beliefs and viewpoints, as different from one's own (Cheney and Seyfarth 152–59). A baboon mother may wade through a river with her infant clinging to her belly, not noticing that the baby has a different experience until it drowns. Similarly, humans only begin to understand, at 3–4 years of age, how others might have a different perspective (Gopnik et al. 41).[4] Before that, the child thinks the other (an adult or puppet) will know where an object is, although the object was moved while the child was watching and the other was not present. Infants as young as 14 months will cry along with another child, showing emotional contagion and gestural mimicry, yet also helping behaviors as the child matures, with mimicry decreasing as helpfulness increases (Preston and de Waal 1). This reveals a continuum of self-other identity in humans prior to the full development of self-recognition: with an ego image in the mirror around age two, then the further development of internal "metamind" representations (involving the right hemisphere and prefrontal cortex) by age 3, and then perspective taking about others' beliefs, including false beliefs, at age 4 to 6 (Barr and Keysar 272; Bogdan 201–2). Such animal-to-human developments in children enable a further sense of communal identity through empathy and generosity.[5]

Preschoolers may actually be more conscious of *current sensory experiences* than older children and adults, with more cholinergic transmitters in their brains and less inhibitory circuits, for greater awareness of the present moment, like adults in meditation (Gopnik 119–20). More like our animal ancestors, preschoolers have an "attentional lantern"—rather than a "spotlight" of consciousness—with less focus on an autobiographical past and future (125, 145–48, 153). That lantern, as a wider staging/screening of consciousness, includes a sense of spirit animating the world. At 5, children around the world develop a vitalist theory of biological causes, like the Chinese sense of *chi*, with a life force in us and in all living things, which weakens with sickness and death (36). It thus seems to involve positive and negative spirits, with us or against us.

Such emotional alignments also develop through *group identifications*—altering our inner theaters. Around the age of 3, children form in- and out-groups, focusing more empathy toward those like them, even those wearing the same color of T-shirt, whom they see as nicer and better to play with (Gopnik 219). An early ethics develops through the theatricality of Self and Other, with our in- and out-group projections. "One-year-olds understand the difference between intentional and unintentional acts, and behave in genuinely altruistic ways. Three-year-olds have already developed a basic ethic of care and compassion" (204). Starting around age 7, girls respond more to babies than boys do, but "boys direct their nurturant impulses toward a broader range of targets, such as animals" (Baumeister 132). Yet parents, teachers, and peers shape the development of children through their expectations—like handlers unconsciously influencing the performance of lab rats when assuming them to be smart or not (McLeod; Rosenthal and Babad). According to this Pygmalion or Golem Effect, the Other's view of us, through like and unlike identifications, with in- and out-group projections, is contagious.

Tests with rats and monkeys demonstrate the development of emotional contagion and empathic helpfulness in our mammalian heritage, plus *self-sacrifice* and a sense of *fair play*. A rat will retreat to the corner of its cage, cowering motionless, while seeing and hearing the distress of another rat given an electric shock (Preston and de Waal 1). But a rat that sees its friend suspended in a harness will press a lever to lower it and then stay close to help reorient it. Rhesus monkeys will refuse to pull a chain giving them food when they find that pulling it also gives a shock to another monkey. In the test, two-thirds of the monkeys pulled a non-shock chain instead, although it gave them only half as much food as the shock chain. The other third refused to pull either chain for a period of time, one for 5 days, another for 12, starving themselves to avoid giving pain to the other monkey. In the wild, rhesus macaques live in a very interdependent, hierarchical society, so these monkeys may lessen their own contagious distress by stopping that of the other. Yet in the experiment, they caused pain to themselves in this way, although more often "between familiar than unfamiliar individuals" (de Waal, *Primates* 29).

Capuchin monkeys have larger brains than macaques, showing independence within groups, but also cooperative, food-sharing behaviors. In laboratory tests, they will throw their food and refuse to do a task—when other monkeys are given a better treat for the same work (de Waal, *Primates* 48).[6] These experiments, with capuchins and macaques, show that a sense of fairness and yet self-sacrifice (as an extreme form of targeted helping) may have evolved through the monkey lineage of our primate ancestors. But fairness expectations have also been found with wolves, coyotes, dogs, and ravens.[7] *Vengeance* has been found in chimpanzees, too, as the dark side of primate expectations of fairness (de Waal, *Our* 204–9), but apparently not with human spitefulness (Jensen et al.). Perhaps unlike "wild animals," we find *pleasure in punishing others*, as shown by an experiment involving monetary punishment that activated the pleasure networks in the brain's striatum (Ariely, *Upside* 124–27; Fehr and Camerer). Likewise, righteous or "altruistic anger" activates that "reward center" in the brain (De Quervain et al.; Goleman, *Social* 131).

By the age of 15 months, human infants, like capuchin monkeys, expect fairness in food distribution tests. And yet, one-third of those tested demonstrate "altruistic sharing" of a favorite toy with an adult, when another toy is also in hand (Schmidt and Sommerville 5). Regarding adult behavior, economist Daniel Ariely calls this difference in priorities, between fairness and generosity: "market norms" versus "social norms" (*Predictably* 67–69). He gives the example of offering to pay one's mother-in-law for her work, cost, and "love" at the end of a lavish Thanksgiving dinner—thus offending her by replacing the social norm of generosity with the market norm of fairness. Ariely and his researchers set up an experiment with a boring task (dragging a circle to a square on a computer screen repeatedly). They found that if it was introduced with a market norm, a payment of five dollars or just fifty cents, participants would do more when paid more (70–71). But if it was introduced with a social norm, "as a favor," then participants dragged more circles than when paid either amount. A different version of the experiment showed that a small "gift" still fits within the social norm of doing the job as a favor (72–73). But if participants were told the cost of the gift, the results went back to market norms and less work for a less expensive gift.

Ariely refers to another experiment showing that people can be primed to work longer on a puzzle, without asking for help and without giving help, if "high-paying salary" is mentioned obliquely (in a scrambled sentence) prior to the puzzle work, because market norms are suggested (*Predictably* 74–75). They were then more "selfish and self-reliant," wanting more time alone, selecting more individualistic tasks, and sitting farther away from coworkers. Ariely also points out, with another real-world example, that when a market norm collides with a social norm, the social norm goes away and it is difficult to reestablish it (77). He says that businesses, especially recently, have confused social and market norms, advertising themselves as "a good neighbor," without recognizing how this can backfire when, for example, a late payment is fined and not just forgiven (78–79). But social norms with positive emotions are also important for employees—building toward loyalty, trust, creativity, and commitment beyond market norms (83–84). In some careers, this might even reach the level of *self-sacrifice*, for the sake of pride and duty beyond salary, as with police officers, firefighters, and soldiers—or potentially in education, with a "sense of purpose, mission, and pride" (84–85).

It might be argued, though, that market norms are particular types of social norms that evolved for efficiency in capitalist systems—with the money exchange abstracting the gift giving and product or service exchanges of tribal cultures. This changed a friendly debt, or relationship of generosity, to a monetary sense of fairness, objectifying the seller and consumer bond (through left cortical networks of the inner theater). Such market-norm abstraction can be taught to other animals, too, such as capuchin monkeys. They then demand *equity* with tokens of work. But they also demonstrate how a primal sense of "unfairness," regarding objective rules and tokens, can provoke rebellion or revenge, as Self becomes distinct from others, yet with equal expectations, in mammalian brains.

Rats and monkeys show the possible evolution of altruism, self-sacrifice, and fairness in our ancestors, from right-cortical social emotions to left-cortical market abstractions. Likewise, rebellion, sacrificing others, and vengeance, in the shadows of such virtues, are key elements of the vampire, werewolf, and beast-people movies explored in later chapters here. These fantastic creatures onscreen, with their physical transformations, hybridity, and bloodlust, *reflect the uncertain mirrors* of shifting identities and contagious emotions in our animal-human theaters, as highly social mammals with complex, fragile egos.

MAMMALIAN EMPATHY, MIMICRY, DETACHMENT, AND ACTOR-SPECTATORSHIP (LEFT AND RIGHT)

Psychologist and primate researcher Frans de Waal has defined various levels of empathy and imitation, from animal to human, showing the levels of interpersonal connection in our evolutionary heritage. At the most primal level is *emotional contagion and motor mimicry*, found in mammals and birds,[8] with startled herds or flocks, for example, or shared crying among human infants. At level 2 there is *sympathetic concern and the coordination of shared goals*, found with particular mammals such as dogs, cats, and primates, plus the *active consolation* of an injured individual, shown by birds and apes, or reconciliation after a fight. Level 3 involves *perspective taking and targeted helping*, along with a "true imitation" of the other by the self, as exhibited by apes—although targeted helping has also been found with dolphins, whales, and elephants ("Putting" and "Russian Doll"). Basic empathic imitation can be seen in various primates with contagious yawning or scratching, resumption of eating when others do, even when an individual is full, and copying a peculiar walk ("Putting" 287). Although increased by positive relationships, empathy in apes and humans may be suppressed with strangers and defectors (291–92). It may even lead to schadenfreude and vengeance.

de Waal also defines, within the targeted helping of apes and humans, *three kinds of altruism*: a spontaneous altruistic impulse, learned altruism, and intentional, predictive altruism, perhaps involving reciprocal expectations ("Putting" 281). Altruism has internal benefits for the performer, such as oxytocin release in the reward system of a mother's brain when she nurses her child, which may encourage learned and intentional altruism. Yet, if there are intrinsic biological (short-term mood change) or extrinsic social (long-term pay back) rewards for the altruistic performer, they are not certain in advance. Spontaneous empathy and imitation come first, according to de Waal, from the deeper levels of our brain's animal heritage (281–82).

Jaak Panksepp and his colleagues offer a related model of animal legacies for a variety of human emotions (see Table 1.1 in Chapter 1). The seeking system can be found not only in the reptilian and fish brain, but also in the invertebrate brain, as shown with crayfish, who seek areas in their experimental tank paired with drug reward experiences: amphetamine, cocaine, and morphine (Huber et al.). Reptiles also exhibit the "primary-process" emotions of rage, fear, and lust (Davis and Panksepp 1948). More distinctive, mammalian, social emotions include *caring, sadness, and joyful play*—with human spirituality at a higher level of feeling (1948–49). This list of invertebrate, reptilian, and mammalian emotions might also be compared with Aristotle's mention of sympathy (caring) and fear, regarding theatrical "catharsis," and with the ancient Indian *bhavas* of anger, fear, love/happiness, sadness, humor, courage, awe, and disgust, plus the ninth *rasa* of peace, as the cathartic goal of mindful attentiveness to such feelings through theater.

Rodent playfulness and cooperation show the potential for *emotional contagion in our mammalian brain*. Mice treated with morphine are imitated by other mice that never received the drug but simply observe their highly active, pleasure-exhibiting behavior on multiple days (Panksepp and Lahvis 1873). Rats also "have a preference for socially positive partners and appear to modulate their play behavior accordingly when the behavior of their partner is discordant with their own." A rat will work with another to pull a baited platform toward the cage—even with an unfamiliar colleague in an adjacent cage—if the rat has been helped by another, but not necessarily the same colleague, in the past (Rutte and

Taborsky). This has been called "generalized reciprocity" and pleasurable social feelings may contribute to the effect.

In experiments with humans, too, emotional contagion and motor mimicry can be shown as the *unconscious* basis for sympathetic concerns, shared goals, and then reciprocal helping. Subjects react to happy or angry faces with similar feelings and facial expressions, even when shown such a face for just 30 milliseconds, faster than it can be consciously perceived, and when it is masked for a longer time by a neutral face, consciously seen (Dimberg et al.). Fearful, full-body gestures, by actors in still images (taken from videos), activate brain areas for a similar feeling in spectators, suggesting a mechanism for fear and flight contagion in humans—as well as the *vicarious* experience of emotions when watching actors onstage or onscreen (de Gelder et al.).

Also, being mirrored in facial and bodily expressions increases how much one *likes* the mimicker and one's prosocial behavior toward others, measured by money donated to a charity (Baaren et al.). With such mimicry and kinship feelings in ordinary life, not only the visual, motor, and intuitional signals of mirror neurons may be involved, but also *audio* mirror neurons, which activate when a monkey hears a sound for a certain action or sees it or does it (Kohler et al.). These studies suggest a neural mechanism for primal, often unconscious communication between humans, built upon the mother-infant ties of facial expressions, gestures, and nonverbal sounds—which also relate to *cinema* with its big screen faces and emotional soundtracks. In other words, the backstage agents in our brain theaters are *signaling* to one another, much of the time, without our consciously knowing.

So that we do not "lose ourselves" in such emotional, mirror-neuron mimicry, *self-other distinctions* are made by the brain. These are crucial for higher levels of empathy, as shown by the rat that is not just overwhelmed by a friend's pain but feels separate enough to help the other. This separate sense of Self, as a "first-person" perspective, involves your somatosensory (body-sensing) cortex, plus the frontal and inferior (lower) parietal lobes (Ruby and Decety, "How" 995). It then enables not only a "second-person" point of view, but also a "third-person" perspective. This goes beyond simulating the other's goals and feelings, like a stage or screen actor. It means observing both self and other, as if from the outside, like a spectator or narrator, (or the "critic/scripter" described in Chapter 5), which involves more of the left temporal lobe.

As infants, we begin to recognize ourselves in the mirror, during the first 2 years of life (Lacan, *Écrits* 75–81). But we also become *alienated* from that mirror ideal, which inverts and reduces us (by about 50%). It shows us in a more stable form than the multiple selves and urges within us. Our *internalized mirror ideal* and the various reflections of Self given by others, along with our mass-media identifications, repeatedly betray us—as our face and body age, or we fail to live up to a screen ideal, or we realize a new context in others' views of us, through gossip and conflict. Also, the mirror image itself can be overlaid with an inner, negative image, as happens with anorexia.

Striving to glimpse the other's view, and reflections beyond it, like a spectator of self and other, may be crucial to our personal development from primal empathy to detached self-awareness and a more complex sense of compassion. Through stage, screen, and life experiences, this might produce cathartic changes in the networks between people. Such a *catharsis* (and *rasa* refinement) in theater or daily life would involve each brain's frontal and left-cortical sense of a proper narrative Self, right-cortical self-other relations, limbic and temporal-lobe emotional drives, and the parietal lobe's (somatosensory) mapping of bodily perceptions, simulations, and actions.

Brain scans show that when one imitates the other, activity increases in the left inferior parietal lobe, but being imitated by the other activates the corresponding area in the right hemisphere (Decety and Chaminade 586). Taking a first-person perspective also involves the *left* inferior parietal lobe (Ruby and Decety, "Effect"). A third-person view, or perceiving actions as caused by the other, activates more of the *right* inferior parietal as part of the brain's "Who system" (Farrer and Frith). Joining the two viewpoints, a "shared neural network" of *self and other representations* involves the right posterior temporal and inferior parietal lobes, and their ties to the prefrontal cortex (Decety and Sommerville). Perspective taking, as a "Theory of Mind" (ToM) *evaluating the other's* beliefs, desires, and knowledge, activates the right temporal-parietal junction (TPJ). This integrates thalamic circuits (at the top of the brainstem) with visual, auditory, body sensing, and emotional, limbic areas, through reciprocal ties with the prefrontal cortex (Apperly et al.), especially the right PFC, with its reasoning about the other's feelings (Decety 151). Stimulation of the right TPJ can also create an "out of body experience" (Blake et al.). In these and other ways, the brain's *left* cortical hemisphere stages the Self (the ideal ego), as "interpreter" or war-room "General," and the *right* cortex stages various relations of self to other or an otherness beyond the self, involving body, emotion, cognition, and perhaps soul (Gazzaniga; Ramachandran and Blakeslee).

Psychologist Paul Ekman distinguishes between "automatic appraisal" of our relation to others, when an emotion is triggered as a rapid evaluation with a physical reaction, and "reflective appraisal," when one becomes cognitively aware of emotion, often in a more ambiguous situation (24–32).[9] Knowing our "hot triggers," and imagining scenes when they might occur, can lead to more reflective appraisal and emotional choices at such times, through our inner theater (32–34). "We can, in our own minds, rehearse and try out other ways of interpreting what is occurring, so that it doesn't fit our usual hot triggers" (34). This relates to two types of memories, remembering as a first-person *actor* or observing oneself as a *spectator* (73–74). The latter, with more perspective taking, along with imaginative rehearsals of reflective appraisal, may increase *attentive mindfulness* (as in Buddhist meditation practices): "we are able to observe ourselves during an emotional episode . . . and consider whether or not our response is justified" (75).

The ability of the inner theater to interpret real-life dramas, through perspective taking, reflective appraisal, and attentive mindfulness, as aspects of emotional catharsis (or *rasa* refinement) between left and right hemispheres, hinges a great deal upon relations to the mother and other primary caretakers *at an early age*. Such family relations set up the networks of Self and Other within the brain, involving personal "affective styles," though these may change through later life experiences, such as therapy and meditation (Davidson and Begley).[10] Research projects with monkeys and with humans, following them from infancy into adulthood, show that parental nurturing in both species affects later "behavioral problems" more than biological nature (van der Kolk 160). A long-term study of Minnesotan families on public assistance revealed that "the combination of vulnerable infants and inflexible caregivers made for clingy, uptight kids. Insensitive, pushy, and intrusive behavior on the part of the parents at six months predicted hyperactivity and attention problems in kindergarten and beyond. . . . The children who received consistent caregiving became well regulated kids, while erratic caregiving produced kids who were chronically physiologically aroused" (160–61).

We move in our lives from a self-other continuum as newborns to the infant's mirror-stage identifications with and alienations from an ideal Self, through preverbal, emotional mimicry with the (m)other—especially with her and others' expressions of love, playfulness, joy, weariness, pain, anger, or depression. While carrying within us particular *self-other experiences*, as networks in our brain's theater, we move to new social contexts, gaining various first-, second-, and third-person viewpoints, through inner simulations and conceptual theories of others' minds. In the theater of everyday life, this becomes a fractal *hall of mirrors*, with multiple reflections of inner and outer stagings. We take various actor and spectator positions to glimpse how our image of Self is like and unlike others' views, along with our differing views of many things. We imagine actions in our brain's theater, as actors and spectators, testing the desires of Self and others (and the demands of the collective Other) against our current bodily state—and against our actual efforts, in play or work (Blakemore and Decety; Decety and Grezes 9). Thus, each person's actor-spectator views the matches and mismatches of that inner theater through the many mirrors of life's outer theaters and its media representations.

HUMAN ANIMALS WITH MORE IN PLAY

Young mammals practice and refine their social skills through "rough-and-tumble" play, in a safe zone with adult feedback (Siviy and Panksepp 1828). Males engage in mock battles to rehearse for more serious competitions later, to find their *fitness role* in the social hierarchy. The mammalian brain's play circuitry involves many areas, including the amygdala's fear system and the ascending dopaminergic seeking system, plus endogenous opioids and cannabinoids (inherent versions of opium and cannabis), which modulate play in positive ways (1821). But humans *extend* such play-work of inner and outer, physical and emotional feedback networks, in a safe ludic zone with adult monitoring, to many arenas in and beyond childhood, from neighborhoods, schoolyards, and parks to gyms, videogames, and professional amphitheaters.

Our bio-cultural evolution has extended youthfulness or "neoteny," in the human physique as well as behavior, beyond that of other primates (de Waal, *Our* 240–41).[11] Like younger apes, adult humans have round heads and flat faces, while often showing a more playful curiosity and creativity. de Waal even states that our neoteny is more like that of the promiscuous bonobos, which retain into adulthood the rounded skull, high-pitched voice, and playfulness of an infant chimp—in female-dominant hierarchies, unlike other ape species.[12] Bonobos, like human females but unlike other apes, have frontally oriented vaginas (241). They use them, as well as other body areas, for brief, sexual exchanges throughout the day, every 90 minutes on average: female to female, female to male, male to male, and infant with adult (89–106). They also use tongues for kissing and engage in various forms of masturbation (which other apes also perform). "For bonobos, erotic contact mixes freely with everything else they do"—with sex used for reproduction, pleasure, appeasement, and affection (90, 106). Humans, however, *sublimate* such promiscuity into private or public aspects of romance, games, sports, politics, and the arts, according to a particular culture's social and market norms. With much less instinctual ordering of brain circuits and behaviors, humans learn through *extended play and spectatorship*, in various halls of mirrors on stages and screens, what their social roles and class interpretations might be, regarding gesture, speech, and action, as legitimate or perverse performers.

Whether rough-and-tumble, sexually promiscuous, or creatively productive, the human brain's inner and outer theaters—of play, learning, and work—involve all three levels of empathy in de Waal's model. Level 1, emotional contagion and mirror-neuron mimicry, happens mostly unconsciously, for performers and spectators in all such interactions. Level 2, sympathetic concern, coordinated goals, and acts of consolation (or reconciliation), occurs between players, teams, and fans. With social contests, bounded by game or work rules, sympathy and coordinated behavior are channeled into team alignments, for and against each other. Yet, consoling feelings may occur across that divide when a loser is displayed in sports, social media, or tragic drama. There is also coordination between opposing sides, in following rules and offering a challenge for the other. Level 3, perspective taking and targeted helping, occurs in team play. But it may also occur across the friend-enemy divide, with a spectator's empathy for the other side, through a third-person perspective that involves an inner simulation and yet self-aware distance.

This might relate to the ancient notion of "catharsis" in theater, but *not* in the sense of just venting or purging emotions. Instead, watching performers in a play, movie, or game, competing within its limits, may evoke a greater awareness of fears and desires (and a refined *rasa* of various emotions), "clarifying" the collective dangers of empathy, rage, and vengeance—in the spectator's own behaviors. This involves all three levels that de Waal describes, from mammal to primate to human. *Rasa*-cathartic "perspective taking" at level 3 also produces changes in the animal legacies and early life influences of our inner theater's circuitry, affecting our social interactions and, to some degree, our culture's future evolution.

CROSS-CURRENTS OF EMOTIONAL IDENTITIES

Empathy with struggling performers evokes a sense of pride, duty, sacrifice, teamwork, and fairness. But empathy for victims can also spur vengeance against others. The adult play and work of professional sports rehearses such in-group and out-group divisions, sometimes on an international scale. Tens of thousands of people in a stadium, or hundreds of thousands, perhaps millions, watching on TV, identify with players performing individually or as one team against another. Mirror neurons fire, fueling emotional contagion, sympathetic cheers, vicarious coordination of goals, and the targeted helping of noisemakers, signs, costumes, makeup, and elaborate chants or dances (even by fans watching through the TV screen). But this may also involve rage against opponents, fans fighting in the stands, and plans for revenge, rhetorically and physically, within the rules of play or beyond. How do empathy and antipathy flow between us—in the theaters of sports, mass media, or everyday life?

Animals often perform collectively for coordinated work, safety, or hunting, in what we call hives, swarms, schools, pods, flocks, herds, and packs. Songbirds, for example, flock together, moving like one organism, for alertness in flight and safety against predator birds. Mating also requires togetherness. When a male songbird is perched alone, the low opioid levels in its brain trigger dopamine-mediated reward-seeking behaviors, such as repeated vocalizing for the female, like the calls of an infant mammal in separation-distress mode, seeking its mother (Riters 1842).

In humans, too, the lack of social contact can trigger a degree of panic, drawing on the neural traces of specific experiences, especially in childhood, of ego alienation and fragility. But that can be remedied with communal or romantic activity, stimulating an opioid (pain reduction) rush within the brain. Yet, the bottom-up networks in the human body and brain, for emotional sharing and physical connection (or mimicry), are also inhibited by top-down controls from the prefrontal cortex (Decety and Lamm 1159–60). This may involve negative criteria for selective attention, through perceptions of unfairness, ugliness, bad reputation, or enemy attributes—restricting empathy for pain. Or there may be a top-down enhancement of *seeking and communing circuitry* when the other is idealized as beautiful and good, or felt to be extended kin as fellow fan or countryman, regarding survival and reproduction drives.

Many of us enjoy being alone. But even when we are the only person in a room with the doors closed and windows covered, we are performing externally and imagining ourselves internally (like I imagine this "we" now)—in relation to others in our lives and the Other of our social norms, as if God or a video camera might be watching. This involves memory traces in our brains left by prior social interactions, especially the foundational circuits pruned by early childhood experiences.

Yet some people panic when others are around. Such *agoraphobia*, like its opposite, may be a part of all of us at a primal level: in our infantile experience of the (m)Other's presence as overwhelming at times for our fragile, budding ego. In adulthood, the Self may oscillate between being overwhelmed by contagious emotions involving crowd pressures, especially through our seemingly omnipresent screen media, and a painful, low-opioid loneliness, even in a crowd.

From childhood on, we develop "defense mechanisms" (involving the left ventromedial prefrontal cortex) to bolster a fragile ego against the threats of emotional contagion and isolation. Even at just a year old, some babies show a top-down neural inhibition of primal instincts, avoiding contact with the mother after separation from her, "to defend against the presence of a caretaker who . . . evokes unpleasant emotions" (Cramer 640). Researchers categorize babies as "avoidant" when they ignore the caregiver's presence or absence, as "anxious" when they get distressed at the caretaker's absence and are difficult to console, as "secure" when they attach to a caretaker and react to her absence, yet are quickly consoled when she returns, or as "disorganized" when they are inconsistent in such reactions (Gopnik 181–83). A culture's stress on *independent* egos or *interdependent* communal connections may influence this. German babies tend to be more avoidant than American ones (182). Japanese babies tend to be more anxious. But such terms also reflect a Euro-American bias toward independent egos. As researcher Alison Gopnik puts it: "An avoidant baby might do well on the playing fields of Eton, where hardly anyone expresses much closeness, and an anxious baby might thrive in a small African village where hardly anyone is ever alone." Such research also relates to the mammalian instincts of panic and then depressive grief, of anxiously seeking the parent with cries for help and then avoidantly hiding from predators—a defense mechanism shown by the rat pup when separated from its mother, as considered in Chapter 1 here. (The four attachment styles are also considered further in Chapter 7.)

Each family is a subculture with degrees of *ego defensiveness and communal attachment*, producing positive or negative effects at different levels of developing personalities. Thus, mothers may unconsciously repeat the "attachment style" instilled in them as children, or repeat the abuse they suffered, with their own children (Cramer 640). Research shows that children who perform with "extremely positive [attitudes] . . . are often denying or defending against an underlying sense of imperfection." Yet a child using ego defenses after a traumatic event may be "protected from psychological upset." According to various studies, defense mechanisms increase with status crises or identity threats, especially when central to self-representation. Women with "a strong feminine gender identity" are more likely to use typically feminine defenses, such as "turning against the self," whereas men with a strong masculine identity use male defenses, such as "turning against the object."

According to the *Diagnostic and Statistical Manual of Mental Disorders* (DSM 4), ego defense mechanisms range from highly "adaptive" (with humor, altruism, and sublimation) to "image distorting" (with idealization, denial, projection, and rationalization) to apathetic withdrawal and passive-aggressive behaviors, to "dysregulation" with delusional distortions (Cramer 642). And yet, *image-distorting projections* can be evoked in anyone through specific experiments (explored further in Chapter 5). Simply touching a subject's face, while she sees a stranger's face being touched in the same way at the same time, produces a "self-other merging" when she is asked to rate herself and the stranger on a series of issues, involving inner states, attraction, closeness, positive feelings, and conformity behaviors (Paladino et al. 1203).

Along with this *social* overlap of self and other, through the "projection" of self and the "anchoring" of one's behavior in the other, *body* merging also occurs (Paladino et al. 1206). After a simultaneous touching of faces, the subject judges the stranger's face as more like her own than before—in a computerized video spectrum that gradually mixes self and other faces, from 100% self to 50/50 to 100% other (Tsakiris). Thus, the self-other overlap "is typically experienced with close others and in-group members," but it can arise "from a purely sensorial experience" (Paladino et al. 1206). Mimicry, through "embodied processes," forms the basis for "self-other overlap and social identification." Adolescent brains, in particular, struggle to mix various identifications with others into "a stable personal identity" regarding social mores and roles (Wexler 127). But for adults, too, this involves "parts of the self," stemming from early identifications in childhood, "projected out onto the world" (128). The world is then "perceived and/or experienced as if it had these properties."

Projection of one's facial features and social attributes on the other, while also incorporating the other in oneself, occurs with *positive*, in-group bonding. But projection can also be defensive, with *negative* aspects of oneself, often repressed and unacknowledged, attributed to the other—in order to protect the ego when the other becomes an uncanny reminder of one's own flaws.[13] An inner chorus of memory traces and imagined supporters may form a protective in-group for the ego, as it disassociates from the other as a threat. Groups may do even worse as they *stereotype and scapegoat* certain people—while projecting characteristics on them that come from their own self-disgust, in a *hypocritical* attempt to purify the in-group.

Research by Daniel Ames shows such positive and negative projections in the process of "overgeneralized similarity" and dissimilarity. When one person finds similarities with another, she readily expects that they have "most everything (including a wide range of beliefs, desires, and feelings) in common" (164). Dissimilarities are assumed as well, from one to many, producing "empathy gaps" (Van Boven and Loewenstein). "When such a gulf opens between self and other, stereotypes are likely to rush in" (Ames 164). Experiments reveal that unconscious stereotypes even apply to nonhuman objects (Reeder and Trafimow 107)—though this may be influenced by cultural contexts. A study in 1944, during World War II, found that subjects presented with a short film of a large triangle moving near a circle saw it as a "bully" that was "chasing" it. The spontaneous projection of motives onto another person also affects initial impressions, trait attributions, and predictions about that person—from obedient to helpful, for example, or selfish to unhelpful (109–10). Even *unconscious*, "implicitly activated" goals can be projected on the other, as well as "explicitly assigned" goals (Kawada et al. 556).

Psychologists Glenn Reeder and David Trafimow distinguish "simple projection"—involving automatic, "egocentric" attribution of one's own physical needs, motives, knowledge, or goals to the other—from conscious, "effortful perspective taking" (114–15). Children as young as four are able to perform effortful perspective taking. Yet even as adults, we often rely on *automatic projections*. "The evidence suggests that an effortful version of simulation may not be the default process for inferring motives but instead takes place only under ideal circumstances" (115). Also, while *effortful perspective taking* "is certainly a more sophisticated and potentially accurate way to infer motives," it can be *biased* by preconceptions or by the perceiver's own goals (116, 119). For example, "naïve realism," which is more apparent in children before age four, may persist into adulthood, with egocentric projections biasing the effortful perspective taking (119). "If others fail to see things similarly it is because they are lazy, irrational, or biased by ideology or self-interest." Thus, even knowing about simple projections, as biased by ideology or self-interest, can fuel such projections—of one's own faults—on the other. "Naïve realism suggests that when another person (or group) disagrees with our opinions we tend to see the motive of self-interest at work ('They have an axe to grind')." Such a projection was found in a survey of Americans and Canadians about whether they supported the 2003 war in Iraq: both sides, for and against the war, saw the other as guided by self-interests rather than ethical principles (119–20). But projections can also be revealed by ironic twists in beast-people movies like those considered later in this book.

Biased attributions of motives can happen especially when "identity concerns and self-esteem needs" of the individual or group ego are evoked (Reeder and Trafimow 120). For example, "self-esteem needs may lead us to attribute positive motives to a subordinate who flatters us, whereas an observer (who overhears the flattery) might attribute an ulterior motive to the ingratiator." I would add that theater, in its various stages and screen forms, may help us to observe such flaws in our animal and cultural natures, particularly with the more *complex, effortful perspective-taking evoked by tragicomic ironies,* shifting spectators' projections toward a greater awareness. Gradually, this might change our frontal and limbic networks with cathartic "mindfulness"[14] about the dangers of projections in real life—although the stereotypes evoked by melodramatic theater may reinforce them instead.

OTHER AND SELF DECEPTION, FORMING THE SELF THROUGH THE OTHER'S VIEWS

With dating and married couples, researchers find that "egocentric" projections of similarity, even when wrong, lead to happier and more stable, romantic unions (Murray et al.). Each person in the couple may think she or he is concerned with the other's desires, but actually projecting one's own on the other. This may be an example of what primatologists call "Machiavellian intelligence" in our animal heritage—deriving from an allocentric (other centered) mechanism as well. Within the hierarchy of primate groups, the survival and reproduction of a certain genetic pattern benefits from the individual avoiding fights with higher-ups, deceiving others when food is found, and finding sneaky ways to mate outside the sightlines of controlling alpha males, by false signaling or withholding information (Hauser 114). Primate intelligence evolved toward the tremendous expansion of the human brain through such social complexity: by manipulating others for personal gain via "deception," as well as by "helping and co-operation" (Byrne and Whiten 12).

Machiavellian intelligence may be a primate legacy in our brain's theater because "functional deception," such as false signaling, was selected and perpetuated as a successful competitive behavior (Hauser). It was reinforced as a learned behavior in our mammalian or earlier invertebrate ancestors simply through trial and error. This evolved toward "intentional deception," which is guided by conscious beliefs and desires, including the "misrepresentation of belief states" in more intelligent primates (113). Thus, the *self-deception* of our ego defense mechanisms, such as repression and projection, developed through the increasing social complexity of large primate groups. According to biologist Robert Trivers, self-deception evolved because the truth is less likely to "leak" out unconsciously when an individual deceives oneself along with others (Miller 322). Types of self-deception in humans include denying responsibility for harming others, exaggeration of helping others, illusion of consistency from past to present, perception (or projection) of assumed relationships, self-defensiveness, and self-satisfying vigilance (Trivers 418–20).

On a group level, such egoistic deception, bolstering the self through trickery, relates also to the "protean" uncertainties that evolve in social hierarchies (Miller 322). If an alpha male in a primate group punishes others at a consistent threshold of their misbehavior, then subordinates might misbehave (such as grooming his females) with increasing success just under that level, tricking the alpha and creating alliances against him (323). But if the alpha adopts a trickier "Mad Dog" behavior, punishing others whimsically—like a human despot—then the terrible uncertainty of punishment functions to control underlings and increase their stress, keeping the alpha in power (323, 328).

Such "social proteanism" in our primate ancestors may have evolved not only through alpha male anger, but also female *lust* (Miller 324). The tactic of females mating with several males while ovulating, along with the biology of *concealed ovulation*, creates uncertainty in males as to which offspring are theirs and decreases the likelihood of *infanticide* (of a male killing the female's current offspring to get an evolutionary advantage for his own genes when he mates with her). "Most female primates in multi-male groups seem to use either protean promiscuity or concealed ovulation to protect against infanticide," but humans use both (Buss).

And yet, there are evolutionary benefits for females, as well as males, in *long-term pair bonding*: frequent sex and greater emotional and physical security. The latter is especially important for reproductive success in our big-brained species because of the prematurity and helplessness of the infant at birth (Linden 149–52). "[M]onogamous, mostly recreational sex has two effects: it gives a high probability of accurately knowing paternity of the resultant offspring and it helps to reinforce a lasting pair bond, both of which promote continued care of the offspring by both parents" (151). There are also dangers to promiscuity, especially for females: not just pregnancy, but also sexually transmitted diseases—rendering a third to one-half of women permanently infertile in certain African cultures where premarital and extramarital sex is encouraged (Baumeister 355–56). In both tribal and modern societies, women often restrain one another's sexual behavior, so that sex is not readily available to men and its "price," in gifts and commitments, will remain high (358).[15]

Collectively, as with couples, a tension evolved in our evolutionary ancestors between *trickery and loyalty*. Unpredictability in generosity, grudges, or forgiveness might have lessened freeloading and cheating, increasing group cooperation and a variety of personality types (Miller 324–27). But it may also have inhibited learning, blocked cognition, and overloaded memory. With group competition in our primate relatives and in many theaters of conflict today, the collective benefit of social stability provides a counterbalance to protean uncertainty and its egoistic tensions. This is because "selfish individuals beat altruistic individuals [yet] . . . groups of altruists beat groups of individualists" (E.O. Wilson, *Social* 243).[16] Dictators, freeloaders, and rebels might compete, within a group, for selfish, genetic gains. But with larger and larger human groups competing against one another, during the course of our bio-cultural evolution, *altruistic cooperation within groups* also survived and was creatively reproduced in various ways.

Indeed, outside threats to a group may bind individuals toward more cooperation—as in the "United States" after 9/11—though also toward the sacrifice of individual rights within the group, as more power is given to its controlling leadership against the common enemy. With the extremes of a Mad Dog dictator, the unpredictability of beneficial relations can lead to friction in the social order and alliances against him—as seen in the recent "Arab Spring" revolts in Tunisia, Libya, and Egypt. Ironically, oppressive social controls may spawn *greater creativity* for those rebelling, as when theater artists work against censorship to communicate in new ways, or people use social media to coordinate political demonstrations.

Competitive, deceptive behaviors in our primate ancestors, within protean, hierarchical groups, may have been a "crucial randomisation mechanism" in the "evolution of human creativity" (Miller 329). Sexual selection probably played a key role as well—with human females choosing mates with greater creativity, intelligence, and novelty displays, while males used such skills to woo or control them, as well as control and trick one another, and then such traits were reproduced in their offspring (330–32). Yet, *protean flexibility* in social orders, propelled by the "neophilia" (love of newness) in mating choices, also involved an interplay with "ritualisation," offering communal patterns of behavior and altruistic stability (332). Nonhuman animals have instinctual rites of behavior. But with our human ancestors, Machiavellian deception, neophilic inventiveness, and ritual orders of transcendent meanings had a "runaway" effect. The bio-cultural evolution of our big brains and social groups involved a tremendous variety of performance practices through our species' reproductive success in the theaters of sexual and environmental selection around the globe.

Research has found "signs of deceit in children's everyday social life," *as early as age two and a half* (Whiten 155), involving right frontal cortex activation (Keenan 226–27). If children are put in "the role of the deceiver" in a story-telling experiment with puppets, they will "remove tell-tale cues, and even lead false trails, to mislead a protagonist about the location of some hidden reward" (Whiten 155). At three, they will make accurate predictions about a character's lack of knowledge about a hidden object—if they are "*actively involved* in the deceiving role." As mentioned earlier, similar experiments show that this realization about the other's perspective does not occur until age four if the child is in a more passive role. So deception of others seems key to the development in children of a "Theory of Mind" about the other person's distinctive viewpoint and perhaps to our evolution of that ability as apes.

At age three, children are not as good at using deception for personal rewards as they are when a bit older. They are likely to deceive, by laying false trails, even when they could benefit from truthfully informing a partner (Whiten 155). They seem to act with "pretense," like actors on a stage, rather than with the intent to deceive. Three year olds appear to "create an as-if scenario and invite others to join in with them" (Sodian 398). Even 2-year-old infants "can indulge in pretend play, yet not grasp [the other's] false belief" (Whiten 156). This may be akin to other apes who exhibit "true pretense," regarding food or sex when a competitor or superior is watching, perhaps as an "implicit theory of mind" (156, 161).[17] Studies show that as early as age one infants may have an "intentionality detector" attuned to the desires and goals of others (prior to an awareness of their own), like non-human primates, forming the basis for later aspects of "mindreading"—in primate evolution and human development (164–65, 168). The Other's desire shapes Self-awareness in the neural feedback of mimetic mirroring between the infant and its caretakers, especially in the right frontal cortex (Keenan). As adults, too, we have a "sociometer," in the brain's theater, monitoring *self-esteem* as "relational value" through others' apparent views (Leary 374–79). The cracks and shards of such mirroring between brain theaters, from child to teen to adult, through the evolution of human proteanism, trickery, and sacrifice, are shown with various types of beast-people onscreen, as considered in Chapters 4 and 6.

OXYTOCIN TRUST AND SOCIAL CATEGORIES

With an inclination toward *pretense*, at play with the Other's desires from very early in human life, one's ego identity becomes built upon Other and Self deceptions, through nurturing experiences and alienating illusions. But *trust* is crucial to the balance between stability and proteanism—in our inner and outer theaters. Oxytocin, a neurotransmitter and hormone, acts in the brain and body to increase maternal functions in women, such as labor contractions and breast milk secretion, while activating dopamine pleasure networks in the midbrain (Zak). Oxytocin also increases trust levels in both women and men, and is known as the "cuddle hormone" because levels rise during sexual intercourse (89–90).

A related protein, vasotocin, which evolved in fish 100 million years ago, helps with reproduction by reducing the female fish's *fear* of the male when she is ovulating (Zak 89). Vasotocin evolved in mammals as oxytocin and the related hormone vasopressin. In two species of vole (a type of rodent), oxytocin and vasopressin are concentrated in distinct areas of the brain—in the midbrain with prairie voles, which mate monogamously and have fathers caring for the young, or elsewhere in the brains of montane voles, who are promiscuous and solitary (89–90). With humans, studies show that inhaling an oxytocin nasal spray increases trust between adults, even strangers. Oxytocin may have evolved in humans to facilitate our extended years of child care (more than double that of chimps), plus attachments beyond kinship to mates, friends, and social groups—or to places and objects, such as our homes and cars today (91). The anterior cingulate, nucleus accumbens, hypothalamus, and amygdala (areas above the midbrain) are involved in the release of and response to oxytocin. Studies also show that the amygdala, your brain's fear and rage center, is activated when you are judging the *trustworthiness of someone's face* (Grèzes et al. 5502). The superior temporal sulcus, anterior cingulate cortex, and orbital frontal cortex are active in experiments with judging *social deception*—when an actor holding a box shows how much it weighs through nonverbal cues without facial expressions.

Testosterone increases in men, though not as much in women, when they feel *distrust* from another, showing the greater tendency of men toward an "aggressive response to being distrusted" (Zak 92). Also, the female hormone estrogen increases the uptake of oxytocin by the body's tissues (95). Such complex interactions with other hormones and neurotransmitters suggest that "life experiences may 'retune' the oxytocin mechanism to a different 'set point' and thus to different levels of trust throughout the course of life."

In judging another person's trustworthiness, we often apply a "social category," which then becomes "a lens" that molds impressions and distorts memories, affecting unconscious mirroring and reactive behaviors (Freeman et al. 673). For example, activating the social category "Elder" *primes* a subject to walk more slowly. The category "Black" can cause nonverbal hostility. The category "Professor" can increase achievement on a general knowledge test. Such social categories may be triggered unconsciously through perceptual cues and "bottom-up" brain mechanisms (674). Once prejudices are applied—prejudging the behaviors of self and other—we tend to "anchor" perceptions of those *like* us in projections of our own characteristics and split off the opposite expectations for those not like us (Cadinu and Rothbart). Thus, *in-group and out-group binaries* form through a need, from childhood onward, to frame our fragile ego identities with social categories, often projecting unacknowledged problems in one's own group on others as scapegoats.

But social categories, such as race, may also be perceived in varying degrees, along a *spectrum of group distinctions*, rather than as all or nothing—through a "cascade" of partial associations (Freeman et al. 674). In this "dynamic continuity account," neural networks of social categories compete for a dominant framework, as in Baars's cognitive model of the brain's theater. Thus, "multiple social category alternatives (e.g., Male or Female) are simultaneously and partially active, continuously competing for activation while perceptual evidence for alternatives is gradually mounted." Our animal ancestors survived and reproduced, especially as emotion-driven, social-learning mammals in hierarchical groups, by identifying others around them as male or female, kin or not kin, friend or foe, and of higher or lower status. Binary categories are stressed in today's *melodramatic formulas* of movies, television shows, and videogames, with heroes versus villains. But behind such stereotypes, there are many alternative associations that may be less deceptive and more complex—in the fractal, emotional hall of mirrors between our brain theaters.

Even the basic, evolutionary distinction of cooperative partner or competitive opponent has many shades in love, war, sports, or tragicomic, stage and screen theaters. Lovers may idolize each other, as they cooperate toward shared goals, yet also at times oppose each other—with both aspects involving self and other deceit. Warriors may demonize and objectify the opponent, but with admiration, envy, and fear (or compassion) leading to other projections as well. In sports, teammates may bond together, like warriors on the battlefield, yet compete against each other for playing time or starring roles. In various theatrical media today, and in the theaters of everyday life, different degrees of *cooperation and competition* are involved, through shifting social categories and their projections. This will be explored in the next chapter through specific films with a lone vampire or werewolf—or with packs of them—which illustrate, in supernatural ways, the mammalian heritage of our brain networks.

MERGED IDENTITIES, MIMETIC DESIRES, AND THE WHITES OF OUR EYES

How does the hall of mirrors between brains and human groups shift from cooperative idealizing and communal merging, with deceptive yet positive projections, to competitive demonizing and sacrificing of the other, as an enemy or scapegoat, as a threat to one's body, ego, and group survival? For over a half century, social psychologists have explored "behavior matching" in the mirroring of actions, when "the observer feels a sense of shared identity with the actor," regarding the "self-concept [as it] guides and develops behavior" (Goldstein and Cialdini 402). Even earlier, over a century ago, sociologist Charles Cooley gave the term "looking-glass self" to reflective appraisals, with *self-definitions shaped by others' judgments and responses*, at least as imagined by the actor. Current researchers Noah Goldstein and Robert Cialdini have used this legacy to study the "spyglass effect" of perceiving *one's Self through the Other's views and reactions to similar circumstances*, with the Other as spectator and actor, in the vicarious "merging" of self and other identities. They find that subjects not only project their own attributes on others like them, but also take on the characteristics that the "merged other" is modeling, such as helpfulness or "self-sacrificing actions," especially when given "attachment-related cues" to evoke kinship (406). This suggests how individuals also *merge with a group identity*, changing their distinct perspectives to conform with the collective Other's, while projecting negative categories on scapegoats or outsiders, as threats to group unity.

I would argue that in complex forms of theater, combining the ideas of Aristotle and Bertolt Brecht (although Brecht called his theories anti-Aristotelian), the looking-glass and spyglass effects of judgment and behavior matching might involve a further stage of oscillation. Through shifts in the spectator's "shared identity with the actor," a cathartic, alienation effect might be evoked through the *merging* of Self and Other, yet also an awareness of such mimicry at critical moments of *distance* between them. As Goldstein and Cialdini found in their psychological experiments, subjects' perceptions of the "merged other's attributes" were *not* stable over time and context, as previous research had assumed. Instead, Self perceptions "change as a result of observing how merged others behave in response to specific situations they encounter" (414).

Another group of researchers, basing their work on that of literary anthropologist Rene Girard, describe an inherent "mimetic" mechanism in the human species. "Imitation is at the basis of social harmony and rivalry. It is a more powerful drive in humans than in any other animal because humans are not dominated by instinct" (Anspach 131). For us, "the other is already a constitutive part of the self" with the further evolution of our mirror neuron systems via cultural learning (Gallese, "Two Sides" 102–3). Nonhuman apes, through emotional contagion and functional deception, might react to others' motivations, perceptions, intentions, goals, and knowledge, but "they do not understand that others have mental representations of the world" (Kruger 117).

Humans, at a young age, "ape" more than other apes. Young chimpanzees will copy an adult's use of a new tool, emulating the goal with it, but leaving out or changing unnecessary steps. Yet a child will faithfully copy each step, including ones that are mechanically irrelevant, over-imitating the ritual "even when doing so is costly" (Lyons et al. 1158). This indicates the distinctive source of our species' "artifact-centric" cultures, from tribal to modern and mass-media varieties (1166). It also shows "the powerful motivation of children to be with others in a special way—to act and think and feel with them, even if the short-term gain is not evident" (Kruger 117). Nonhuman apes do not enter into "joint attention" like children, who show a toy to an adult for the joy of experiencing it together, or perform for the attention of the Other, even by misbehaving, thus demanding that a parent watch. This suggests that "the action itself is not satisfactory, or even real, unless the other is sharing it"—in a theatrical way.

The *theatron* of gaze following and performing for the Other, with some degree of kinesthetic awareness, evolved at least 2 million years ago, in the "mimetic" stage of our bio-cultural evolution (Donald). Today, humans are the only primates with no pigmentation in the sclera (white part) of the eyes. Our sclerae are also three times larger than in orangutans, with which we share a common ancestor 14 million years ago (Tomasello et al. 315). The white sclera increases our ability to follow others' gazes, even at a distance, showing the significance of human eye contact, joint attention, and gaze communication in our ancestors' survival (Kruger 120). It will be significant in our exploration of recent *Planet of the Apes* movies in Chapter 6.

Human children share food and information, and join in cooperative ventures, more readily than young apes.[18] As our ancestors evolved to live in larger and more complex social groups, the adaptive advantages of cooperation extended ape "social learning" toward human "cultural learning." This involved an expert and novice, or even novices together, actively teaching and inventing through a "communion" of viewpoints, with a "representation of the other person's perspective internalized by the learner" (Kruger 120–21).

There is also a dark side to our mimetic drive. We like to think that merged identities, from teacher and student to sports teams to romantic lovers, always lead to better cooperation. But mimetic theorists, citing various experiments, argue that the mirroring of desires between individuals may produce *rivalry*, not just harmonious goals, as when two men desire the same woman (Anspach 130). When rivalry produces injury and then revenge, *reciprocal violence* can occur between the parties, back and forth like a family feud—until a substitute victim is found and a scapegoat ritual instituted (132–35). Such mimetic violence is mobile and contagious. It tends to "propagate from one actor to another," shifting its target (142). Like other civil hierarchies, religions may prohibit violence, but they also channel it, from primal sacrificial victims to symbolic communal rites and overt "holy wars" (132). Whether religious or secular, groups are susceptible to "enmity contagion" with a splitting of "us versus them" and the emergence of a leader who hypnotizes members into devoting their lives to the fight against other humans as a threatening pseudo-species (Konner 164–68). Even when rivalry within a group is reduced through a common goal like this, violence may erupt against outsiders, through insecure egos and their in-group ideals.

GROUP VIOLENCE THROUGH MIMETIC, OXYTOCIN RIVALRY

In Stanley Milgram's experiments of the 1960s, subjects were ordered by an authority figure in a lab coat to give electric shocks to someone unseen and most did as they were told, even when hearing the victim's cries, when the victim claimed heart trouble, and when the victim pleaded to be released (Konner 168). Peer encouragement increased the shocking obedience—although many subjects resisted or expressed discomfort when complying. In another experiment in 1971, Philip Zimbardo had two dozen male Stanford University students, who were selected from seventy as the most psychologically healthy, play the roles of prison guards and prisoners. They performed so fully, merging with their abusive and rebellious roles—as prisoners rioted on the second and subsequent days, and the guards responded with physical and psychological punishments—that the experiment was ended early, after six days, because of the brutality. A female graduate student whom Zimbardo was dating (Christina Maslach) objected on moral grounds, persuading him to stop the experiment, but she was the only person out of 50 observers who did so (Zimbardo).

Mimetic theorist and neuroscientist William Hurlbut argues that group aggression is self-perpetuating for at least two reasons: (1) it gives individuals a sense of social approval, which "becomes the dominant imperative of personal fulfillment" and (2) "aggression feels good" via oxytocin and dopamine systems in the brain (187–89). In nonhuman mammals, "natural mechanisms of submission" (such as vasopressin) inhibit the acceleration of aggression, but humans do not show such "preestablished programs" (189). Yet patterns contributing to human aggression can be found. A cross-cultural study revealed that the more violent human cultures tend to segregate men and women, whereas less violent ones are matrilocal or have strong husband and wife relationships (Konner 163). It can thus be argued that "the social dynamic of male groups fosters violence," as with urban gangs and terrorist organizations.

Likewise, chimpanzees form all-male "power coalitions" in the wild (Boehm 345). They go on regular patrols of their territorial boundaries, "attack and kill neighbouring males or females with infants, acquire females ready to breed, and bluff against enemy patrols." But chimp societies are mostly organized in a male "social dominance hierarchy" (361). Humans, on the other hand, have many different types of societies, sometimes creating "stable, permanent whole-group anti-hierarchical coalitions" in egalitarian systems that use "ridicule, ostracism, and execution as tools of counter-domination" (355). Mimetic desire and rivalry[19] can lead to reciprocal vengeance, but it can also be a means to dislodge rulers, through anti-hierarchical alignments—such as workers' strikes and street riots.

Oxytocin, which functions as a "cuddle hormone" in the physical bonding of lovers and of parents with children, can also contribute to *mimetic rivalry*, especially in the competition between groups for dominance. An Israeli study found that individual players of a simple videogame, who were losing or winning against an imaginary opponent, rated their feelings of envy or schadenfreude (gloating at the other's misfortune) as higher if they were given a nasal spray of oxytocin (Shamay-Tsoory et al.). The researchers proposed a more basic effect than trust, envy, or gloating: "the oxytocinergic system is responsible for modulating the salience of social agents in social contexts. As a result, the administration of OXT [oxytocin] may provoke a wide range of emotions and behaviors related to social behavior and parenting, such as trusting collaborators, attacking potential intruders, and competing with rivals" (870). A review of various studies found that a high level of oxytocin in subjects' blood plasma was associated with trust, supportive communication, positive parenting styles, and yet interpersonal distress and major depression (Bartz et al. 307). Thus, oxytocin may have a downside, involved with prosocial behavior, yet also "interpersonal sensitivity." Context and personality indeed matter in how this social-salience neuropeptide affects the nervous system.[20]

By whetting mimetic meaningfulness, oxytocin increases the perception of fear in another's face (Fischer-Shofty et al.) or of happiness or anger, even unconsciously, when the other's face is presented in a fraction of a second (for 18, 35, or 53 milliseconds) and "masked" by a neutral face before and after it (Schulze et al.). Another study showed that oxytocin increases the "mind-reading" of emotional states and subtle social cues by male subjects, just in seeing the other's eyes (Domes et al.). It promoted *in-group favoritism* and out-group derogation in a test of ethnocentrism in Dutch males, regarding German and Middle-Eastern others (De Dreu et al., "Oxytocin Promotes"). And yet, in another experiment, oxytocin increased the willingness of males to align with others who look highly threatening, but may be useful as strong teammates (De Dreu et al., "Oxytocin Modulates"). It also decreased the aversion of males to angry faces in a financial reward, associative learning game (Evans et al.).

With non-human mammals, oxytocin increases *territoriality and aggression* (De Dreu et al., "Oxytocin Promotes" 1262). In male humans, it reduces "the willingness to sacrifice in-group targets to save a larger collective but not the readiness to sacrifice out-group targets" (1264). Another study showed that oxytocin increases the male "tend and defend" impulse of self-sacrifice, or "parochial altruism," involving in-group love and trust, plus defensive aggression toward out-group threats (De Dreu et al., "Neuropeptide" 1408, 1411). It also promotes the "tend and befriend" impulse of women, but this might relate as well to maternal aggression in animal studies (Bosch et al.; Campbell 7). Thus, the same neuropeptide that promotes romantic love and parent-child bonding may increase in-group kinship and out-group aggression, through happiness, fear, and anger contagion—involving emotional mind-reading, camaraderie, and self-sacrifice, yet also envy, gloating, and rivalry with others. People who experience neglect and abuse as children may also gravitate in later life toward gang loyalty, authoritarian leadership, and violence, through the oxytocin system in our brain's mammalian heritage (Pedersen 107). Such paradoxical extremes of oxytocin effects can be seen in the nurturing yet destructive drives and animal-emotion scripts of beast-people (and their creators) onscreen.

LIKING OR WANTING, SACRIFICE AND ADDICTION

Neuroscientists make a distinction between liking and wanting[21]—which are deeply connected in our experiences of love, camaraderie, addiction, and sacrifice. *Liking*, in the double sense of consciously enjoying subjective pleasures and feeling kinship with another person or group, involves oxytocin systems for mimetic, social meanings. The brain network for "liking" includes the ventromedial prefrontal (a.k.a. orbitofrontal) cortex, "which mediates planning, evaluation, impulse control, and empathy" (Hurlbut 184). It also involves the insula, which is "crucial for body representation, regulation, and subjective emotional experience"—as well as the "mirror neuron-like connection" of external and internal experiences regarding Self, admiration of others, and shared social emotions.[22]

The insula is key to the "disgust" reaction (one of the *rasas* of ancient Indian theater theory), when the brain responds to images of contamination and mutilation. This is the opposite of "liking" and may relate to the fragmentation of identity and bodily perception that lies behind the insecurity of Self, especially as the ego initially develops, through sociality, in the "mirror stage" (Lacan, *Écrits* 75–81). As Hurlbut puts it, with reference to Girard who was influenced by Lacan: "This triangular connection between self, other, and images of mutilation may mediate the personal and social transformation from chaos to order effected by sacrifice of the sacred victim described by Girard in his theory of the scapegoat mechanism" (184). Beast-people movies often focus on such odious scenes of *mutilation*, involving victims of the monster, but also the monster itself as a scapegoated victim of personal, social, and physical transformations, between order and chaos, like and unlike us.

Wanting, as "active anticipation," involves the brain's dopamine network (and Panksepp's seeking system), which turns a mere bodily sensation, produced by a repeated, habitual or ritual action, into a "motivational magnet," confirming "reward learning" through attention and arousal systems (Hurlbut 184–85). This is most obvious in the "dominating and destructive power of addiction," where the *compulsive wanting* involves anticipated reward, through personal rites of drug use or another attachment, even as the pleasure from it lessens and it is *disliked* more and more (185). Such compulsive, self and other destructiveness is driven by "an overwhelming sense of defect and deficiency." It involves a feedback loop of remembered liking (or loving) with a drug, person, or group; then disappointment at the loss of oxytocin pleasures; and yet *habitual*, dopamine anticipation of reward, in an impossible return to first-time pleasure. "Addicts generally hate their addiction, just as those caught up in mimetic rivalry seem compelled toward tragedy they neither choose nor understand." This may also involve alienation, depression, and shame through the panic/grief of prior attachment problems (Solms, Pantelis, and Panksepp). Vampire and werewolf films show such tragic mimetic passions, reflecting different wants, fears, and addictions in various decades of cinema history, like the early *Dr. Jekyll and Mr. Hyde* films (1920, 1931, and 1942), with transformations from civilized human into wild animal, sometimes involving self-hatred and disgust as well. But craving, anger, and delusion may evolve into something more mindful (Goleman, *Destructive* 198–232) through *rasa*-cathartic effects in the viewers of these monsters onscreen.

There is *a tragic cycle* to many human relationships: liking others, bonding with them, wanting what they want, competing in mimetic rivalry with them (and with outsiders), becoming disappointed or disgusted with the lessening pleasures that the idealized Other offers, and yet sticking to the habit and its reward anticipations, or repeating the cycle with a new object of desire—whether in drug use, romance, or group camaraderie. Such a tragic habitual cycle also produces collective suffering, with crime, abusive parenting, hate groups, and scapegoating movements. It shows a dangerous dimension in the development of our human ego beyond instinctual rules, through mimetic desires and sacrificial drives. This may be a tragic flaw in our animal to human shift from biological to cultural values, *demanding more meaning to life than just living together*. But such a flaw in our big brains could also be the basis for a further development in each of us. This might alter the animals of our inner theater, beyond the priorities of the "selfish gene" (or "intentional" genome) and the ego's related game of survival and reproduction in a mass, social media age.[23]

BIO-CULTURAL MIRRORS, MASKS, AND LENSES BETWEEN US

We start our lives, like our animal ancestors, in a continuum of self with other, before the development of ego awareness. But later, too, we share feelings with others, often unconsciously, through emotional contagion, while simulating their actions through mirror neuron networks in our heads. Our egos and in-groups become defensive, territorial, and deceptive through survival and reproduction drives. With tragic blind spots, we reconstruct memories and project expectations of good or evil, happiness or loss, onto other persons and groups, which refract as fantasy figures in the hall of mirrors between our brain theaters.

At the lowest level of animal-human empathy and imitation, *emotional contagion and motor mimicry* derives from our mammalian ancestry, but also relates to birds, with the collective reactions of herds and flocks. This is reflected in the infectious bloodlust and radical transformation scenes of humans into werewolves or vampires into bats, especially when they form packs and clans, in the films considered in the next chapter. At the second level, we have *sympathetic concerns and shared goals*, like certain mammals and primates. These elements develop in vampires and werewolves, or their slayers, when they form teams to save one another and to battle more perverse members of their super-natural species. At the third level, humans practice *perspective taking and targeted helping*, as do apes, dolphins, whales, and elephants. These empathic dimensions also emerge with vampires, werewolves, and their slayers onscreen, especially as such characters change their views, and perhaps spectators', too, through tragicomic twists in the melodramatic taking, saving, or avenging of victims. Fairness in market norms versus generosity in social norms, even to the point of altruistic sacrifice, as with capuchin and rhesus monkeys, become ironic themes with vampires and werewolves also—as they pay for, take power over, and yet suffer more than their victims.

Mimetic *rivalries* often arise, even with those "like" us, when we "want" what they desire, yet we strive for ego distinction. This might be seen in the vampire's taking of blood from its victims to perpetuate its immortal illusions. We are fueled by oxytocin bonding pleasures and dopamine reward systems in our mammalian brains and bodies, while also seeking "meaning" beyond biological instincts and cultural constraints—through our awareness of Self, Other, and mortality. We cooperate with others through reciprocal altruism or with moral principles that elevate one's sense of Self. But we will also betray others, even sabotage them, if we can *rationalize* this in a different framework of meaning. For example, in one study, subjects were allowed to give hints to help or hinder a friend on a test after taking the same test themselves (Gilbert 167). When the test was described as a game, subjects helped their friends. But if it was explained as an important test of intellectual ability, they actively hindered them.

Despite our unconscious connections with others, we become *egoistic* in at least three ways, as psychologist Daniel Gilbert points out (230–31). First, the way we know ourselves is special because we can only infer the experiences of others. Second, we find meaningful identities, beyond biological values, by joining certain groups, but we also insist on our own uniqueness: "research shows that when people are made to feel too similar to others, their moods quickly sour and they try to distance and distinguish themselves in a variety of ways." And third, we see ourselves as unique because "we tend to overestimate everyone's uniqueness."

Yet we also find meaning in *groups*, a value for Self through the Other. We might rescript a sour memory of group oneness as blissful camaraderie, even if it was painful. Through *hazing* rituals in sports teams, college fraternities, and the military, individuals submit to painful experiences, bonding with one another and earning the right to belong—even identifying with those who cause the suffering. A study showed that volunteers experiencing severe electric shocks, as an initiation rite, valued the club they were initiated into more than those getting a mild shock (Gilbert 180). When people "believe they are suffering for something of great value," they actually feel less pain. In another study, subjects received negative feedback after taking a personality test, while others took the same test and watched the criticism, but were not criticized themselves (182). The victims of the criticism liked the psychologist who gave it to them more than the watchers did. Likewise, beast-people (and their scientist-creators) may be "liked" onscreen—if the viewer identifies, at a safe distance, with the suffering victim in the monster's embrace, as being transformed through super-natural infection into an elite group. This may involve ironic *rasas* of sympathetic sorrow, eroticism, disgust, and awe, sometimes with tragicomic insights.

Against expectations, intense suffering may "trigger our defenses," so that we give a positive spin to the experience (Gilbert 183). We find egoistic meaning in the suffering—even while joining a group or liking a mean authority figure. Summarizing his research, Gilbert defines us as "*the only animal that thinks about the future*," radically affecting its state of happiness (4). But I would expand that to say: we are the only animal that *seeks meaning*, with a sense of gratitude and purpose in life, *beyond* just the pleasure of living or being happy. We do this with others of our kind (and our pets) *by representing* the future, present, and past—through reflective, inner and outer theaters, involving feelings, imaginings, and symbolic cultures.

Vampire and werewolf films, when they shift our sympathies, perspectives, and sacrificial identifications about individuals and groups, may move us toward new meanings: beyond automatic emotional appraisals toward attentive mindfulness about our own outbursts, which make us monstrous at times. The cognitive reappraisal of emotions, through cathartic twists and *rasa* refinements in such films, changes the viewer's inner theater, altering circuitry between the left-cortical, egoistic, narrative Self and its right-cortical, more holistic awareness. This relates to the first- and third-person viewpoints of our memories and dreams, as well as the playful *hall of mirrors* in our everyday outer theaters, involving ego defenses and yet communal attachments with body-distorting projections, which are also reflected in the vampire's immortal ego and the werewolf's pack-oriented transformations.

Machiavellian intelligence evolved in our highly social, primate ancestors, with the deception of others and oneself for individual survival and reproduction benefits against group norms and hierarchies, but also with long-term, oxytocin-fueled, group and pair bonding. This might be seen in vampires and werewolves as well. They reflect the social *proteanism* of our animal heritage, with a tension between trickery and loyalty in their passions and mutations. They also show key elements of our ape social egos explored further in Chapter 5: attraction, trickery, and dominance (sometimes with a female vampire as alpha, like in bonobo troops).

Specific feelings about others depend a great deal on one's personality, family history, and social *context*. Western (European and American) cultures often encourage individualism while many Asian, African, and Native American societies promote collectivism.[24] The Judeo-Christian and Islamic traditions advocate a sequential (more left-cortical) sense of time and relationships, even in a cosmic theater of this life and the afterlife, with each person having a single soul. But Hinduism, Buddhism, and various indigenous belief systems cultivate the holistic (right-cortical) awareness of cyclical life and death experiences. Each lifetime is part of many others, with current relationships, also involving animals and objects, being more spiritually intermixed. Such an alternative, cyclical, holistic, more right-cortical view of life and death is represented as horrific, yet attractive in the Christian cultural context of vampire and werewolf films—with left-cortical slayers or inner restraints of those figures protecting society from their infectious passions.

Even secularists, as conservatives and liberals, differ in valuing competition or cooperation more, through notions of responsibility and freedom, territoriality and kinship,[25] drawing on more rule-based or intuitively flexible, left- and right-cortical networks. Such cultural, religious, or political *stereotypes* are not true to the same degree for everyone in each category. Yet they shape how we feel about—or remember feeling and then define—Self and Other, especially regarding gender and ethnicity, happiness and anxiety.

Gilbert points to various experiments that show common beliefs about men and women, or about different cultural backgrounds, influencing *how people see themselves* (207–8). When volunteers were asked to remember how they felt a month before, and were prompted to think about gender, women recalled more intense emotions and men less. But a similar group of men and women remembered emotions of equal intensity when not primed to think about gender. In another study, males and females played a game and reported similar kinds of emotions right afterward. But parallel subjects, when asked a week later, recalled stereotypical feelings. The women remembered sympathy and guilt, the men anger and pride. A study of females keeping diaries with daily ratings of their feelings for four to six weeks showed that their emotions did not vary in accord with their menstrual cycles. But when asked to reread the daily entries, they remembered more negative feelings on the days they were menstruating.

Regarding cultural stereotypes, Asian Americans tend to believe they are less happy than European Americans because Asian culture does not stress happiness as much (Gilbert 208). But a comparative study showed, by signaling subjects to report their happiness level at random times during a one-week period, that Asian Americans were slightly happier. And yet, when asked to remember their happiness, they reported less of it than European Americans. Likewise, Hispanic and European Americans reported about the same level of happiness during the week. But Hispanics later remembered higher levels of happiness. Anxiety, too, can be misremembered through a stereotype. Students who did well on an exam recalled being more anxious (and those who did badly, less anxious) than a similar group reported when taking it.

Such stereotypes may be the *lenses* we use, and *masks* we wear, to stabilize the multi-reflective, hyper-theatrical "hall of mirrors" between us, as we act upon our memories, dreams, and plans—while imagining one's Self in the Other's view. Through emotional contagion and mirror neurons, we absorb the feelings and simulate the actions of others, often unconsciously, with our inner theater. And yet, we cannot see ourselves, in the theater of everyday life, as others see us. The workers backstage, in each of our minds, signal to others like them, across the gaps between us, without our conscious actors knowing. We often play roles, externally and internally, deceiving others and ourselves, while projecting faulty recollections, predictions, and stereotypes of happiness, pain, and enemy threats.

The vampire and werewolf films investigated in the next chapter expose such stereotyped projections and deceptions by getting under the skin of mutating monsters, behind their masks and lenses, while also invoking a fear of their emotional and physical contagion. They thus become popular mirrors of our everyday *behavior matching and group merging*, with looking-glass and spyglass identifications of Self through the Other. The tragicomic cracks in such mirrors create alienating insights for viewers through sacrificial passions, in a Brechtian and Artaudian sense, involving playful happiness, tragic pain, and various other *rasas*.

With the evolution of consciousness and migrations across the globe, our animal-to-human ancestors freed us from instinctual rules in a limited environment. But our remnant drives and mammalian emotions, with a greater awareness of aging and death, propel us to seek personal meaning in group beliefs.[26] We try to present the right character of Self in different social situations—whether fitting in or rebelling. Yet each *mask* of Self, when presented to others, is incomplete and illusory to the actor wearing it, like one's Theory of Mind about what is onstage in others' inner theaters and the image of Self in one's own. So we often use stereotypes as *lenses* to remember our-selves in the past, project a Self into the future, predict how others might act or view us, and feel a greater wholeness of *meaning* through group alignments and moral beliefs about good versus evil. But such prejudiced views of Self and Other, mixed with emotional drives about nurturing and threatening forces in the environment, may produce dangerous fantasies about others or parts of oneself, *projected* onto leaders and scapegoats with sacrificial results.

The next chapter explores such projections onscreen with vampire and werewolf transformations, involving primal lust, hunger, seduction, abjection, and violence. These horror scenes serve as collective "nightmares" about the animals in our brain's theater becoming visible externally—through specific cultural developments. They reflect the uncanny hall of mirrors between us, including biased perspective-taking through trust or distrust, loyalty or trickery, with stereotyped lenses of gender, class, and race. But they also evoke emotions and desires that disrupt our conscious (inner stage/screen) controls. They sometimes offer new views, not just escapist entertainment, with cathartic, *rasa*-refining insights about our inner theater's sacrificial flaws, regarding our animal-to-human evolution. Yet that depends, too, on how we watch the films and reflect on their influences—as with people and events in our daily lives. For we each play a role, to some degree, in shaping the vast rippling effects of emotional contagion, mirrored actions, and belief systems that move through us, as uncanny neuro-theatrical waves, which some may perceive as supernatural forces.[27]

Chapter 4

Vampires and Werewolves Onscreen

What are the current caves for projecting our inner fantasies, in dream-sharing rituals, like our prehistoric ancestors tens of thousands of years ago? Movies today, watched in a public cinema, or on television at home, or on various portable screens, have become the most popular way to mirror the inner theaters of our brains (along with other showings on such screens). Horror movies, especially those with animal-human monsters, reveal timely conceptions of our body-brain networks, our shape-shifting illusions of identity, and our out-group scapegoating. Such hybrids may be akin to the images considered in Chapter 2, left by Paleolithic people in certain caves, with figures they experienced as moving in torchlight, through a profound darkness, deep within the earth. Yet each human-animal chimera on the movie screen today reflects a specific history, a lineage of figures in the brain theaters of the past, expressing neural structures that we share with them and with each other now—but also cultural and personal differences. Vampires and werewolves will be considered here, first with an exploration of philosophical meanings, folklore traditions, related neuroscience issues, and specific film classics, then with a survey of recent trends onscreen. Further animal-human hybrids will be investigated in Chapter 6.

TO DRINK OR NOT . . . ?

As babies in the womb and then outside it, we experience a "we-centric space" through emotional contagion and unconscious mirroring (as neuroscientist Vittorio Gallese puts it ["Roots"]). This becomes the basis for our developing ego-centrism in particular social environments, while recognizing our membership in certain groups. We move from the choral oneness of the womb and its extensions in early life (the crib and mother's arms, along with the breast or bottle) to performing for and with others as individual actors on various stages in our lives and through rehearsal scenes of possibilities, memories, and dreams inside our heads.[1] The Other that mirrors the Self involves that lost choral oneness—revived by transitory moments of joyful and painful ecstasy (*jouissance*), as in the rapture of mystical experience, sexual orgasm, or intense communal belonging. Monsters in horror movies evoke such primal thrills, especially the vampire with its bloodlust and contagious bite.

According to philosopher Paolo Virno, this primal *intersubjectivity* should change how we understand politics today (175–79). As humans, we each move far beyond the pre-individual, sub-personal, we-centrism of our mirroring primate brains (178). Through language we build up a sense of Self and particular categories for others—*objectifying* members of our own species. Virno gives the example of a Nazi officer treating an old Jewish man cruelly, as not even human, although his mirror neurons are mimicking the abject suffering of the old man, sending signals within the officer's head as if he had similar needs (182). We use language to contain such super-natural evil in our big brains, but language also enables a primal alienation of Self from Other—and the objectifying of others as subhuman. This produces individual and collective wildness, with humans lacking natural instincts, far beyond the "wild" orders of other species.

The vampire figure, from its undead origins to its Dracula nobility and recent teenage permutations, embodies this paradox of a monstrous hollowness in being human, of intersubjectivity along with self-alienation and other-objectification. (It may also reflect an ancient fear of infectious rabies, as considered with the werewolf figure later in this chapter.) Like each of us when in the womb, dependent on the mother's blood to survive, we bear intersubjective passions throughout our lives. But they are twisted and reformed by language, social rules, foundational experiences, and our continued performances for others. We are horrified by and attracted to the vampire, as fanged, bestial, immortal idol and corpse, because it represents unconscious currents of our own deadly, transcendent bloodlust—in various historical and mythic forms. Life feeds on life. And humans feed, in many ways, on each other.

The word "vampire" is a Slavic term dating back to at least 1047, when a monk wrote it in the margin of a Russian Orthodox text, meaning "alien" or "estranged uninitiate" (McClelland 187–91). This probably referred to pagans, witches, and heretics who continued to practice pre-Christian, Dionysian, ritual feasts, involving animal sacrifices and the excessive drinking of wine, sometimes mixed with blood, plus music that was offensive to Christians (xii, 43, 191). Thus, long before it acquired its modern meanings, the term "vampire" was tied to the theater and wine god, Dionysus (40–41). Ecstatic, drunken, orgiastic rites, involving possession by Dionysus, emerged initially in ancient Thrace (now Bulgaria) and the Balkans, spreading south to Athens where the art of theater developed.

At the start of Euripides's play, *The Bacchae* (circa 406 BCE), set in ancient Thebes, a Greek city-state north of Athens, Dionysus himself appears and tells the audience he is from Lydia and Phrygia, just south of Thrace, to the east of Greece across the Aegean Sea (in western Turkey). During that play, the god possesses the bacchae, his female followers in Thebes, who wear fawn skins and "snakes with licking jaws" (line 697). Their rites in the woods, as described by a shocked shepherd, involve tearing apart cattle (*sparagmos*), eating them raw (*omophagia*), and nursing wild gazelles and wolf cubs at their breasts (693–746). They also invade villages, stealing babies and wounding men "with some unknown power" (751–63). Near the end of the play, one of the bacchae, Agave, queen mother of Thebes, brings onstage the disembodied head of her son, King Pentheus. Still in a trance, she proudly shows it to her father, Cadmus, as the head of a lion cub she hunted and killed with her "bare hands" (1238). Even more tragically, yet perhaps therapeutically, Cadmus brings Agave out of her Dionysian trance to see, with the audience, what she has done.

The last part of the play is lost to us, its ultimate catharsis obscured. But Nigerian playwright Wole Soyinka rewrote the play in 1973, as *The Bacchae of Euripides: A Communion Rite*, with Yoruba and Christian imagery.[2] He ends it with the head of Pentheus mounted on a thyrsus (a fennel staff, symbol of Dionysus) shooting a red liquid from its orifices. The blind shamanic seer, Tiresias, dares to taste the fountain and proclaims that it is not blood, but wine.

The early Christian community, in the first century after Jesus, celebrated his "Last Supper" with the apostles in a different way from later Roman Catholics, Eastern Orthodox believers, and many Protestants. This is shown in a text known as the *Didache*, dating back to the first or early second century, as old as the New Testament gospels (Tabor 44). As a handbook for Christian converts, it describes the Eucharist as a thanksgiving meal involving bread and wine, with references to Jesus as stemming from the vine of David. But there is no account of Jesus's words about the bread as his body or wine as his blood. According to biblical historian James Tabor, it is "inconceivable" that Jesus, as an observant Jew, would have associated the supper wine with his blood or the bread with his body (151). Jewish law forbade the consuming of blood or the eating of meat from which the blood was not drained. The early followers of

Jesus after his death, led by his brother James, followed this rule strictly, as shown in Acts 15: 19–20. The focus of Catholic and Orthodox ritual on the reenactment of Jesus's words—about eating the bread as his body and drinking the wine as his blood—was added later, with those words put in Jesus's mouth in the gospels due to Paul's influence. This consuming of the god shows the influence of ancient Egyptian and Greek culture, especially the mystery cults of Osiris and Dionysus (150–51, 258n23). But it also relates to modern myths about the perverse vampire who drinks blood—like chimps have been observed to do after killing a rival in the wild, as mentioned in Chapter 1.

The association of heretics and pagans with vampires, which must be killed again and again after death, reveals an ironic anxiety about the resurrecting body and return of the repressed in Slavic Christian cultures, extending eventually worldwide through our mass media. Pre-Christian blood sacrifices were subsumed within the Judeo-Christian theater, with the priest performing the role of Jesus and the congregation becoming not just spectators but also actors, consuming the god. But the vampire's bloodlust for humans *inverts* this rite of divine cannibalism, with a demon consuming human blood as his Dionysian wine.

And yet, each of us starts out as such a demon, feeding on our mother's blood inside the womb. The vampire frames a monstrous darkness at the edges of human life—fetal feeding in the womb and being eaten in the earth—with the vampire sleeping in a coffin during the day and feeding on others at night. He (or she) becomes the abject, horrific Other, as an adult-size fetus and zombie, dead and alive, wild and civilized, animal and human. The vampire thus expresses the primal emotions in our vertebrate-mammalian heritage of feeding, lust, and nurturing. But it also shows the monstrosity of our distinctively human brain with its right-cortical holistic intuitions and left-cortical, abstract categorizing functions. Even the elite in human culture can become a vampire, like Count Dracula, because all of us depend upon the globally aware "Devil's Advocate" or anomaly-detector in our right cortex to glimpse the big picture (Ramachandran and Blakeslee 136; Stephen Smith et al.). But the *slayer*, with his tools of language, cross, and stake, represents the left hemisphere's linguistic, rational, objectifying, linear repression of the right hemisphere's *vampire* with more intuitive, emotional, sensual, cyclical (life-death), limbic and subcortical ties to the animal brain (McGilchrist). Such left and right cortical functions will be explored further in Chapter 5, regarding neuroscience, and then related again to other beast-people in Chapter 6.

According to Virno's political theory, human language turns us into monsters. It causes our abject alienation from nature and creates the potential for radical evil with the objectifying of others as non-human (as in the Nazis' treatment of the Jews). Yet it also enables a moral containment of such evil. Vampire stories often involve a slayer figure as this poison-cure: an infected hero, like human language itself, as communal vaccine (or "katechon") akin to the evil that it represses. The best vampire and werewolf films explore the inner theaters of such beast-people and their slayers, showing right and left cortical conflicts *within* each character, as well as between them. But this tragic complexity, of outer battles reflecting inner conflicts of villains and heroes, as kin, also appears in earlier folktales about these creatures—even in a melodramatic, good versus evil framework. Such tragic kinship relates to primal drives of bonding, pleasure, and play in all of us, from childhood into old age and death, exemplified by the vampire's immortal lust.

SCAPEGOATS AND SHAMANS, RIGHT AND LEFT

In Balkan village life since medieval times, a professional vampire hunter might be employed when mischief or disease occurred and was blamed on a dead person's return as the threatening Other (McClelland 24). With a plague hitting the village, the first dead victim was usually blamed as the vampire. That corpse was then staked and burned to keep the other dead victims, as vampires also, in their graves (Bunson 200–201). Vampires were scapegoated as paradoxes: human and antihuman, alive and dead, community members and yet threats to communal life requiring further ostracism beyond the grave. But like them, vampire *seers and slayers* were connected to pre-Christian rites, akin to shamanic healers (McClelland 12). As with today's shamans (and perhaps prehistoric cave artists), they used rites, spells, and trance visions to cathartically cure a communal illness by interacting with super-natural, human-animal beings in a cosmic theater (6, 12). The seer-slayer thus bore a kinship to the vampire as a "quasi-supernatural" *vampirdzia*, or Gypsy *dhampir*, the son of a widow, thought to be the son of her dead husband as vampire, or as a *sabotnik*, a dog or a man (rarely a woman) born on a Saturday, with similar demon-detecting talents (59–60). Saturday, as the Hebrew Sabbath, was also tied to Jews as antithetical to Christians and associated, by Russian Orthodoxy, with "unclean" Dionysian dances (62).

According to Balkan folklore of the nineteenth to twentieth centuries, vampires were unlucky corpses, transformed by a cat or a dog jumping over the body before burial, or a chicken flying over it, or an object exchanged over it, or a handshake, or a live person's shadow passing over it (McClelland 53). The space above the corpse was thus a transitional threshold, because the soul was separate yet attached to it until 40 days after death (53–54). The weeks after the winter solstice, from Christmas into the new year, were considered "vampire days," or "undead days"—with the sun rising again after being at its nadir in the sky (57). But the resurrection of the vampire's body, *unlike* that promised by Christ's, meant a devilish lustful hunger rather than angelic destiny with God. The vampire represented a dangerous, Dionysian sense of earth and sun returning to life, of nature's transformations, and of the animal drives within the human spirit. Also, the seer-slayer and vampire show an embattled kinship inside our brain theaters, across the corpus callosum as a threshold between left (civilized) and right (primal) hemispheres.

The animal theaters of our brains create identity projections, fantasy/reality screens, and performance masks—in a hall of mirrors between us—as we interact with each other, aligning ourselves with and against certain groups. Yet, the stagehand and audience networks within our heads continue to communicate *between us*, through gestures, facial expressions, mirror neurons, and emotional contagion (Goleman, *Social* 15–19). This happens especially through limbic and right cortical networks, below the level of our conscious awareness, as it is staged by the left cortex focusing a rational spotlight that limits, yet draws on the right's holistic senses.

Vampire folklore, from past to present media, shows the outer social theaters of many inner brain theaters interacting in specific cultural contexts. The spirits of Self and Other, as animal drives and human personalities, extend beyond each body, in social feedback networks. Positive, nurturing aspects of this are represented by saints and angels. Negative, threatening aspects are shown with vampires and demons. A corpse that appears incorruptible might be viewed as a saint or a vampire, with divine or animal elements, depending on the communal interpretation of that person's spirit—as it persists in limbic memory traces forming the inner theaters of the living. Scientifically today, we know that an exposed, recently buried corpse often has blood in the organs, as well as blood on the lips and a bloated stomach, from bacterial gases inside, plus long nails and exposed teeth, from shrinking tissues around them, all of which might have contributed to vampire lore (Kaplan 140–42).

In North America, from colonial times into the nineteenth century, mysterious diseases such as tuberculosis were blamed on vampires (Bunson 88). Corpses were exhumed to see whether the heart was rotting away (89). If it contained blood, the dead person was thought to be a vampire, feeding on others, and the heart had to be burned. In Bulgarian folklore, the vampire takes on flesh, but not bones, reincarnating as it grows up: from an invisible spirit through amorphous shapes to animal and human forms—with increasingly aggressive power (McLelland 66).

Young vampires feed on their own corpse, or on animals, until they grow stronger, appear like humans, and feed on them, too. Before that, as an amorphous shape, a young vampire is often characterized as a puffed up bag of skin and blood, with a sharp snout for sucking more blood, yet also vulnerable to puncture (66–67). Or it might appear as the head of a pig, buffalo, or ox, rolling along the countryside, thus recalling ancient, pre-Christian traditions of animal sacrifice (67). Even a good person might become a vampire, if an animal passed over his unlucky corpse.

In the nineteenth century, British Romantic writers further transformed the vampire into a grotesque, yet aristocratic and sublime figure, a greedy rich man, a "social parasite" (McLelland 148), more like the subsequent cinematic depictions of Nosferatu and Count Dracula. He became associated with the ancient mysticism of Eastern Europe, the Near East, and Egypt—with Turks, Arabs, and Muslims threatening the "decorum of British society." In 1819, John Polidori published a novella, *The Vampyre,* initially attributed to his friend, the famous poet Lord Byron, with a blood-hungry Lord Ruthven as its Byronic monster.[3] In 1872, two decades prior to Bram Stoker's fictional Count Dracula, Irish author Sheridan Le Fanu created "Carmilla," in his novella of that title: an exotic female vampire, with lesbian eroticism, appearing also with the anagrammatic characters Mircalla and Millarca (as right-brain Imaginary/Real perversions of the left-brain Symbolic order).[4] Thus, noble, bisexual exoticism became a key element of perversely attractive, monstrously sublime, Romantic vampires in literature and of their eventual offspring, male and female, on the silver screen. They show, like the earlier, agrarian vampires, and even like the young amorphous bag of blood with a biting snout: a mix of feminine and masculine, life and death, erotic reproductive and immortal survival drives. They also reveal a potential intertwining, in our animal-brain theaters, of evil and good, lusting and feeding, rage and nurturing, panic and play, fear and seeking networks (and corresponding *rasas*)—through subcortical, limbic, and right-cortical shadows, disrupting the left's rational illuminations of civil consciousness.

LOVE, LUST, AND RAGE: PLAYING BEYOND THE ANIMAL SCRIPT

As considered in the last chapter, aggression feels good to us at times, via oxytocin and dopamine systems in the brain (Hurlbut 187–89). We lack some of the instinctual scripts for submission, which other mammals have with neuropeptides such as vasopressin, to inhibit an acceleration of aggression between individuals and groups. Vampires in folklore, cinema, and other media depict the dangerous human *transcendence* of instinctual scripts by showing a living-dead, animal-human creature lusting aggressively for blood. A vampire exemplifies in the extreme how the nurturing neuropeptide oxytocin may combine with our (left) brain's addictive, reward-anticipating dopamine system, plus limbic fear and rage networks—in repeated infantile tantrums, biting and sucking at another's body.

This primal scene of childlike vampire passion, with intense maternal attachment, involves panic at potential loss, yet also *playfulness*. Sigmund Freud observed a "fort/da" game that his 18-month-old grandson played, tossing a reel on a string and then pulling it back, while saying, "*Fort . . . Da*" (away/here), as a way to rehearse separation from and attachment to the mother. Likewise, Melanie Klein, in the object relations tradition of British psychotherapy, theorized a playful, yet vicious cannibalism in the child's relation to the mother's breast and womb, as good and evil objects, with fantasies of biting and enclosing them, or being consumed and re-enclosed—which relate to vampire fangs and coffins in folklore and popular media today (Creed, *Monstrous*).

Neuroscience offers evidence for a connection between primal play and horror, between the brain's care and grief/panic (separation-distress) systems. Both involve the brain's nurture chemicals, oxytocin and prolactin, which rise in the mother and infant during nurturing behaviors or with adult companions when together. They decrease with separation, along with lowered opioid levels during panic, as when the mother of a rat pup is not near it, evoking the emotion of "seeking" in both of them (Panksepp and Biven 100, 295–98). Opioid pleasure can be raised again with the mother's nurturing return, or other experiences of loving affection later in life, and intensified in addictive ways by opiate drugs (Solms and Turnbull 121, 130–32). During mammalian play, there is a widespread release of opioids in the brain, especially in the preoptic area (POA), which governs maternal and sexual behaviors (Panksepp and Biven 371). Thus, the bloodlust of vampires represents a primal need in us, too, for the pain-soothing, natural drugs of *love and play*, or the addictive potential of such powers, through opioids in the brain and bloodstream, along with oxytocin as both a neuromodulator and hormone, in the brain and body.

As neuroscientist Jaak Panksepp and psychotherapist Lucy Biven point out, the formalized play of adult athletics involves aggressive passions (171–72). The notion of a "killer instinct" mixes the seeking system's predatory feeding drive, in a battle for social dominance as alpha player, with the rage network in the brain. Sports, videogames, and other entertainment media (including "reality TV") play with various vampire-like emotions: grief in potential losses, nurturing through supportive teams and rabid fans, and voracious pleasure with vengeful, aggressive victories. These media, like the vampire's bite, make the viewer into a player, with addictive illusions of immortal power—taking hold and transforming the brain's inner animal theater. Today, our many mass and social media, including TV shows, movies, Facebook pages, Tweets, and consumer websites, act like vampires: holding us with exotic, erotic, and violence-laced playfulness, sucking from our lifetimes to feed a global culture's pleasure, identity, and money network.[5]

Whether playful or not, vampires embody the fate of all of us, to some degree, with our need for nurturing contact outside the womb and for blood earlier within it. Like other carnivorous mammals, but unlike prey animals that live in flocks and herds, human infants are born very immature and dependent on the caretaker (Panksepp and Biven 300–301). Birds and various ungulates, such as sheep, are able to flee from predators soon after birth and have a short period for intense bonding between mother and infant. But predatory animals, especially those that work together to hunt, such as wolves, have much longer bonding periods, with offspring that are socialized in complex ways with others in the group, especially through play.

Human babies show different *attachment* temperaments (as mentioned in the previous chapter): secure, avoidant, anxious, or disorganized. In older children and adults, "panic attacks" arise when the brain's separation-distress system is triggered. This occurs in the inner theater through neural networks of fear and grief, shaped by prior experiences of loss (Panksepp and Biven 340–41). Vampires thus exemplify a "radical attachment disorder" with their anxious lust for human contact, sucking blood like breast milk, and an extreme sense of grief and loss as the undead—through addictive illusions of immortality, at Imaginary edges of the Real (involving right cortical, mother-infant ties). Today's vampire-like mass media extend this attachment disorder of addictive virtual realities, through and beyond instinctual animal scripts, as the pop-cult norm for each new generation of players, devices, and sites.

Specific traumatic experiences in human childhood can carry over to parenting styles in adulthood, crossing generations within certain families through external behaviors and internal epigenetic markers (Panksepp and Biven 342–43). Vampires show this, too, when they infect particular victims with their blood need and suckling style, creating vampire offspring. As ancient "immortals," they are elderly parasites with a child's passions, preying on the life and bloodstream of younger mortals. Their perverse, infectious eroticism may thus represent the psychological contagion of pedophilia and other traumas that repeat across generations (with some victims growing up to be perpetrators), as well as the potential spread of blood-borne diseases, such as HIV–AIDS.

But vampire lore also points to recent discoveries about the bio-power of youthful blood. Experiments with mice show that replacing old blood with young blood can improve mammalian memory networks in the brain (Jha; Villeda and Wyss-Coray). Likewise, bathing an old, overly enlarged heart in GDF-11, a protein from youthful blood, can heal it, with similar effects also on other muscles and the spinal cord (Colen). Of course, older humans have long envied the vitality of younger ones. And youth have often feared the creepy affection of those afflicted with aging bodies, yet immortal desires. Ironically, though, vampires onscreen have become younger and more familiar in recent decades,[6] from the foreign counts, Nosferatu and Dracula, to the American teenagers that Buffy slays and Bella loves. They live amongst us now, reflecting the immediacy of our mass media, where they also appear, feeding on our limited mortal moments, but also entertaining us.

RAT-MAN ONSCREEN (THE THREE NOSFERATUS)

According to folklore, vampires can turn into various animals: butterflies, fleas, snakes, frogs, dogs, cats, bats, and rats (Bunson). Such animals represent a transformative wildness at the edges of human culture, even when appearing tame. Indeed, some of the mammals in this list (dogs, cats, and rats) have brains close enough to ours to be used, along with those of primates, for neuroscience experiments when human brains cannot be used for ethical reasons. Some people, however, question the ethics of implanting electrodes in, or killing and dissecting such animals, *because* their brains are similar to ours. Rats have been especially useful in Panksepp's research on the "intense experiences" that all mammals have when "ancient networks of their emotional brains are directly manipulated" (Panksepp and Biven 391). And yet, rats evoke in many of us, along with blood-sucking fleas, a specific fear of contagious disease, relating to the Black Death that spread throughout Europe in the Middle Ages. A film near the start of cinema's history, F. W. Murnau's *Nosferatu* (1922), relates vampires to rats and plague, with further associations to insects and even plants as showing the perpetual cruelties of Mother Nature.

The silent German film *Nosferatu* is based on Bram Stoker's 1897 novel, yet changes the names and locations. Count Dracula becomes Orlok, Mina becomes Ellen, and Hawkins and Renfield are condensed into Knock. There is Hutter instead of Harker and a professor named Bulwer but no

Van Helsing. London becomes Bremen. Even with such changes, Stoker's estate sued for copyright infringement and a judge ordered all copies of the film destroyed. But several were later discovered and the film came to life again like a vampire. It continues to haunt us today, reflecting how capitalism itself is a vampire, as Karl Marx suggested in 1867,[7] preying on workers but infecting them, too, as consumers with endless desires—especially through the mass media.

Like Count Dracula in the novel, Orlok is a consumer of territory, bringing the young lawyer Hutter to his Eastern European castle for the purchase of land in Bremen. As he sits at a table and examines the contract, he wears a hat that covers his pointed bat-like ears, which are revealed later in the film. While setting up a standard motif for screen vampires, this also draws on stories from prior centuries, when bats became associated with vampires after news reached Europe about real "vampire bats" in the New World, which fed on sleeping mammals by cutting their skin and sucking (or licking) the drops of blood. Indeed, the anticoagulant in their saliva was eventually labeled "draculin" by modern scientists (Wasik and Murphy 86–87).

Yet Orlok's hunger for property also becomes bat-like when he clutches the contract—as when Hutter, in another table scene, cuts his thumb while slicing bread and Orlok leaps at him to suck the precious blood. Orlok looks longingly at Ellen's long neck, pictured in an amulet that Hutter drops on the table. Beauty, blood, and the legal promise of land excite this human-animal figure, attracting him to Hutter, to his fiancée, and to the German city of Bremen. Orlok exemplifies both bloodlust and legal conquest as a primal and elite, mother-child and patriarchal figure. This perverse count thus reflects the inner (and inter-cranial) theater of limbic and right-cortical networks threatening left-brain, symbolic orders, with animal drives extending through human culture.

Hutter writes to Ellen that he found two mosquito bites on his neck, after sleeping the night at Orlok's castle. But the film viewer sees that the bite marks look like fang punctures, akin to the Count's teeth. Hutter also finds Orlok sleeping in a coffin, with long nails, sharp teeth, and pointed ears. Later, he watches the Count put many coffins on a cart, get in one himself, and leave—with horses pulling it and no driver. The Count's further kinship with carnivorous Mother Nature comes when dockworkers inspect his coffins, before putting them on a ship to Bremen. They open one, find it full of dirt, and turn it over; then rats emerge from the soil. One bites a worker's foot and he beats on them with a board in revenge.

The film juxtaposes scenes of a madhouse cell containing the land-dealer Knock (who initially sent Hutter to visit Orlok), grabbing a fly to eat while exclaiming, "Blood is life," with Professor Bulwer showing students a Venus flytrap eating its insect meal. The professor points out "Nature's mysterious ways" and says to his students, about the meat-eating plant, "Like a vampire, is it not?" He also shows them a microscopic creature, which he calls "a phantom almost." These card titles, between the silent film scenes, tie the meaning of the vampire to the human fear of Nature's mysterious cruelties—exemplified by predation and epidemic disease. Indeed, the sailors on the ship with Orlok and his coffins start dying of disease en route to Bremen. None survive when it arrives, but the rats do, like in medieval times, spreading plague throughout the city.

Another title tells the audience that vampires draw strength from the accursed earth in which they were buried. Murnau's 1922 film eerily evokes an obsession with blood and soil, with power through territory, which increased in Germany as the Nazis fetishized an "Aryan" *Blut und Boden*, like vampires. Ironically, though, the film reflects Nazism in the opposite way also. Its vampire Count (played by Max Schreck) has a large nose, bulging eyes, and big ears, and is associated with rats like the Nazi depiction of Jews in later propaganda movies.[8] Thus, the film illustrates a projection of inner panic and potential evil within the German people upon an outer scapegoat, evoking *rasas* of fear, rage, disgust, and yet awe.

In Stoker's novel, too, the description of Dracula suggests a Jewish stereotype, with his "beaky nose and a black moustache and a pointed beard," along with his hoarding of gold and money in various European currencies (Gelder, *Reading* 14; Stoker 172). The bloodlust of the vampire relates to another Christian and white supremacist stereotype of Jews, from medieval times to the present, as sacrificing gentile babies and mixing their blood into ritual bread (Armstrong 80).[9] Again, this exemplifies a warping tendency of the human brain's inner theater to project the dark side of one's own ideals—Dionysian wine-blood mixed with Christ's bread-body—as a perverse caricature of the stereotyped scapegoat.

Unlike the novel with its all-male "Crew of Light" triumphing over Dracula, in *Nosferatu* Ellen alone saves Bremen with her beauty and self-sacrifice. She reads about and performs a cure for the vampire's epidemic lust—by attracting him to her bed and offering her blood. This delays him until dawn. He fails to heed the cock's crow. When hit by the sunlight through Ellen's bedroom window, Orlok vanishes in a puff of smoke. Ellen thus becomes a Christ-figure, offering her body and blood to bring in the sun's light and power. But she also represents the future fetishizing of romantic racial beauty by the Nazis—as justifying the slaughter of those deemed grotesque like Orlok with infectious, bestial blood. Such scapegoats included Germans with mental deficiencies, political opponents, and a certain ethnic group that the Nazis associated with rats, hyper-sexuality, infectious blood, and super-natural danger, as shown in the 1940 propaganda film, *The Eternal Jew*. There, along with a montage of rats coming out of the sewer juxtaposed with Jewish crowds in the streets of Poland, the narrator warns that Jews have the uncanny ability to look like their "human" hosts.[10]

In Werner Herzog's 1979 film, *Nosferatu*, rats are also shown bearing the animal evil of vampire contagion. And yet, more sympathy is evoked here for the Count and his progeny, by a postwar German director (who also made the Chauvet cave documentary several decades later). At the start of the film, actual mummies are shown from Guanajuato, Mexico. This reflects the movie viewers in an uncanny way: an audience of the undead, as Other to the living onscreen. Herzog restores some of the Stoker names and characters (with Lucy getting Mina's role as Jonathan's wife) but keeps the Count's destination as Germany, shifting the port to Wismar where the original film was shot. Thousands of rats were used in filming the ghost ship's arrival there, whereas only a few were shown in Murnau's film. A bat also flutters at the window curtain of Lucy's bedroom, earlier in Herzog's version, in a montage that juxtaposes her screaming feverously in bed and the long-nailed, bat-like Count approaching Jonathan in bed during his visit to Transylvania. The rat and bat aspects of Nosferatu's limbic, right-cortical threats become more visceral and dreamlike here, especially as movie viewers watch in the dark from the coffin-like space of the auditorium (or at home with today's technology).

After his arrival in Wismar, the Count (Klaus Kinski), like the voyeuristic movie viewer, peers through the window at the beautiful Lucy (Isabelle Adjani). When he visits her home, he tells her his tragic fate: he cannot die and is in abject pain without love—or with a profound, immortal, *attachment disorder*, which might explain why he likes to sleep in a coffin and brings it, along with his rats, to Germany. He says he wants the love that she has for Jonathan, but she refuses the pathetic, large-eared, sharp-toothed, potential lover. Later, in a book her husband brought home, she reads about Nosferatu, the undead vampire, who hunts his victims in the form of a wolf at night, but must follow Nature's laws and can be destroyed by sunlight if distracted from the cock's crow by a pure-hearted woman. She warns Van Helsing to destroy the coffins from the ship, but he refutes her as a man of science. So she acts on her own, crumbling a holy Host, as the divine Body, around a chair where Jonathan (Bruno Ganz) sleeps, because he is now a vampire with a white face and sharper teeth, after a bite on his neck in the Count's castle.

This prevents Jonathan, when he wakes, from stopping Lucy's rendezvous with the Count. She lets Nosferatu suck her neck for a long time, like a baby, while she lies in her bed, making the Count expire with the sunrise, as in Murnau's film. But here, Lucy's sacrifice is greater and perhaps futile, due to the ignorance of legal authorities and the dutifulness of a servant. Van Helsing finds her lifeless form on the bed, drives a stake into Nosferatu, and then is arrested for murder. Jonathan gets the maid's help in sweeping the Host crumbs from the floor and, in a tragicomic twist (or *rasa* mix of awe and humor), leaves on horseback to continue the vampire contagion, saying he has "much to do." Maybe this is a warning from a postwar director (born in Munich in 1942) that ignorant authorities and a dutiful audience can mean the extension of Holocaust brutalities into our own time, in Germany and elsewhere.

E. Elias Merhige's *Shadow of the Vampire* (2000) recreates Murnau's filming of *Nosferatu* with American actor John Malkovich in the historical role of that director. Willem Dafoe plays the actor Max Schreck, playing Count Orlok—although he may be a *real* vampire according to the film. Dafoe is more sympathetic than the original Schreck, or Kinski or Ganz in their film, as an abject immortal. Today's audience knows that he is just an actor onscreen. But he evokes our recognition of the vampire potential in each of us, involving animal drives in our brain's theater.

The crew making the film view Schreck as an extreme Method actor, who sleeps in a coffin and lusts after the lead actress, Greta, like the Count after Ellen, as a way to discover and convey the bestial truth of his role. Likewise, Malkovich's Murnau becomes increasingly obsessed with capturing the vampire-actor on film. Even after his principle cameraman is killed by the Count's (or Schreck's) wolf-like passion, the director makes a deal with him to continue filming and explodes with primal rage toward his colleagues and Orlok.

When the climactic bedroom scene is shot, the director injects Greta with a sedative because she screams at seeing Orlok/Schreck's lack of a reflection in the mirror (an *aphanisis* more in keeping with the *Dracula* than *Nosferatu* tradition). Here, the director exemplifies left-cortical technological control while the vampire-actor shows his right- and sub-cortical animal ambiguity, without a mirrored identity. Indeed, the left brain (and VMPFC) of Orlok/Schreck can no longer restrain his lust and blood-seeking drive. He feasts at Greta's neck and kills two more of the crew. Yet Murnau takes control of the camera and taunts the vampire to persist in his passion, calling him: "Death of centuries! Moon-chaser! Blasphemer! Monkey! Vase of prehistory." He warns him to stay in the camera's frame to be immortalized onscreen. "If it's not in frame, it doesn't exist." This film reflects how another film, early in cinema's history, continues *as a vampire* today, along with its contagious progeny, drawing the lifeblood of audiences in each new generation—and extending the screen immortality of its actors and creators.[11] Together they reveal how strange the human chimera becomes, with hyper-reflective brain theaters, with left-cortical frames screening right-cortical/limbic passions, fashioning bodily survival and reproductive drives into ego ideals of transcending death—through sacrificial bloodlust.

GENTLEMEN PREFER BLOOD (IN DRACULA FILMS)

Tod Browning's 1931 American film *Dracula*, based on the successful stage melodrama by Hamilton Deane and John Balderston, depicts the Count as different from Nosferatu, with a charming, sly, animal-human nobility, in Bela Lugosi's reprising of his Broadway role (Holte 35–36). This time, instead of Harker or Hutter, Renfield travels toward the castle. He gets a warning from the "mountain people" about Dracula and they put a chain with a crucifix around his neck for protection. Then the movie viewer becomes a *seer*, getting glimpses before the hero of his destination: a mountain road, a castle on the mountaintop, and several coffins in a crypt. The lid of a coffin opens and a hand emerges. An opossum is shown next to another coffin, then a female hand creeping out of another, and a bee crawling out of its own tiny coffin. One of Dracula's three brides rises fully out of her sarcophagus and an opossum climbs into another coffin from which a skeletal hand protrudes. The opossum looks like a rat, yet perhaps offers an uncanny, non-mammalian twist, as a native American marsupial associating with European vampires. The bee and its coffin likewise suggest Dracula's bat transformation and flying talents, but with a hive mind and stinging potential.[12] Such animal, skeletal, and female forms introduce the figure of Dracula, who then appears wrapped in a cloak and staring at the movie viewer.

After Renfield arrives, in a carriage led by a flying bat, shots of the castle are shown, with bats flying outside a broken Gothic window frame and two armadillos clambering out of a dark interior space. Here, the bats and Gothic window frame suggest a primal cave and medieval church, but the armadillos also indicate American (maybe Texan or Wild West) connections. Then, from a staircase above him, Dracula welcomes Renfield to the castle, opening his arms with a candle in one hand. Unlike Orlok, this count has a fully round face and is fashionably dressed, with a white shirt and vest, bow tie, and dark suit. Yet he also wears a six-pointed medallion on his chest, hinting at the European fear and scapegoating of Jews, which became extreme in Germany in the decade this film was made.

Lugosi's Dracula, with a charming Hungarian accent and grand gestures, introduced with an opossum, a bee, armadillos, bats, and multiple wives, also represents a more general American fear of erotic Mediterranean males with animal potency. He is a *perverse* Rudolf Valentino (Auerbach 115), pleased with the wolves he hears howling: "children of the night, what music they make" (a quote from Stoker's novel). When Renfield clears a cobweb in his way, and a spider is shown in close up, Dracula says, "The spider spinning his web for the unwary fly; the blood is the life." He calmly philosophizes about Mother Nature's balancing acts—while his own bestiality is barely glimpsed.

In this film, Renfield sucks his own blood when he cuts his finger, because the Count is scared away by a small crucifix. Thus, left-cortex *orthodoxy* inhibits the right and its ties to deeper animal appetites (McGilchrist). Yet Renfield faints at the sight of a bat outside a door that he opens in the next scene. The three brides then approach Renfield, but Dracula, coming from outside where the bat was, waves the women away. The film also moves politely away, cutting to another scene as the Count bends over his victim—not showing the *homoerotic* contact with Renfield.

On Dracula's ship, bound for England, rats and dead sailors are *not* shown (aside from a shadow of the captain tied to the helm). But the Count's effects on his new servant are stressed. A crazed Renfield opens his master's coffin when "the sun is gone." He stares up from the hold at the dockworkers, wild eyed, when the ship arrives at port. A newspaper then states that he is in the care of Dr. Seward and eats flies and ants for their blood. Yet this is *not* shown, just suggested for the inner theaters of audience brains.

Dracula charms Lucy when he meets her at the theater, in her private box with Seward, his daughter Mina, and her fiancé Harker. Later, doctors in a medical operating theater discuss, with others watching, how many blood transfusions the ill Lucy needed. But the Count's theatrical charms meet their match in another alien, the Dutch vampire expert, Van Helsing. He notices that Dracula is not visible in a small mirror under the lid of a cigarette case. When he offers the mirror to the Count, Dracula rashly knocks it away, showing his hidden violent nature and fear of mirrors—or of left-hemisphere containment, inhibiting the right's animal passions. Despite his suave demeanor, Dracula's undead ego cannot face the hollowness that we all bear, behind the mask of Self, from the mirror stage of infancy onward: a dark uncertainty or lack of being, masked by left-brain abstractions of identity.

The Count politely excuses himself and leaves Seward's home. Soon, Harker sees a large dog crossing the lawn (not shown to the movie viewer). Van Helsing explains that it is Dracula, transformed into a wolf, and that he can also become a bat. Like the devil, "his strength" comes from human disbelief in him. Later, Van Helsing proves his will and wisdom as a vampire *slayer*, resisting the Count's hypnotic gaze by holding up a crucifix, "more effective than wolfsbane," again showing the *left* cortex's categorical power to contain right-cortical complexity. Van Helsing also saves the infected Mina, staking Dracula in his coffin—another act of violence not shown onscreen, but representing the left brain's linear, analytical control of the right's cyclical paradoxes. Van Helsing frees Mina from her coffin, for a happy ending in Harker's arms. In this early talky, the explicit horrors of Dracula are pushed into the shadows beyond the frame, evoking further *play* in the chambers of the viewer's brain. Yet this reflects a deeper horror in the vampire's bloodlust and abject immortality: mirror-stage illusions of a transcendent ego, arising from animal drives and childish attachments in the spectator's particular inner-theater networks—and then vanishing in the light of day.

Thousands of vampire films followed, emerging in various forms (Gelder, *Reading* 86).[13] Lugosi's Dracula appeared in a comedy in 1948, joining Abbott and Costello, plus Frankenstein and the Wolf Man, in *Abbott and Costello Meet Frankenstein*. Christopher Lee played Dracula in the British Hammer Films versions, starting in 1958 and then in seven sequels until 1973. (One more Hammer Dracula was made without him the next year.) In the 1970s, Hammer also produced a female vampire series, increasing the lesbian eroticism and overt sexuality of the vampire's predation on his female victims, naked and in orgasm at his bite (98–99). Dracula porn films also developed, expressing sexual liberation ideals, such as *Count Erotica, Vampire* in 1970 and *Bite Me, Darling* in 1972 (Day 26). As a more romantic figure, Frank Langella reprised his Broadway stage version of Dracula onscreen in 1979.[14] The Count was again given comic twists by filmmaker Roman Polanski in 1967 and Mel Brooks in 1995. But the most spectacular, meta-theatrical display of the vampire's powers appeared in Francis Ford Coppola's 1992 film, *Bram Stoker's Dracula*, which added a pseudo-historical framework to the novel (with a screenplay by James V. Hart).

The origin of Dracula's lustful rage is shown at the start of this film, narrated with a voice-over by Anthony Hopkins, who plays an Eastern Orthodox priest in 1462 and then Van Helsing 400 years later. First, a cross falls from a dome, symbolizing the fall of Constantinople (today's Istanbul), the thousand-year capital of the Eastern Roman (Byzantine) Empire, to the Ottoman Turks in 1453. Their march across Transylvania then appears on a map (in a film made soon after the United States felt so threatened by another Islamic state, Iraq, that it started the First Gulf War). And then, in 1462, a Romanian knight in red armor, Vlad, son of Dracul, kisses his wife, Elizabeta, and goes outside to fight the infidels. (Her name echoes the myth of Elizabeth Bathory, who reportedly killed hundreds of virgin girls, bathing in their blood as a fountain of youth, although she lived a century earlier.) This opening is both retrospective and prescient, evoking the sympathy of a Western audience toward the Christian knight, in his crusade against Muslim invaders a half millennium ago.[15] Instead of being the alien Jew or Eastern European, as in *Nosferatu* or *Dracula*, this Count is one of us—or of the Christian American majority in the Hollywood film audience—battling the anti-European, non-Greco-Roman other.

With his long spear, Vlad the Impaler (representing left-brain categorical certitude) kills many enemies, as shadows on the battlefield, and bodies are shown perched on stakes stuck in the ground. Hopkins's voice-over tells about the "vengeful Turks" shooting an arrow into the Count's castle with a note giving the false report of his death. In despair, Elizabeta flings her body into a river and dies. Vlad returns and finds her corpse on a chapel floor with blood dripping from her mouth. He tells a priest (played by Hopkins) that he will rise from death in vengeance. Then he stabs a stone cross in the chapel. Blood flows from a breast-like bulge at the center of the cross, where Vlad's sword sticks—as a *right* cortex, cyclical paradox. Then he drinks sacred wine, the Blood of Christ, from the chalice on the altar and more blood flows, like a fountain across the chapel floor and Elizabeta's body. In this operatic, super-natural spectacle, the animal passions in Vlad's brain theater twist from reproductive love and communal duty, in kissing his wife and fighting the Turks, to immortal rage at God, at fate, and at the Church, perverting the Christian symbol of Jesus's sacrifice, which replaced earlier blood offerings.

The scene then shifts to London in 1897, with Renfield (Tom Waits) in a madhouse, showing his animal passion by catching and eating a fly. Jonathan Harker (Keanu Reeves) also describes, in a letter to Mina, his impression of leaving the West and entering the East, as he travels by stagecoach through the wolf-infested landscape of Transylvania. When he is left on the roadside for a transfer, the *reptilian* hand of the masked driver of his new coach touches Harker's shoulder to urge him inside. As they reach the castle, the coach passes through teeth-like gates, again suggesting the animal edges (or fish-reptilian, subcortical structures) of the character's and audience's brain theaters, interacting onscreen.

In this version, Count Dracula (Gary Oldman) purchases ten houses scattered around London—with a large map of that city on a wall behind the vampire, as capitalist consumer. He appears now to Harker as an old, very pale, yet vigorous man, with his white hair tied in two large buns on top of his head and a long braided ponytail behind it—like breasts with a phallus. When Harker cuts himself shaving, the Count licks blood from the razor, delicately and yet voraciously. His shadow on the wall acts independently of his body, which also does not appear in the mirror. Like Schreck's *Nosferatu* and Lugosi's *Dracula*, this Count is a pre-mirror-stage monster, a blood-addicted, adult fetus, behind the *mask* of civilized aristocracy.

Harker tells Mina in a letter that he is a "prisoner" of the Count's castle, as his perverse, *external womb*. This becomes even more apparent when Harker explores forbidden rooms. Against the Count's order, he lies in a hexagonal bed and apparently falls asleep. He experiences (or dreams of) three beautiful, bare-breasted, yet fanged women, one with Medusa snakes for hair, emerging from the satin sheets to fondle, kiss, and bite him. As they approach his chest, the metal cross on it melts. Harker is clothed in this scene, but naked animal contact is evoked, at the edges of the frame and in the viewer's brain theater, with one of the female vampires between his legs, her mouth at his crotch. The women also kiss each other, again showing the right hemisphere's alternative drives, against the left's patriarchal symbols.

Dracula appears suddenly and forces the women, with a wave of his arm, to scatter backwards on all fours. Two of the women become attached to each other like a large insect. Dracula speaks a foreign language to them, with subtitles translating his claim on Harker as "mine," because he can "love" also. But the borderline of this attachment disorder is then revealed: Dracula's bride-brood ask if they are "to have nothing tonight." He produces a naked crying baby, which they gather around, apparently (though the child is hidden) to *feed* upon, not nurture.

A hint of bisexual perversity is also shown on the human level, back in London, with Lucy and Mina kissing in a storm, playing in a garden maze, and giggling about Lucy's three suitors, including the American cowboy Quincy with his long bowie knife. The other suitors are Arthur, a British aristocrat, and Dr. Jack Seward, who is in charge of Renfield and injects a clear liquid into his own vein, trying for the madman's insights. The movie viewer sees, too, that Mina looks very much like Vlad's Elizabeta, with both being played by the same actress (Winona Ryder). This encourages, in the spectator's inner theater, a dopaminergic anticipation of a climax to the Count's 400-year passion for his reincarnated wife, although she is currently engaged to Harker.

After the death ship arrives in London with Dracula in his dirt-filled coffin, plus 50 boxes of "experimental earth" (according to Hopkins's voice-over), Lucy, in a sheer red nightie, is drawn to the garden on a moonlit night, followed by Mina in blue. There, Mina watches a hairy wolf-man with a bat-like face move between Lucy's legs and bite her neck while she writhes in erotic pleasure on a stone bench or tomb. But unlike the movie viewer, Mina as voyeur interrupts this primal scene. The animal-human hybrid looks toward her and says, "No, you do not see me." Then he (or it) vanishes. The camera moves likewise, later in the film, in quick leaps across the night landscape, as if showing the vampire-wolf's point of view, as it seeks its prey—indicating that this monster of vengeful lust resides in each of us, potentially, as voyeurs and actors.

Dracula's kinship with wolves becomes apparent again when he meets Mina as a young stranger on a London street. She leads him, at his request, to a cinematograph, an early multiplex theater (in another scene added to the novel) with tinted erotic images onscreen and soldiers in shadow play.[16] He comments about there being "no limits to science" and speaks to her in a foreign language. Elizabeta *within* Mina starts to recognize Vlad, who says he "crossed oceans of time" to find her. She submits to his bite, but he hesitates. Then a white wolf disrupts the cinema, scattering the audience as a thrill that is all too real. Yet the wolf nudges Dracula like a pet. The Count encourages Mina to stroke it, charming her even more with his civilized wildness. "He likes you," Dracula says, as if speaking for her brain's subcortical and limbic, vertebrate and mammalian heritage. "There is much to be learned from beasts."

Later, the film intercuts glimpses of the Dracula wolf-man feeding again on Lucy with scenes of Mina marrying Harker in Romania, in an Eastern Orthodox nunnery, after his escape from the Count's castle and brides. Mina writes the Count a "Dear John" letter. This causes the monster to weep, evoking the viewer's sympathy for a malignant, erotic, and yet tragically romantic character (Kane 98–102).

Van Helsing also becomes a perversely sympathetic *anti-hero* in this film, as (im)mortal enemy to the anti-villain Dracula. He is introduced in a medical teaching theater, showing his students, on wooden benches above him, how a vampire bat needs blood, offering some from his own finger—and joking about civilization and "syphilization" advancing together. Eventually, he teaches Lucy's three suitors that she is a vampire, observing her strange illness, wanton passion, sharp teeth, fang wound in the neck, and mausoleum undeath. He tells them they are fighting "not one beast but legions . . . age after age." This encourages *rasas* of fear, heroic vigor, and awe in the movie viewer, but also perhaps disgust at his hubris and cavalier teaching methods.

Van Helsing's Crew of Light, including suitors as slayers, confronts the Lucy vampire as she returns to her coffin with a child to consume. Then she vomits blood at Van Helsing's face, like the demonically possessed girl in an earlier horror film, *The Exorcist*. Yet the scientist-slayer also appears possessed at times, with perverse *jouissance* (Lacanian sexual-mystical pleasure-pain) in finding his longtime prey, his super-natural mate in the cosmic battle of good and evil. He hugs Quincy, for example, bouncing him homoerotically, chest to chest, before leaving with Jack. In a restaurant scene, Van Helsing's steak is juicy with blood, soon after he and his crew have finished off the undead Lucy, by stabbing her corpse and cutting off her head. With such slayer and vampire parallels, the film mixes *rasas* ironically, showing a kinship between the opposing, yet complimentary functions of left and right cortical networks, also involving our animal emotions and drives.

When the youthful Dracula seduces Mina again, with a bite on the neck, she asks him to make her a vampire, to free her from death—despite what happened to her friend, Lucy. So then (in a scene taken from the novel) Dracula cuts his breast with his claw-like fingernail and Mina sucks the wound, giving him an oxytocin rush of male-mothering pleasure. Yet the Crew of Light disrupts this perverse, anti-pieta scene and Dracula turns into a huge bat-devil. He first tries for sympathy as an abject, fallen angel appealing to Lucy's former suitors, "Look what God has done to me," suggesting also that a divine spectator may be watching.[17] But when the Crew shoots their guns at the Dracula bat, he dissolves into many rats, which scurry past them—reflecting, perhaps, the multiple animal selves of emotional contagion in the watching audience. Likewise, jostling red blood cells (like the microscopic creatures in *Nosferatu*) are shown at various points when the Count lusts for and feeds on a victim, suggesting the many animals in each of us, moving with the movie.

The film's final act also reflects the vexed relations between different aspects of Self in the brain's theater, with left-cortical inhibition of, yet inspirations from right-cortical animal drives. Van Helsing rescues but is almost seduced by the vampiric Mina, while Harker, Quincy, and Arthur race on horseback, trying to catch the Count's stagecoach, as it heads back to his power base in Transylvania—showing the repressive and expressive powers of cortico-limbic and limbic-subcortical networks. Mina's teeth turn into fangs and she attacks Van Helsing's neck. But he breaks away from her seductive body and vampire kiss, burning a host then onto her forehead, and thus restoring her prefrontal cortex sense of self-control. Yet she demands to know whether he will drive a stake through her heart and behead her like he did to "poor Lucy." Van Helsing has vowed to protect her, however, and does so, even when the Count's demonic brides approach their campfire like wild animals and attack their horses.

Mina, with kinship now to the Count and with witch-like powers over Mother Nature, summons clouds to cover the sinking sun, in order to aid the fleeing Dracula. When the men catch him, after subduing his "gypsy" allies, then slit the Count's throat and stab him, Mina points a shotgun at them to protect him. She helps him into the chapel and Harker prevents his friends from pursuing her saying: "Her work has just begun." Thus, Mina's limbic nurturing drive shelters a subcortical living-dead monster, via a left-cortical (patriarchal) weapon and sanctuary. Her husband, resonating emotionally with her, provides a right- to left-brain acknowledgement of her "work." But various *rasas* might also be evoked in the working inner theaters of movie viewers—angry, fearful, erotic, wondrous, disgusted, courageous, and sorrowful tastes—through tragicomic twists of melodramatic sympathies.

Instead of a villainous Count dissolving into ashes, as in the novel, Mina expresses her reincarnated love for him, "stronger than death," as he lies on the chapel's stone floor. Dracula even paraphrases Jesus's words on the cross: "Where's my God? He has forsaken me. It is finished"—making him an ironic, villainous-heroic, Christ figure. He asks Mina for "peace," so she plunges the knife deeper into his chest. She then kisses the old man's lips, pulls out the knife, and slices his head off.[18] The film ends here—without clarifying whether Mina becomes a vampire or is healed and has a child with Harker, as in the novel. This ending challenges the movie viewer to reflect on various potential identifications with and against Dracula: (1) as perverse father/mother figure to Mina, Lucy, and his three brides, (2) as heroic warrior whose passion for the cross and his wife turned him into a monster, or (3) as fallen angel, purely evil and deceptive all along.

Such tragic/melodramatic ambiguity may provoke the audience to rethink their feelings about malignant aliens (like Schreck's Nosferatu or Lugosi's Dracula) and their attraction to romantic shape shifters. It may also affect the spectator's inner theater of feedback loops (especially in the VMPFC) between frontal cortex and limbic system, or between left and right hemispheres. While invisible in mirrors, Dracula himself is a mirror to the film voyeur, as he was to Mina, Van Helsing, Lucy, and others. He represents the Other side of desire within each of us: the animal passions fueling our highest hopes for love and immortal meaning.

SYMPATHETIC BLOOD LUST (*THE HUNGER* AND *INTERVIEW WITH THE VAMPIRE*)

As Coppola's *Dracula* illustrates, the undead human-animal became a more sympathetic creature for movie viewers in the last decades of the twentieth century. Vampires have become associated not just with foreign aristocrats challenging an Orthodox Church and unjust God, but also with marginalized groups in our current patriarchal society, such as lesbians, gay men, AIDS victims, and abused children (or rebellious teens).[19] Vampire fans are attracted to this new breed of the undead, as "an outcast who both rebels against the society it haunts and is marginal to it." Thus, according to British theorist Milly Williamson, the twenty-first century vampire in literature, film, and television is "speaking to our culture's unacknowledged pain, dramatising the desire to transcend the socially imposed definitions of what it means to be human" (190). She also finds in the vampire: "a creature of capitalism . . . imbued with ideas of individualism and a focus on the self so central to modern society, even as it poses the problem of the lack of individual significance" (186).

In the psychoanalytic view of various commentators, the vampire represents, from folklore to today's films, the dryness of the corpse, lacking the fluidity of living energy, with the victim's blood akin to breast milk (Copjec; Creed, *Phallic* 81–83; Dundes 20–29). Thus, the corpse, whether a mourned family member or a threatening foreigner, lusts for blood—like all of us needing Mother Nature's nutrients and feeding at the breast or table from infancy onward. I would add that this bloodlust intensifies with Western culture's stress on individualism and the media's stimulation of unsatisfiable desires. Massive social networks prey on us, like vampires, drawing our time, money, and ideals toward competing idols. But they also turn us into little vampires, attached to certain brands and their latest products, as we seek collective validation of our individual significance through cinema, TV, and newer virtual screens, as our consumer "boob-tubes."

A stylish example of such sympathetic vampire bloodlust appears in *The Hunger*, a film from 1983, based on a novel by Whitley Strieber. It was Tony Scott's first feature film and led to his later "recasting" of several popular genres with a "postmodern temporality," as with various scenes showing billowing curtains in this one, which suggest "the screen-like nature of memory, the ways it both obscures and reveals the past and its relation to the present" (Phillips 253–54, 260). As Barbara Creed points out, this film also updates the traditional vampire from folk monster and aristocratic alien to postmodern urbanite in New York—with no fangs, bats, wan virgins, or gliding predators, and not even the term "vampire" being used (*Monstrous* 68). Yet the film juxtaposes, in alluring and horrific ways: glamorous youths and aging bodies, artistic beauty with musical instruments and scientific monstrosity with animal experiments.

Elegant French actress Catherine Deneuve performs the role of Miriam Blaylock, an "archaic mother" (Creed, *Monstrous* 72) who makes others into her vampire children as well as victims. She incestuously drinks their blood, yet also offers her blood to some, with the lure of long-term youth, though at the cost of sudden and immortal aging. Her mate at the start of the film, John Blaylock, is played by the glam-rock megastar David Bowie, who became famous in the 1970s with his androgynous, guitar-playing, alien come to earth, Ziggy Stardust persona onstage. Miriam and John initially appear in a nightclub, which has cage-like wires, picking up their victims as sexual liaisons. When they bring a young woman and man to a private room, she dances erotically against a bare wall and he relaxes on the couch. Soon, they are French kissing with Miriam and John. But this sensual scene is intercut with glimpses of a macaque bearing its fangs and viciously biting and eating another monkey. An ankh, the ancient Egyptian symbol of eternal life (looking like the Christian cross with a loop at its head) is shown in close-up with a dagger tip. It is used to cut a human victim. Then two of the ankh daggers are washed in a sink and bodies in black bags are burned.

The ankh blade appears at other points, later in the film, when John and Miriam use it—instead of vampire fangs—to kill victims and drink their blood. Perhaps it symbolizes the "hunger" (of the film's title) in nature and culture, as initially juxtaposed with a monkey's fangs, performing a similar bloodletting. The survival hunger of our animal ancestors fueled the immortal dreams of ancient to modern artworks, and of today's mass-media technologies, which exploit the young through lures of fashionable pleasures, benefitting the elite. Tellingly, this film offers classical music being played in Miriam's New York mansion, by her on the piano, by John on bass, and by a neighbor girl, Alice, on violin—after the initial nightclub scene. Flashback memories are also shown of John with a white, Enlightenment wig, suggesting a darkly passionate (or right-cortical) side, hundreds of years before, during the Age of (left-cortical) Reason.

In the film's present timeframe, John notices in the mirror that he is aging quickly, with crow's feet at his eyes and hair coming out in clumps. He visits a clinic to meet with Dr. Sarah Roberts (Susan Sarandon), who wrote a book on "Sleep and Longevity," which Miriam purchased. John ages even more suddenly while in the waiting room, as Dr. Roberts discusses with her colleagues how they sped up the inner, biological clock of monkeys in their experiments, apparently causing the vicious attack seen earlier. Thus, the left-cortex focused scientist, as non-nurturing female professional, makes her fellow man (or vampire) wait and suffer, like an overly bureaucratic healthcare system.

Likewise, the alpha competition in ape troops, as a more primal bureaucracy of survival, mostly among males but with an *alpha-female* at the top in bonobos, is shown in Miriam's home. The aging John reveals a final burst of abject power, with reduced front-cortical restraint, killing his young friend, Alice, while she plays the violin for him. Thus, with Miriam away from home, a youthful female beauty cannot control a savage male beast—despite her musical talent. But later, with perpetually youthful power over the grotesquely aging John, Miriam carries him upstairs, like a mother taking a child to bed, and places him in a coffin in her attic, with other elderly offspring, also in coffins. In another scene, Miriam seduces Sarah in her home, feeds on her, gives Sarah her own "not human" blood, and teaches her to be a vampire. Miriam, as archaic mother, as super-natural, incestuous bisexual, offers love "forever" to Sarah, as she did to John.

However, Sarah finds a way to change the alpha pyramid. She uses the ankh dagger to stab her own neck while kissing Miriam. This somehow awakens the mummy children, who meet their mother when she carries Sarah to the attic. They (and John) crowd around her, with excessive immortal love, as dried walking corpses, zombie-vampires, touching and kissing her. This, along with the doves flapping around her, horrifies Miriam and drives her off the balcony ledge. Breaking the banister, she falls to her death below. Sarah is shown as the survivor then, looking young and beautiful on another balcony, becoming the new alpha. Thus, conflicting *rasas* are mixed as this film develops toward its climax: eroticism, sorrow, and disgust (with the youthful-looking vampires seducing their prey and one of them aging suddenly), awe and fear (with the vampire/monkey "hunger," especially as Alice becomes a victim), and then rage and courage (as Sarah and the elderly zombies rebel against Miriam).

The Hunger was not a commercial success, perhaps because it presents an implicit critique of the glamorous film and TV industries as vampire-like. (David Bowie as John watches TV cartoons while aging; Sarah appears on TV selling her book before she is changed by Miriam's love.) It also offers a grotesque reminder of aging and death for those watching, even alpha females. And yet, the film evokes tragic sympathies, too, for its vampires. John asks forgiveness from Alice before he slits her throat—with an old man's sad regret, while driven to repeat a horrible crime, after failing to find a suitable victim in other places. When Miriam comes home, she forgives him for killing their young friend. He then requests death, asking her to "release" him, but she cannot, although she has killed many others. Perhaps her passionate attachment to him, another sense of animal "hunger," prevents her from being merciful. Yet she is also repulsed by his appearance. She cannot kiss him on the lips, though he asks her to remember his youthful beauty. After he falls down the stairs and she carries him upward to his attic crib/coffin, Miriam wails beside John's grotesquely aging body. She introduces him by name to the coffins holding her other aged, immortal children and asks them to be kind to him.

Sarah also draws sympathy from movie viewers, after being made a vampire, even from those who might feel averse to her sexual scenes with Miriam. She kneels on the floor in convulsions, with Miriam's alien blood in her veins, and tries to kill herself rather than killing others due to her hunger. Sarah's earlier role as a scientist studying human aging, with brutal experiments on monkeys, and John's aging while waiting for her at the clinic, also suggest a theater of cruelty that many of us will experience someday—as medical technology allows us to live longer and age more, as consumers being consumed, victims of vampire capitalism.

Likewise, in *Immortality* (a.k.a. *The Wisdom of Crocodiles*, Po-Chih Leong, 1998), a hemophiliac vampire, Steven Grlscz, tries to restrain his bloodlust for Anne, whom he loves. Yet he needs to drink her loving blood to stop his own internal decay. He explains to her that she is in bed with a "horse and crocodile," because of our *mammalian and reptilian* brain structures. She tries to nurture him back to health, as he weakens and ages without drinking blood. But when the crocodile in him gains strength again, Anne flees and then fights back. She stabs his hand with a metal chopstick, even as he saves her from falling off a building, and he is left, ironically, to bleed to death.

Vampires as sympathetic antiheroes are given more historical and theatrical twists in Neil Jordan's *Interview with the Vampire* (1994), based on Anne Rice's popular novel. Like *The Hunger*, this film explores the vicious passion, yet abject hunger of embattled, immortal alphas and newly made vampires. But while Miriam is an affectionate maternal figure to her aged mate, her coffined offspring, and her lesbian lover (and while Anne nurses Steven in *Immortality*), the vampires in *Interview* lack such a positive supermom. They suffer, even more than their victims, from Mother Nature's theater of cruelty.

Present day, San Francisco reporter, Daniel Malloy (Christian Slater), functions as a mediating chorus for the movie audience—to meet and interview Louis de Pointe du Lac (Brad Pitt), who takes us back to the beginning of his vampire career in 1791. He was a 24-year-old human then and master of a Southern plantation. His wife died in childbirth and Louis, distraught at the loss, longs for death, too. The watching vampire, Lestat de Lioncourt (Tom Cruise), notices Louis's death drive and, like a Mephistophelian devil, gives him a choice between getting that wish or immortal life. Biting Louis in the neck, Lestat flies high into the sky with him, in a surprising homoerotic twist for these macho movie stars. This scene also reflects the vampiric lure of the cinema screen itself—with its power to edit mortal time and space, to manipulate our identity projections, and to give us moving, super-natural experiences. Eventually, Louis accepts Lestat's offer to become an immortal youth (to "come" with him), sucking the blood from his wrist—while the heartbeat in the soundtrack quickens, joining the audience to Louis's transformation. But Louis gets pains in his stomach (like Sarah in *The Hunger*), as if with menstrual or pregnancy cramps, and his face becomes very pale.[20] Dying then, he is reborn, seeing the world in new ways, including statues coming alive in the graveyard. Yet Louis becomes even more abject, in his whiter-faced immortality, than he was as a slave master with a dead wife.

Louis's new vampire identity refracts the adoring gaze of Brad Pitt's fans, revealing something uncanny behind the illusions of the silver screen. We objectify and fetishize the bodies of such stars, male or female, through the cinematic apparatus (Mulvey). But we are also enslaved by them, each time we submit a bit of our lifetime to their projected immortality. Here, too, Pitt/Louis summons our sympathy as a victim of Cruise/Lestat, his homme fatale, evoking an ironic awareness regarding Euro-American imperialism. Though a slave owner, the American Louis cannot accept the teachings of the European vampire Lestat, who shows him how he kills two to three victims a night and quips: "it's so easy, you almost feel sorry for them." Outraged at such casual slaughter, yet feeling a similar lust, Louis drinks the blood of his house slave, but then burns down his mansion and frees the others, telling them: "your master is the devil." Hence Lestat, as a white-faced European vampire with a long history of objectifying people, shows a casual, aristocratic aloofness when feasting on his victims. However, Louis, as an American with new imperial power, expresses a contradictory self-loathing, a tragic conflict about his use of and need for colonial subjects—evoking *rasas* of moral disgust and rage in movie viewers as well.

Louis tries to survive on the blood of animals (rats and doves). Yet Lestat taunts him with the pleasure of cruelty, taking his time in killing and feasting on a young woman, even putting her temporarily in a coffin. Her fear of death tickles his will to power and amoral joy. This may elicit an ironic *rasa* mix of fear, awe, dry wit, and playful joy in movie viewers who sympathize with both victim and villain here, altering prefrontal networks toward tragicomic mindfulness.

More mindful than Lestat, Louis refuses to drink the victim's blood or take her life. He wanders the streets of New Orleans at night and finds a child (Kirsten Dunst) at the body of her dead mother, a plague victim. She tells him her father also left her and hugs Louis. He then gets Lestat's help to save the girl from death and loneliness, making her into a vampire. This new "happy family" (as Lestat says) may suggest the legitimacy of gay marriage, at least for vampires. But it also produces a further theater of cruelty around the lack of a compassionate, nurturing mother figure[21]—because Louis's misplaced empathy toward the little girl, Claudia, is not maternal enough.

During the plague in New Orleans, with bodies in the streets, Lestat tells Louis that "God kills indiscriminately and so do we." This dark view of the Creator is reflected not only in the vampires' plague of random serial killing, but also in Lestat's gift of eternal life, as a kind of anti-Christ. He gives blood from his wrist to the little girl, Claudia, like he did with Louis, to keep Louis from leaving as he fears. Claudia sucks voraciously at Lestat's wrist—like an infant at the breast—so that he becomes her male mother, like Louis, giving birth to her as a vampire, although Lestat is more ruling and *paternal* otherwise. Claudia then gains blonde curls and a very white face, like a porcelain doll. Throughout the film, when Lestat and Louis feed, they also get white faces, ironically looking anemic after drinking blood—or more human when hungry. They represent the dark powers of a perverse, white God, with their elitist, Euro-American hubris, acting like a superior species, but with bestial appetites.

Lestat teaches Claudia to feed and she, being a quick study, becomes more ruthless than he: killing an affectionate maid, a seamstress working on her dress, a piano teacher on the bench with her, and a matron whom she lures to comfort her as a child without a mother. Thus she turns her pre-vampiric mourning into a trap for parental figures. Eventually, she hates her vampire-God, her creator Lestat, for making her an eternal doll, a child monster who can never become a woman, so she tries to murder him. Vampire family bonds and attachment cravings turn into a female Oedipus complex—with a girl killing her father figure (Lestat) and escaping with her mother figure (the long-haired Louis).

In a Nietzschean act, beyond good and evil, Claudia lures Lestat into sucking the blood of a dead boy, one of the twins she presents to him as a peacemaking gift, pretending they are a drunk and just sleeping. Having feasted on "dead blood," Lestat becomes incapacitated. Claudia then slices his throat and he transforms into a rotting corpse. Getting further revenge against their Creator, she and Louis dump him into a swamp. But he proves to be more an immortal force of Mother Nature than they thought. He returns to them later as a piano-playing, undead monster, who has fed on the river's snakes and toads. So Louis sets him on fire, along with their home and neighborhood—before he and Claudia escape on a ship bound for "the Old World."

In Paris in 1870, in search of the source and meaning of his immortal existence, Louis meets the sly, alpha vampire Armand (Antonio Banderas), who boasts that he is the oldest of their kind in the world, the clownishly wicked Santiago (Stephen Rea), and a host of cynically sadistic, theatrical vampires. Playing the figure of Death with a scythe onstage, Santiago teases the human audience of his vampire theater about the whimsicality of mortality, dispatching quickly with fictional lovers on a bench and a monk at a well. But then he produces a real victim onstage: a girl who screams that she does not want to die. A female in the audience screams to him: "Take me." But he tells her to wait for her turn. This again reflects the voyeuristic, fetishistic, and sadomasochistic gaze of the movie viewer, seeking an intimate identification with the screen's artful projections, which objectifies the viewer even more than the performer. Here, however, an element of real horror appears—in contrast to the prior parody.

The movie viewer is lured into a "theory of mind" about the victim onstage, in a *hall of mirrors* with the watching audience in the vampire theater onscreen. Those spectators onscreen look aghast at the stage, as if speculating about how real the fear of death is inside the victim's mind as she stares at them seeking help. This sets up a "mind reading" (by the real movie audience) of minds (in the audience onscreen) reading a mind (the victim onstage) reading their minds in the vampire theater, or at least an imagining of the others' inner theaters by the actual movie viewers.[22] In reality, the actress offers her body, too, to the others' desires, although her character resists, as Santiago unveils her breasts. Then Armand enters, takes over the scene, and disrobes her completely. He shows a greater seductive power, calming the girl and offering her a youthful death with "no pain," instead of having her "pink flesh . . . become gray with age." Her naked body collapses in his arms and he lifts her, over his head, to a swarm of vampires behind him. Then the movie audience is given an overhead, God's eye view, as the anonymous vampires crawl bat-like around the girl to feed and she disappears under their black capes. This might again evoke fear, awe, lust, disgust, joy, and sorrow in viewers, toward *rasa* refinement and cathartic awareness, especially about the mortal passions of aging and feeding.

At first, Louis and Claudia enjoy the show from their box seats, with the avant-garde performance of "vampires pretending to be humans pretending to be vampires" (as Louis says). But then, like the human audience in the vampire theater, they become *disgusted* at the sacrificial spectacle—perhaps honing this *rasa* in movie viewers toward multiple moral reflections. Later, Louis and Claudia follow Armand on a tour of the vampires' lair, in the catacombs of the cathedral-like theater. Santiago reads Louis's thoughts about Lestat and warns him that the worst crime for a vampire is to kill another vampire. But Armand takes a liking to Louis, calling him beautiful as a Zeitgeist, "the spirit" of his age. Homoerotically, Armand offers to the teeth of Louis the hand of a human boy he keeps at his side. Claudia sees Armand and Louis forming a new couple—and feels even more the lack of a mother and of her own womanhood, as she foresees her male-mother leaving her.

Claudia demands to have a woman, a victim of Louis, turned into her vampire-mother, before he leaves her. The woman is willing and Louis complies. But then a horde of theater vampires captures them, gleefully putting Claudia and her new mother in a deep well (like the one that Santiago made the fictional monk fall into onstage)—where the sun roasts them. Louis, trapped by these left-cortical slayer-vampires in a coffin behind a brick wall, is then saved by Armand. But Louis is too late to rescue Claudia and her newly made mother, who holds her as they both turn into ash sculptures. Here again, the movie shows its peculiar misogyny, with a perverse patriarchal horde (including a redheaded female) lacking a nurturing mother figure, yet sacrificing Claudia and hers, like they sacrificed the naked woman earlier onstage. In both scenes, Armand functions as a corpus callosum, mediating between the left-cortical competitive predators and right-cortical, more cooperative and nurturing audience or Louis/Claudia/mother figures.

Louis gets a spectacular revenge against them, burning some in their coffins, slicing Santiago into pieces with his scythe, and burning down the vampire theater of cruelty, with his own twists of right-cortical sorrow and left-cortical vengeance. Armand escapes and even compliments Louis for his alpha act, telling him to rise above his "regret." Yet Louis refuses to do so, retaining that tragic human flaw. A century later, in New Orleans in 1988, Louis meets a weakened Lestat and again refuses to join him. And yet, while apparently surpassing Lestat, Louis still feels himself a failure when the reporter, Malloy, wants to join him as a vampire. His story (and perhaps the film) has appealed to his audience too much as a melodramatic thriller, not enough as a tragic warning. Enraged at Malloy, Louis squeezes his neck, lifting him to the ceiling, but then disappears.

Driving away from the interview room in his red convertible, Malloy thinks he has escaped with a great story. But then he meets another master and potential father-mother-Creator for his desire. Lestat appears in the car seat behind him, bites his neck, and takes over the wheel, promising him "a choice" he never had. This ending leaves the movie audience, in their seats, with a choice, too: is the film a melodramatic thriller or a tragic lesson?

Other vampire films have offered historical fantasies (such as *Jesus Christ Vampire Hunter* in 2001 and *Abraham Lincoln: Vampire Hunter* in 2012) or theatrical reflections of past and current media (as with *Queen of the Damned* in 2002, a sequel to *Interview* with a rock-star vampire). Many, such as *Blacula* (1972) and *Scream Blacula Scream* (1973) with a former African prince as black vampire, *Def by Temptation* (1990) with a black female vampire, *Vampire in Brooklyn* (1995) with vampires from the Caribbean, and the *Blade* series (1998–2004) with a black vampire hunter, put racial and class twists on the vampire myth.[23] Even the recent *Batman* film series (2005–12), with its monster/slayer antihero inspired from comic books (like *Blade*), might be seen as a tragicomic twist on the vampire bat-Count myth (Pizzato, *Inner* 210–12). But *Interview* offers direct insights about the perverse impulses of the human animal through historical, theatrical, racial, and class contexts—depicted in immortal, yet tragically flawed vampires. Humans are extremely successful as a highly reflective, creative, and experimental species, colonizing and changing much of the globe. Yet our aspirations toward immortality, with increasing technological powers, have challenged the balancing acts of Nature. Indeed, some of the worst cruelties of our species have occurred through "white" cultures colonizing or expelling others, as enslaved subjects and genocidal objects, in the last half millennium.

Akin to the white-faced vampires in *Interview*, many consumers today of various races are tempted by the pleasures of upper-class wastefulness, especially through the Western-led mass media. But like Louis (and eventually Claudia) more than Lestat, Santiago, or Armand, we might become regretful about the tragic errors and immortal temptations of our brain theaters—so as not to be so wastefully destructive, even if we perceive Nature or God that way. Otherwise, our hunger for pleasure, meaning, and fame, or for vengeance, beyond balanced animal instincts, may lead to more suffering and sacrifices for ourselves, our offspring, and others around us. A new generation of vampires and their werewolf relatives, emerging in the late twentieth century, showed this, too, with supernatural teenage appetites, reflecting an insatiable, youth-fetishizing, mass audience and the "slayer" needed for self-control.

RABID TEENS WITH PACKS AND SLAYERS

The heroic slayer in earlier *Dracula* films becomes a weak reporter in *Interview with the Vampire*, learning about modern vampires with his notebook and tape recorder, without the proper book knowledge and pest-fighting tools. Yet both types of characters, the slayer and journalist, represent distinctively human, left-cortical, language and abstraction processes encountering the right-cortical, limbic, and subcortical animal energies of vampires. As with Malloy's intimate interview with Louis, and further contact with Lestat, sympathy and terror were evoked by werewolves and vampires in various films from the 1980s to 90s—showing them as angst-ridden yet feral adolescents.

Even prior to the youth trend with these cinematic monsters, early depictions of adult werewolves evoked sympathy for their split personalities. Wilfred Glendon (Henry Hull) in *Werewolf of London* (1935, dir. Stuart Walker) and Larry Talbot (Lon Chaney Jr.) in *The Wolf Man* (1941, dir. George Waggner) were each Jekyll and Hyde characters, transforming painfully and then becoming horrified at the results of their killer instincts, especially toward a beloved female. Likewise, Phyllis Allenby (June Lockhart) feared that she may be a "Wolf-Woman" causing the brutal violence around her in *She-Wolf of London* (1946, dir. Jean Yarbrough). These werewolves struggled against their dark identities, unlike the Counts in early vampire films. But ties were also made to the 1931 *Dracula* with Bela Lugosi playing the werewolf (named Bela) who infects Larry Talbot in *The Wolf Man* and with sequels, like *House of Frankenstein* (1944) and *House of Dracula* (1945), featuring both kinds of monsters, plus a lab-created zombie. All three monstrous types reflect the horrors of fascism, war, and the holocaust in the 1930s–40s, yet also harken back to much earlier fears of humans and animals, even pets, turning rabid—as if with an undead fury.[24]

Weeks, months, or even years after being bitten by an infected animal, the rabies virus may enter the brain of its victim (if the modern vaccine is not given) and monstrous symptoms result (Wasik and Murphy 7). Pain usually returns at the wound, even if healed by then. Fever and fear of water follow, along with extreme thirst (8). In the last stage of the disease, cries of agony may turn into animal-like barks, with frenzied biting, hallucinations, and hypersexuality (9). Yet, as with early werewolves onscreen and the reflective vampire Louis, lucid moments occur, allowing the rabid human to realize the horror of his or her transformation (10).

Werewolf of London suggests this, too, with British doctor Wilfred Glendon being bitten in Tibet, yet trying to cure himself with a "mariposa" plant he brings back to London, which blooms in moonlight, as the only known antidote to the rabies-like "epidemic" that infects him and others.[25] In a more recent film, *The Howling* (1981, dir. Joe Dante), a doctor administers a rabies vaccine to the victim of a wolf bite. But later in the film, the doctor turns out to be the infected leader of a "colony" of werewolves. In *Wolf* (1994, dir. Mike Nichols), a book editor, played by Jack Nicholson, becomes infected with lycanthropy, hunts a deer, expresses a feral power at the office, kills his wife, and infects his girlfriend and competitive protégé—with parallels of animal-human lust in the woods, at home, and at work.

Tales of werewolves go back at least to ancient Rome (Frost 5, 50). The word, meaning "man-wolf," occurs in an 11th-century Anglo-Saxon church ordinance, referring to the devil (5). Yet the now-familiar mythic elements—bite infection, full-moon transformation, rage to kill a lover, and death by silver bullet—did not become consolidated as a stereotype until 20th-century cinema. Rabid youth joined the cinematic paradigm in 1957 when *I Was a Teenage Werewolf* (dir. Gene Fowler, Jr.) extrapolated the bestial characteristics of adolescent bodily changes and male testosterone surges to wolf hair, teeth, and ferocity. Two more teen-monster movies were released in 1957 (both directed by Herbert L. Strock): *I Was a Teenage Frankenstein* and *Blood of Dracula*, with a male teenager turned into Frankenstein's monster and a female into a wolfish vampire—by an unethical teacher/experimenter/hypnotist. Such films may be a response to postwar, Rock and Roll gyrations. But they also foreshadow the increasing adult fears of a baby-boomer generation coming of age in the 1960s during the hippy, antiwar, feminist, and civil rights revolutions. As Miss Branding, the teen vampire's teacher, says in the latter film, "I can release a destructive power in the human being that can make the split atom seem like a blessing." These films also reflect the split self in each of us, especially during the wildness of our teen years, when the wiring of the brain's frontal-lobe controls are not fully insulated (myelinated), allowing more impulsive behaviors than a decade later in life.

A teen werewolf came to the screen again in 1981 with *An American Werewolf in London* (dir. John Landis). The film begins with two American boys, Jack and David, being attacked by a werewolf in rural Yorkshire, after leaving an unfriendly pub and getting lost on the moors. This may show a fear of Old World folk passions infecting modern American individuals in the postcolonial era of a global superpower, also threatened by Iranians students as terrorist kidnappers. With comic twists to the horror, Jack dies on the moors but comes back as a grotesquely decayed ghost haunting the infected David (mixing *rasas* of fear, awe, disgust, and humor). David recovers his health in a London hospital and falls in love with his nurse, but also suffers transformations that turn him into a killer. His London victims haunt him, too, counseling suicide like Jack, who says that his boring "limbo" will not end until David dies as "the last" werewolf. There are witty references to the werewolf movie tradition and David follows the call of nature back into a human trap, sleeping in the wolf cage in a zoo. He wakes there, naked and scared, but escapes by climbing over the fence, showing his primate brain's alternate evolution, as he glibly tells his fellow wolves to "take any calls." Yet the film also reflects on the dangers of cinema as an erotic, objectifying medium. It climaxes with David attacking fellow viewers in a porn movie theater, then others in Piccadilly Circus, before being cornered by the cops, offered the love of his nurse again, and slayed.

More purely comical versions of teen werewolves appeared in the 1980s. In *Full Moon High* (1981, dir. Larry Cohen), Adam Arkin's character is bitten while on a family trip in Transylvania and then tries to keep his transformations secret while back at his American high school. In *Teen Wolf* (1985, dir. Rod Daniel), Michael J. Fox finds his inherited wolfishness useful for high-school basketball, becoming a super-primate also. In the sequel, *Teen Wolf Too* (1987, dir. Christopher Leitch), Jason Bateman gets animal power as a boxer in college. In these sports films, each teen eventually chooses to compete as a human, rather than a super-natural "wolf," even while trying to win a pretty girl. This may suggest the physical and moral dangers of steroid use by athletes, but in a light-hearted way. Such films also set up the darker comedy of MTV's popular *Teen Wolf* series, starting in 2011, with a werewolf trying to keep his animal identity a secret in high school, while finding others like him with super-natural senses and strengths in hierarchal packs.

Another comic-horror film in the 1980s presented an adult female vampire seducing a teen and transforming him. In *My Best Friend is a Vampire* (1987, dir. Jimmy Huston), Robert Sean Leonard plays Jeremy, who finds his sexual initiation is more than he expected, with bodily changes and vampire slayers on his tail. But a wise teacher, Modoc, gives him a guide book and explains that vampires are just a persecuted "minority group." Jeremy's parents also worry that he might be homosexual and promise to help him with that "problem." Yet the adult vampires convert the lead slayer, McCarthy, into one of their kind—even as he tries to free the "soul" of one of the teens. Jeremy then introduces his girlfriend to his parents, reassuring them that he can appear normal, at least in a heterosexual sense.

Such associations of vampires with gays and other minorities reflect an important development in werewolf films as well. In the early films, adult werewolves, Wilfred Glendon and Larry Talbot, became alienated like many teenagers, ashamed of their passionate drives and lack of self-control, even to the point of wanting to die. (This was stressed even more in the 2010 remake of the 1941 film.) Likewise, David in *An American Werewolf in London* is tempted toward suicide by the grotesque ghosts of his victims and his former buddy, Jack. The love of an older woman, his nurse

Alex, is not enough to save him—as a lone werewolf trapped in the city. However, a growing sympathy in the 1980s toward minorities in America, even those with alternative sexual orientations (through greater awareness of an AIDS epidemic), and toward teenagers as a powerful consumer underclass, became reflected in many werewolf and vampire films. Increasingly, werewolves and vampires, adolescent and adult, gained communities onscreen—suggesting changes in viewers' "social brain" networks (Matthew Lieberman), as well as in mass and social media consumerism, from Baby Boomers to Gen-Xers to subsequent Millennials. And yet, such right-cortical variability of sexual, screen-media identifications also increased the appeal of a left-cortical *slayer*, sometimes with female strength and youthful ethics beyond patriarchal oppression and alien fears.

In 1997, *An American Werewolf in Paris* (dir. Anthony Waller), a sequel to the 1981 film set in London, again showed the American fear of Old World infection—but added a female werewolf as slayer of her kin. Initially, the French femme fatale, Serafine (Julie Delpy), appears to commit suicide by jumping off the Eiffel Tower, which causes an American teen, Andy, to "fall" in love with her, as he tries to save her from the fall. But to find her again, he must confront the older male community of her hybrid species. They live in a "villa" with Serafine and they "love Americans," luring tourists into a nightclub and Gothic church as killing zones.

Andy, infected with the werewolf virus, yet maintaining more morality, loses one of his American buddies in such a bloodbath and then comically sees his ghost, plus that of a lusty American girl his werewolf alter-ego killed (mixing *rasas* of fear, sorrow, rage, humor, and eroticism). With the help of another American buddy, bolstering their heroism (and becoming a Christ figure), Andy and Serafine use werewolf fierceness to slay their fellow hybrids, who were cloaked as monks in the church's tourist trap. Ultimately, Andy and Serafine accept their animal curse, as a power tamed by love and marriage. But they still need extreme thrills, like bungee jumping off the Statue of Liberty, to express their super-teen passions as a new balance of right and left cortical networks: with fully present, ecstatic awareness, yet a logical lifeline.

Packs of werewolves and vampires, as bestial consumers, appeared in various films in the late twentieth and early twenty-first centuries—like the shopping mall zombies in *Dawn of the Dead* (1978), but with greater energy, requiring new forms of slayers. Perhaps this relates to a real-life shift after the 1960s, which social psychologists observed, from organized gangs of teenagers to "packs" that were loosely structured, engaging in random violence, "with little empathy for the victims" (Staub 395, citing an unpublished work by Saul Scheidlinger). Reflecting this onscreen, a strange pack of vampire drifters, of various ages from child to adult, appeared in the American southwest, as a bullet-proof, serial-killing tribe in *Near Dark* (1987, dir. Kathryn Bigelow). One of them, the beautiful teenager Mae, lures a local human teen, Caleb, into being bitten and joining their immoral kin group. But eventually Mae joins Caleb in resisting the pack, helping to save his little sister from a lusty little-boy vampire and his reckless, bloodthirsty anti-family. And yet, it also takes sunlight to slay them, as a left-cortical containment of right-cortical misrule.

Similarly, a motorcycle gang of young vampires appears in *The Lost Boys* (1987, dir. Joel Schumacher), with a girl from their group seducing a teenager, Michael, who recently moved with his divorced mother, Lucy, and younger brother, Sam, to the decadent beach town of Santa Carla. They transform Michael and claim him, with blood in a wine bottle, during a party in their cave lair, where they later hang upside-down, bat like. But comical twists occur as Sam joins a pair of preteen, comic-book inspired, overly confident *slayers* in order to save Michael, plus his girlfriend and a little boy she cares for, from becoming fully vampire with the gang. They must also battle a perverse father figure, Max, a vampire leader formerly hidden in the normal adult community, as a video shop owner on the boardwalk—suggesting that vampirism lies under the surface of many aspects of pleasure-seeking consumerism in the Reagan era, including movies.

More recently, *Blood and Chocolate* (2007, dir. Katja von Garnier), showed an adult pack of werewolves in Bucharest as human free runners in the city and wolves hunting a human scapegoat in the woods, with a female in their group rebelling against them due to her love for an American male, as in *An American Werewolf in Paris*. Reviving traditional names and associations, yet with videogame hyperactivity, swarms of bat-vampires appeared in *Van Helsing* (2004, dir. Stephen Sommers) as the brood of Count Dracula and his three wives, along with Frankenstein's monster and an antiheroic Van Helsing.[26] He eventually becomes a werewolf and kills his beloved mate, who had been a slayer with him, and then howls in tragic mourning. *Underworld* (2003), *Underworld: Evolution* (2006), *Underworld: Rise of the Lycans* (2009), and *Underworld: Awakening* (2012) involved modern battles with high-powered weapons and silver nitrate bullets, between hip, black-leather wearing vampires and grungy, street-smart werewolves (Lycans) as slayers to each other in a multi-era class and clan war.

However, the remake of *The Wolfman* (2010, dir. John Johnston), starring Anthony Hopkins and Benecio del Toro, stressed a father and son rivalry, as werewolves, rather than clan warfare. It develops tragic edges and ironic *rasas* with the slaying of two sympathetic beast-people by those who loved them: a father killed by his son in a werewolf fight and that son slayed then, reluctantly, by his girlfriend. (Yet the film failed to make a killing at the box office.) Set in the 1890s, it reflects connections between theater, movie fantasies, and real life, with the son introduced as a troubled Shakespearean actor, suspected of violence because of his tragic roles onstage, and then becoming monstrous, like his father, who was infected by a werewolf bite many years before (also suggesting that werewolves age, unlike vampires).

Kinship is a powerful animal drive in humans, as in other mammals that evolved to nurse their young and survive in extended family groups. Even in current Western cultures that shape many young brains toward an independent sense of Self, the social circuitry of others' views are crucial to one's developing identity. Neuroscientist Matthew Lieberman provides evidence that even with the Western stress on independence, we have a "Trojan horse self" that soaks up social influences "without realizing where these foundational worldviews came from" (235). Like the ancient Greek trick of putting many men into a huge wooden horse, as a gift to Troy that ends a long war, evolution has created a brain theater in us where ideas, narratives, and feelings of Self (or free will) emerge from the networks of mother-infant "connection," child and later life "mind reading" (mentalizing about others' views), and preteen to subsequent "harmonizing" with society, through self-control (11). These three developmental stages also relate to our brain's mammalian, primate, and distinctively human evolutionary structures, according to Lieberman.

Peer groups and counter-cultures, along with mass advertising and social media networks, greatly influence our remnant animal instincts, mammalian emotions, and ape egos—as we break away from early life, family values. In movies with werewolf and vampire packs, subcultures are shown that express the 1960s lure of "free love" (or love-free sex) as a communal fountain of youth, along with free running and other primal appetites. But they also show the abject superficiality of selfish pleasures, especially with left-cortical, dopamine and right-cortical, adrenaline feedback loops fueling our brain's contagious passions, fantasies, and addictions. In 1980s–90s vampire and werewolf films, such a theater of wild desires is often confronted by romantic love, evoking another primal instinct, with oxytocin networks for long-term mating and child care. This gives such hybrids, recast as slayers or as vampires and werewolves with self-restraint, an alternative, social-brain power.

GIRL-POWER GODDESS AGAINST THE VAMPIRE FATHER-MOTHER

While most vampires are male, female vampires have haunted the screen, too, as examples mentioned earlier suggest—involving men and sometimes women in a slaying role. Female werewolves (like Serafine) appear less frequently but also offer intriguing gender twists. In *Dog Soldiers* (2002, dir. Neil Marshall), British soldiers in a war game in the Scottish Highlands become infected by werewolf bites, as experimental pawns of the patriarchal military, through contact with a local family of werewolves. The soldiers coordinate their defenses in the family's empty home, killing the werewolves as intruders into *their* territory, with pack-like camaraderie, some sacrificing themselves for their buddies. But the soldiers are betrayed by a female zoologist, who turns out to be a werewolf, too, like some of the bitten soldiers. Just one of the men survives, along with a pet dog. In the Canadian horror film *Ginger Snaps* (2000, dir. John Fawcett), a teen's start of menstruation, with wild animal emotions, is related to her werewolf symptoms of a tail, fangs, and killing sprees. Her sister tries to cure her with a plant antidote but kills her accidentally instead. (The film had two sequels in 2004.)

Such slaying or curing of male and female monsters for the human community's survival, regarding home territories, became the focus of a comic horror film, *Buffy the Vampire Slayer* (1992, dir. Fran Rubel Kuzui), which spawned a very successful, yet more serious TV series, created by the film's screenwriter, Joss Whedon (1997–2003). Focusing on a blonde, high school cheerleader named "Buffy," the movie and TV series mixed romantic and communal relations of post-feminist pleasure and purpose, with the "Chosen One," her male teachers, and her slayer friends fighting various forms of (or becoming) the undead. Influenced by comic teen and horror romance precedents of the 1970s–90s, *Buffy*, as a film and TV series, showed a millennial generation's right-cortical, limbic, and brainstem ambiguities, mixing various conflicting *rasas*, sometimes with tragicomic insights. It also presented a young woman in the traditional, left-cortical *slayer* role, restoring some degree of order, while joking about saving the human race and facing the tragic, inner conflicts of these brain functions.

Film historian Stacey Abbott finds a turning point in horror cinema with George Romero's zombie films of the 1960s–70s, involving the violent reality of human hunger. She also cites Romero's *Martin* (1976), with its human serial killer who jokes about being a vampire, or may believe he is one, yet uses a blade (like in *The Hunger*), instead of fangs, to kill and drink the blood of his victims, as a sexual perversity (116–21). From then on, the screen vampire, even when super-natural, became like the zombie, according to Abbott, "a despiritualized being, defined by the abjectness of its corpselike body and its hunger for human flesh and blood" (121). But I would argue that such postmodern vampires also reveal a new craving for meaning, from natural drives to super-natural desires, even if not in a fully credible cosmic theater. Vampire and werewolf films from the 1980s to 90s and into the new millennium show teenage and romantic monsters who develop restraint, sacrificing selfish pleasure so they can live more meaningful lives, or even find a mission as *slayers*, stopping the bloodshed by others.[27]

Similarly, but with a broader scope of examples, Emily McAvan has found a "postmodern sacred" in our mass-media rites. Looking at fantasy, science fiction, and horror, in novels, films, and television shows, McAvan finds that

> the postmodern sacred is a paradoxical attempt at accessing spirituality, using the symbols contained in explicitly unreal texts to gain a second hand experience of transcendence and belief. This . . . displaces the need for belief or real-world practice into a textual [and virtual screen] world, requiring little of its consumers . . . [except that they] must return again to purchase another text. (19)

And yet, I would argue, this also shows a new form of sacrifice by the reader/viewer to the media apparatus, as fan of a specific genre, show, or series and its fictional, super-natural gods—through a fetishistic, real-world ritual, even with the denial of full belief.

The consumer of postmodern sacred fiction, with vampires and other super-naturally ambivalent, good and evil creatures, becomes a vampiric subject, craving their virtual, yet transcendent blood, as if it were *ichor*, the mythical, golden fluid in the veins of ancient Greek gods, toxic to mortals. But such consumers also become the victims of this postmodern, mass-media communion, transformed with each bite into an illusory immortal being, identifying with the transcendent figure onscreen, while feeding it their blood by watching. In the weekly ritual of TV episodes, or the binge watching of shows through the Web, or the media events of new movie releases, or the interplay with their videogame offshoots, or the nostalgic watching of earlier films whose sexy stars do not age onscreen, we become the (m)Other to the human-animal super-natures in our media rites—with their immortality feeding upon our brain theaters and mortal lifetimes. They become our perversely ideal super-egos, promising fantasy pleasure and a playful, yet demanding purpose to life today, in relation to mythic dimensions of folklore and religion in the past or sci-fi projections into the future. But sometimes, vampire and werewolf films also show a rebellion of the undead, yet aging audience, as with Miriam's children rising from their coffins to hug her to death in *The Hunger*. Or they show a way to modify our contagious vampirism and virtual bestiality with the "chosen" Slayer and her friends, fighting a new generation of media-infected, social-networking, pleasure-seeking youth.

Buffy, as a film and TV series, exemplifies this playfully ironic, yet dangerously manipulative potential of a postmodern sacred theater, with a mass-media communion between super-natural monsters, the human slayers (as priestly teachers and warriors) fighting them onscreen, and their mortal movie audiences on earth. The film begins with the premise (continued in the TV show) that a young female "Chosen One" has been reincarnated across many generations to fight vampires, with one appearing in each generation, though it is left unclear how that "generation" or its geographical location is defined. Implicitly, this harkens back to Ellen in Murnau's *Nosferatu*, using her female beauty to catch the Count's lust and slay him by sunlight—without the "Crew of Light" from Stoker's novel. Likewise, Mina has a stronger role in Coppola's *Dracula* than in the novel, as both seer and slayer of the Count, though in the novel, as in this film, she helps the Crew to track the vampire with her telepathic tie to him (McClelland 162–67).

In *Buffy*, a postfeminist heroine (Kristy Swanson, doing her own stunts) uses a cheerleader's gymnastic training, plus further action movie skills taught to her by a reincarnated "Watcher" named Merrick (Donald Sutherland), to slay vampires in her Los Angeles high school and local neighborhood. In the TV version, Buffy (Sarah Michelle Gellar) has a new high school, Sunnydale, as "Portal to Hell," and a new Watcher, Giles (Anthony Head). She also becomes more of a third-wave (collectively acting) feminist, not just an independent-minded postfeminist (Levine 170–76), leading her "Scooby Gang" of friends, including converted vampires, as vigilante slayers—against a new monster each week and a "Big Bad" villain each season.

Both the film and TV series involve a typically melodramatic, good versus evil formula, pervasive on American stages and screens since the 19th century. The comical sidekick to the hero, in that 19th-century formula (e.g., the bumbling assistant to Dr. Frankenstein, who was added to the early stage and screen versions of the novel), becomes in the *Buffy* film and TV episodes an intrepid sense of wit in the heroine herself, and in some of her colleagues, including a new boyfriend as romantic focus of the film. Yet tragic edges emerge, too, with inner conflicts, errors in judgment, vexed love affairs, and human-animal-monster transformations involving moral ambiguities, even challenges to patriarchal traditions that are part of the film/TV show's rules.

As Jana Riess has pointed out, the film and its subsequent series involve traditional Christian and vampire-fiction symbols (such as crosses, holy water, and church rites), plus Jewish and Wiccan references.[28] But they are driven more by ideas from Buddhism or New-Age Spirituality: from Zen with a privileging of experience over teaching and Tibetan Buddhism with the premise about a reincarnated Slayer who draws on knowledge from the past (xvi). The film even starts with Buffy in a prior lifetime, during the "Dark Ages" in Europe, with the same actress having red instead of blonde hair, meeting with an earlier incarnation of her Watcher (the same actor as later). A voice-over explains the rule about the "Chosen One" and then a Southern Californian gym is shown, with Buffy and her fellow cheerleaders, during the "Lite Ages."

Such wit is often used in the framework of the film or TV show—and by its young slayers, especially Buffy, as a weapon to startle the zombie-like vampire they are facing, before a sudden kick or punch, and then a stake to the heart (Riess 42–43). This provides not only comic relief for the mass audience, but also a sly connection to the heroes' very American, left-cortical optimism. Their can-do confidence becomes a trickster power (also using right-cortical alertness) against the grave undead—like that of comic servants against their masters in ancient Greek and Roman comedies, or later commedia dell'arte. And yet, for viewers young and old, the immortal youthfulness of the film/TV stars as slayer-heroes, especially of the Chosen One reincarnated in our time to stop evil's return, may reveal an uncanny, vampiric twinning (and *rasa* recipe) of the beautiful and grotesque, good and evil forms of youth onscreen, masking the aging of actors' bodies and of mortals watching in the mass audience.

Much has been written about the traditional archetypes and progressive feminist twists in the TV series, plus other issues in its many plot complications over seven seasons—with several Slayers added to the show, despite its premise about Buffy as the only Chosen One (because she dies twice and comes back to life, and alternate Slayers also die).[29] There is even an academic journal, *Slayage*, devoted to the series. But I will focus here on the original film of *Buffy* as exemplifying the postmodern, animal-human-spiritual, and vampiric lure of screen media today, "both subversive of and complicit in dominant culture and ideologies," as Lorna Jowett says of the TV series (2).

The film's flashbacks show the medieval, red-haired Buffy with an earlier incarnation of the teacher/Watcher and then fighting her nemesis

Lothos (Rutger Hauer) and his fellow vampires, with the latter framed as a dream of the current high school cheerleader, clinging to her teddy bear in bed. These prior Buffy scenes also relate to a modern film tradition of violent female superheroes, such as Ripley, Sarah Connor, and Nikita in *Alien* (1979), *Terminator* (1984), and *La Femme Nikita* (1990), or other high-kicking women on TV that followed the *Buffy* film, such as Xena and *Alias*'s Sydney (cf. Jowett 20). There has also been the rare female vampire or werewolf who aided men in killing her kind, such as Mina (after she was infected by the Count) and Serafine. But there have been few if any human females with a mythic mission to kill vampires, prior to Buffy.

Like the male hero on a folkloric quest, especially in movies based on Joseph Campbell's "monomyth" formula, Buffy at first refuses the call to be a Messianic warrior (Frankel 6–12). But the film foreshadows her inevitable destiny with a visual overlap, crossfading from a cruciform wooden dagger held up by the medieval redhead (who says, "I shall be his sword"), in the initial flashback scene, to a pompom held by the blonde cheerleader today. Later, when her Watcher (Merrick) confronts Buffy alone in the school gym, claiming that she has a mission to kill vampires, she finds him "creepy." Yet he convinces her of her mission by knowing about a mole on her left shoulder as mark of the Slayer (though she already had it removed) and about her dreams of fighting vampires in earlier time periods of history, including their leader, Lothos, whom she also dreams about holding her in bed.

Merrick shows Buffy the urgency of her mission by taking her to the local graveyard. There, as if inspired by Romero's zombie films, two newly dead (a male and female) crawl out of the broken earth of their graves and attack Merrick and Buffy with vampire teeth—until she stabs them with wooden stakes that Merrick brought. Later, Buffy skips a training session, so as not to miss cheerleader practice. Merrick finds her in the women's locker room at school and throws a metal dagger at her head. She catches it with her left hand, without thinking, but then becomes quite verbal in complaining about it. Thus Merrick, as archaic patriarch, threatens Buffy's physical survival at the graveyard—and again with a phallic dagger, triggering her right cortical and limbic reaction: a supernatural awareness, fear, and action response on the left side of her body (controlled by the right sensorimotor cortex). Merrick bypasses her inner theater's left-cortical identity categories of cheerleader, mall rat, fashion snob, and dance planner—shown in earlier scenes—with a sudden knife test. But he also shows a new superego *demand* for her body and time, to be the chosen Slayer of vampires at her school, who attack with fangs.

Initially, Buffy is torn between the demands of her peer group for cheerleader practice or school dance discussions, along with her *desire* to kiss her jock boyfriend, and Merrick's for her Slayer training. But Merrick forces Buffy to realize her talents with the old phallic technology of stakes and daggers, illustrating the main themes of the film and its subsequent TV series.[30] Can today's generation of young women reconnect with a male-dominated heritage and improve it for the future? Can they balance a desire for sexual pleasure and long-term romance with a feminist wariness about being objectified and possessed by the male gaze (or vampire's craving)—becoming active agents of a phallic yet feminine "girl power"?[31] Such a postfeminist reconnection to sexiness and romance reveals a conflict for men, as well as women, in the movie audience, even as older male Watchers, teen girls, and the bestial vampires between them exemplify changes in cultural and psychic pressures, from the Dark to "Lite" Ages.

Aside from the archaic Merrick and vampire-leader Lothos, all the adult males in the movie are fools, including a school administrator, who reminisces about his own drug use while not noticing Buffy's superpower with a tack that she spits at a distant fly, and her father, who just tells her to "stay away from the Jag." Buffy's mother is also oblivious to her boyfriend's name and her Watcher's training program, while more concerned about a new wristwatch not working. Buffy soon realizes that most of her friends are petty minded, too. She goes from the jock, Jeffrey (after he breaks up with her on her answering machine), to a lower-class, automechanic boyfriend, Pike (Luke Perry), who saw his best buddy, Benny, floating with fanged teeth, outside his window. Though others at school do not notice such vampire transformations, Pike joins Buffy in fighting the fanged teens in various scenes. Pike also encourages her to return to Slayer duty after Merrick is killed—when she avoids mourning and mission dangers, buying a dress instead for the school dance.

Merrick's paternal nurturing of Buffy's Slayer talents and Pike's romantic appeal, as the only peer to support her mission (like a comic yet cagey sidekick), both relate to a rival claim on her by the perverse father figure, Lothos, who arises from his crypt to battle and bite her as an archaic mate. He and his vampire assistant, a male with a beard but a female name, Amilyn (played by Paul Reubens, better known in the 1980s as the children's TV star, Pee-Wee Herman) are the source of the fanged teens that appear around Buffy, whom she must kill as evil, though she knew them before as classmates and party friends. One vampire also appears as a player on her high school's basketball team (as in *Teen Wolf*). Thus, Lothos and Amilyn represent an archaic evil, plus homoerotic fears, erupting through the surging hormones, peer pressures, and identity confusions of high school students, indirectly reflecting the AIDS, drug use, and school shooting contagions of the 1980s–90s.

But Lothos expresses a romantic claim on Buffy across many lifetimes, as a perverse patriarchal figure with archaic *left*-cortical predatory rules, awaking from his crypt to meet her postfeminist challenge—while she puts holes in his teen protégés with phallic wooden sticks. "It's time to put away childish things," he tells her, during their final battle together. "You and I are one." But Buffy replies: "One what? Cute couple? I don't think so." Overall, the romantic comedy question of the film involves five potential matches for Buffy: Jeffrey the basketball jock, his buddy who fancies her butt (coveting it when it crosses his chest in Jeffrey's convertible and grabbing it in the school hallway, though Buffy then flips him to the floor), Pike as a better teen alternative, Merrick as dream interpreter and mission giver, or Lothos as dream danger threatening her world with his unreality. These options also reflect various identifications for the Watcher of the movie—the film audience as Other to Buffy—as male gaze, young or old, objectifying her (and Kristy Swanson), yet sacrificing time to her immortality onscreen.

Eventually, Buffy crosses the class and cheerleader elitism of her peer group (mostly white, though one of her girlfriends is black), in a rebalancing of left-cortical norms and right-cortical/limbic passions, through a *new desire* for Pike as the boyfriend who values and helps her as heroine. But she also finds herself reborn into an archaic rivalry between the Watcher and lead vampire as father figures—representing the emotional *drives* of her inner theater and the new *demands* of her postfeminist (and premodern) social context, regarding the *needs* of her and the vampires' bodies for survival and sexual pleasures. In my view, this Lacanian distinction, between need, drive, desire, and demand, relates to (1) the primal needs of our species for food, safety, and sex in the vertebrate brainstem and subcortical areas of bodily maintenance, involving survival and reproduction instincts, (2) our mammalian limbic emotions, as culturally shaped drives, focused on particular objects of pleasure or pain, (3) our neo-mammalian, right-cortical, Imaginary desires for various ideal identifications, and (4) our left-cortical Symbolic demands for limiting and judging the deeper animal passions. But these brain areas, evolutionary legacies, and Lacanian cultural dimensions also relate to a distinctive human need, drive, and desire: for a sense of *purpose* in the given social demands, which Buffy exemplifies in accepting her mission as the Chosen One.

Merrick tells Buffy, during her training, that he "must not interfere" in her fights, he can only train her. Later, when Merrick violates his cardinal rule, raising a dagger to kill Lothos, as the head vampire seems about to bite Buffy, Merrick gets slayed instead. Yet he also taught her that her menstrual cramps are a natural signal that a vampire is near, which she can use as super-natural Seer, as well as Slayer. A father-daughter bond developed between them, as he gave a greater purpose to her life and redefined the pain of Mother Nature's demands on her as a young woman—the instinctual needs of her womb, as vehicle for her genes' immortal reproduction. Merrick's rule breaking and sudden death, out of concern for her, makes her womb pain and its Chosen purpose even more poignant. This produces a new network of inner conflicts between vertebrate-brain needs (survival/reproduction), mammalian-limbic drives (fear/revenge), right-cortical desires (for personal or communal protection), and left-cortical demands (of her Slayer duty)—in Buffy and her sympathetic fans.

Buffy's hand touches her womb and she winces a bit each time she nears the danger of the fanged ones and prepares to slay them, in various scenes that build to her battle with Lothos. In a sense, she is aborting each one she kills, limiting the contagious, bloody womb (or Kristevan *chora*) of Lothos and Amilyn's monstrous fertility. Amilyn loses his arm in a fight with Pike, but continues, although symbolically castrated, to lust, fight, and feed. Eventually, Amilyn is staked by Buffy, as a "gift" from Lothos, who taunts her about the intimacy of that slaying as her first "real" kill. Thus, Buffy's menstrual cramps signal not just the polymorphous perversity of Lothos/Amilyn's fanged offspring, but also her own bloody power. She must develop a new left-cortical game plan to contain the genetic needs and painful pleasures of her body, as well as her confused identifications with different men in her life, replacing the upper-class parents who overlook her. Even her good father/teacher, Merrick, tells her, while he is dying, "Don't play our game."

Buffy wears what looks like a white wedding dress to the Senior Dance. But Pike helps her rip the lower part off, so she can fight the vampire gang outside the gym, before facing off with Amilyn and then Lothos. This suggests she is a virgin bride, fighting to resist a *deflowering* by the various vampires she faces. Such a challenge climaxes with the Droit du Seigneur claim of Lothos (like Count Dracula taking Mina from Harker in Coppola's film). Several times during this film, the gaze of Lothos puts Buffy into a trance, which neither Merrick nor Pike can wake her from. But in her final battle at the dance, Buffy recalls the dying words of Merrick, "when the music stops, the rest is silence" (quoting Hamlet's final words, but perhaps meaning here: when the party pleasures and consumer illusions end, we realize our mortal aging). Merrick's ghostly words give her enough left-cortical distance from her archaic father/lover, Lothos, that she can fend of his samurai sword with a California flagpole and slay him then with the broken leg of a wooden chair.

Early in the film, when resisting her fate as the Chosen slayer, Buffy tells Merrick that she just wants to graduate from high school, go to Europe, marry the movie star Christian Slater, and die. She says nothing about having children. Her menstrual cramps, demanding some recognition of that possibility (through her inner-theater stagehands), are deferred toward the greater purpose of vampire detection and slaying. And yet, once she has staked all the vampire teens around her and beaten the Big Bad Lothos, she turns in a still white dress toward the loving arms of her new boyfriend and appears ready to marry the movie star, Luke Perry, with a new purpose, perhaps, for her menstrual cramps.

In the subsequent TV series, with a different actress playing Buffy, there are many more permutations of the Chosen One. Buffy appears as a morally good and bad, rational and yet emotional Slayer, juxtaposed in various seasons against the rule-following, Caribbean-accented, black Slayer, Kendra, and the wilder, lower-class, white Slayer, Faith (Jowett 45–47, 83–90). For a time, Faith occupies Buffy's body, fooling her friends. Buffy also dies twice and returns to life. She changes her social class, too, dropping out of college for a while and working under the pseudonym "Anne" at a fast-food diner, before returning to her mission and friends. At another point in the series, a robot version of Buffy turns up, a "Buffybot." Romantic relationships change her, too. In the first season, Buffy falls in love with a much older, reformed vampire, Angel, who turns wicked again when they have sex. In later seasons, she develops more perverse passions for another vampire, Spike, who has violent sex with her. But Buffy gains Spike's help in slaying other vampires, with the aid, too, of her human friends and Oz, a teen werewolf. This surrogate family with a super-natural mission, led by Buffy and eventually including her little sister, Dawn (adopted by Buffy, even after she learns that they are not biologically related), becomes a third-wave feminist collective that also involves the bisexual witch, Willow. The vampire-slaying adventures of Buffy's Scooby Gang thus reveal various inner conflicts in the changing social networks and erotic dynamics of friends and foes.

Yet, a key reproductive issue suggested by the original film—will Buffy's womb (and inner-theater stagehands) produce offspring to replace some of the fanged figures she slays?—is not addressed until the end of the seven seasons. In the final episodes of the TV series, Buffy, Faith, and Willow cast a magic spell that produces Slayers around the world as their offspring, while destroying their greatest enemies (with Spike's self-sacrificial death). But this does not involve Buffy's womb as a maternal force.

A novel and film series that emerged a few years later further developed the image of a teen goddess, as destructive warrior, yet creative womb, killing and coupling with vampires, through an alternative family structure again involving a male teenage werewolf. The *Twilight* books and movies (2005–08 and 2008–12) showed a love triangle between an archaic vampire, a Native American werewolf, and a teen human, Bella. She is romantically involved with them, not a chosen slayer. But she also becomes a vampire-fighter with them, heals an ancient rivalry between their species, and produces a new hybrid creature from her womb, tied to both her vampire and werewolf boyfriends. Thus, she becomes a post-postfeminist (or neoconservative) Mother goddess, as life-giving vampire, much different from Buffy as Slayer.

In the *Buffy* film, Merrick gives his protégé a crucifix as well as stake when they go to the graveyard to meet zombie-vampires. After his death, she tries to use a crucifix as a shield against Lothos and he mocks it, lighting it on fire super-naturally. But even with Buffy's "keen fashion sense," combining a can of hairspray with this traditional symbol of Christ's sacrifice to shoot the flame into Lothos's face, the weapon fails to stop him. She must use her own sacrificial body, with her training as a gymnast, kickboxer, and Slayer, plus an improvised wooden stake, to extinguish the source of invasive evil in the school gym—feminizing the archaic, perverse Father. She then dances with the lower-class Pike and rides away with him on his motorcycle, toward a distant, immortal, vanishing point. Buffy thus exemplified, in the 1990s, a teen goddess potential for young female viewers (becoming more complex and popular through the TV series) with Pike as a role model for men. However, with the *Twilight* series came another teen transformation: from virgin-girl to warrior-woman against vampire promiscuity and oppression, and then to an animal-human Mother goddess (akin to the Virgin Mary as the New Eve, stomping on the serpent in Roman Catholic iconography, yet here with Mormon twists), for the next generation of silver-screen worshipers.[32]

BELLA AS TRAGICOMIC GODDESS IN A NEW VAMPIRE-WEREWOLF FAMILY

The hugely popular *Twilight* novels, by Stephenie Meyer, were turned into five films by screenwriter Melissa Rosenberg, working with Meyer, and by four different directors, the first female, but the rest male: Catherine Hardwicke (*Twilight*, 2008), Chris Weitz (*New Moon*, 2009), David Slade (*Eclipse*, 2010), and Bill Condon (*Breaking Dawn, Part 1* and *Part 2*, 2011–12). The first film begins with a narrative voice-over by the lead character,

17-year-old Bella (Kristen Stewart), musing about death as sacrifice, while a deer is hunted and caught in the arms of an ambiguous human figure: "dying in the place of someone I love seems a good way to go." Bella remains the focus for the entire series, but less as a narrator[33] and more as the prime object of desire for the vampire Edward Cullen (Robert Pattinson) and the werewolf Jacob Black (Taylor Lautner), involving sex and death drives, with various sacrificial ideals. Like Buffy, Bella crosses class, but the other way, as she falls in love with Edward, moving from *middle to upper* and from human to superhuman when she meets, dates, and eventually weds him. He is also much older, over 100 years old, though he appears to be her age as a student in her new high school, because he was turned into a vampire at 17. Like a combination of Lothos and Merrick/Pike to Buffy, Edward brings a death-drive passion to his romance with Bella, craving her blood, but he restrains that vertebrate-mammalian lust, in order to purify his desires and hers toward long-term demands.

Bella, a child of divorced parents, has left her mother, who has a new husband, in Phoenix, Arizona, and now lives with her father in the very wet, wooded town of Forks, in western Washington, near the La Push reservation of the Quileute Indians, where Jacob lives. Bella arrives at her new high school feeling alienated, driving an old red truck that her father gave her. And yet, she becomes the "shiny new toy" at school (as an envious girlfriend says), akin to the strange Cullen kids, who are even more white faced. She soon falls in love with Edward Cullen and joins in the family activities, such as baseball, though they play with super-natural speed and strength.

As vampires, Edward and the other Cullens must control their vertebrate-brain's craving for Bella's delicious scent and flesh, through their mammalian-limbic social emotions, humanoid left-cortical rules, and prefrontal-lobe restraint. Aside from Edward, the other four Cullens at the high school are paired, like incestuous siblings, though all, including Edward, are adopted. Alice, who can glimpse the future, often holds hands with Jasper, who has the most trouble controlling his craving for human blood. Rosalie, who is still angry at being made a vampire and easily expresses limbic rage, stays close to the fun-loving Emmett. Their foster parents are Esme and the physician Carlisle. He turned them all into vampires to save their lives, at different points in the past, making them immortal. He leads them now, with strong (prefrontal cortex) self-control, teaching them to be domesticated "vegetarian" vampires who only drink animal blood, not human.

Edward Cullen desires Bella partly because he can read everyone's mind, human and vampire, but not hers. So the mystery of her brain's inner theater fascinates his—as soon as they meet in their high school biology class. Jacob, a Quileute Indian who goes to the reservation school, knows Bella from when she lived in Forks earlier, as a small child, before her parents divorced. Now in rivalry with Edward, he also wants to protect Bella from the Cullen family, because his Quileute tribe is descended from wolves (according to actual legends of that tribe in western Washington). In Meyer's and the film's fiction, the tribe includes members who are "genetic" werewolf shape shifters. That makes them "natural" enemies with the predatory vampires. Yet the tribe has a *treaty* with the Cullens, from a previous century, about living in peace and hunting animal prey in separate territories. Carlisle lived there before, but left and has recently moved back with his new family.

Bella thus becomes drawn toward rival kin groups, through the adjacent territories of Forks and La Push—like Buffy being torn between her cheerleader and Slayer identities, with her high school as a Valley Girl hell mouth. Bella is introverted and less athletic than Buffy, evoking different reader and viewer identifications. Bella does *not* get a sudden call to a mission or train for it as the Chosen One. But eventually, she is chosen by Edward and the Cullens as a family member and trains to fight vampires with them in the final film of the series. By that time, she helps to renegotiate the rivalry between the Cullen vampires and Quileute werewolves, so they and other vampire allies can fight together with her against a larger force of evil vampires in an overall melodramatic climax to the five films.

Even in the first film, Bella's value, as the object of others' desires, increases quickly with friends gathering around her as the new, exotic girl at school and with Edward at first resisting and then, with periodic absences for self-control, choosing to spend time with her. Yet he mysteriously resists her desire for more contact: "What if I'm . . . the bad guy?" Bella becomes suspicious when Edward saves her from an out-of-control car in the school parking lot—stopping and denting it with superhuman strength. He then saves her again by suddenly appearing with his Volvo to drive her away from four men who endanger her in a dark alley. Bella does research online and in books on the possibility of Edward being a vampire, as in the myths she finds from around the world, including Egypt, India, and Peru. She worries that the mysterious deaths in her community might be due to Edward. He thus exemplifies how desire may involve fear and fear may increase desire, especially in romantic attachments, when such limbic passions and right-cortical paradoxes (against left-cortical binaries, such as good versus evil) relate to deeper, reproductive and survival needs.

But Edward also takes a risk with Bella, admitting to her that her research is correct. He takes her to the woods and shows why his species avoids the sunlight—not (like Nosferatu) because it kills, but because it makes his skin sparkle "like diamonds" (or snow as shown in the films). This evokes a further passion in her, but he warns her against the lure of sensual beauty, as a deadly erotic temptation: "I'm the world's most dangerous killer; everything about me invites you in." He likewise finds it difficult to love her respectfully, due to his primal attachment disorder and blood-sucking drive: "You're like a drug to me." Edward appears to be Bella's age, with the actor, Robert Pattinson, appealing to movie viewers with his mature teen beauty.[34] (He was almost 22 when the film was shot.) Edward also combines the elder, elite, and yet perverse charm of Count Dracula, in early vampire films, with the self-disgust, inner conflict, and tragic awareness of more recent blood-drinking hero-villains, such as Louis and Miriam (or Oldman's Dracula)—potentially refining the *rasas* of awe, eroticism, and disgust in movie viewers, through ironic twists in the plot. Pattinson's Edward may thus provide a *warning* to fans of the film series, as well as Bella, especially to the teen girls who made Meyer's novels so successful—as addictive *consumers* of the vampire-lover fantasies they offer.[35]

In the first film, Bella dreams of Edward visiting her bedroom, or thinks it is a dream, because she sees him there and then he disappears. Later, he admits that he sometimes comes in through her window and watches her sleep, which vampires never do. Like the film viewer, Edward is a Peeping Tom, but instead of creeping Bella out, this bonds her to him even more. He then takes her on super-natural leaps between pine trees while she clings to his back like a primate infant. He also kisses her in her bed, while she is just in her underwear. Yet he leaps back then against her bedroom wall—careful not to become too passionate, with the vampire "venom" in his mouth that might kill her (Mukherjea 72). Here, Edward stages a primal erotic scene for the film fan, while insisting on abstinence, despite Bella's (and the fan's) desire for more. He is a poisonous snake-man, with fangs, yet he also evokes the potential for primal bliss (or Lacanian *jouissance*), like the serpent tempting Eve in Eden with a tree of greater knowledge of good and evil.[36]

At the climax of the first film, Edward must save Bella's life again, but becomes even more tempted by her blood. Three nomadic vampires, who were the ones killing humans in Forks, find Bella with the Cullens during their baseball game. One of them, James, desires her as a tasty "snack." The Cullen family tries to protect Bella, using her clothes to mislead the other vampire pack as it hunts her through the woods. But James finds her mother in Arizona, calls Bella, and then lures her to a ballet studio where she took lessons as a child. In the mirrored dance room, James shows Bella a video of her as a child, in ballet tights, telling her mother she "sucks." James takes a new video of her, terrified, when he is about to kill her, saying that such a performance onscreen will "break Edward's little heart" and then "his rage will make for more interesting sport." Thus, an evil vampire, who still consumes human blood, uses mammalian senses and predatory skills, plus human mirrors and video technology, to catch Bella and evoke a temporal-lobe ecstasy in her inner theater, involving emotional memories and present terrors, as homoerotic bait for another vampire—and for movie viewers' various gazes.

Edward arrives to save Bella, but she is bitten by James first. After the other Cullens arrive to finish the fight and destroy James, Carlisle advises Edward on how to remove the evil vampire's venom. When he sucks it from Bella's blood, removing the other male's evil, Edward is tempted to indulge too much in the pleasure of her body—again endangering her life. And yet, he is able to restrain himself, saving her from James and his own animal scripts. Edward turns a villainous hunger and sex drive (with dopaminergic pleasure) into limbic, nurturing (oxytocin) desire, with the help of his foster father's left-cortical categories and prefrontal-cortex demand: "Remember who you are."

This bonds Bella even more to Edward and other vampires. Primal pathways of her inner theater are transformed through the broken mirrors of the ballet studio, with James's terrorism, his destruction by the Cullens (Alice rips his head off and the others set him on fire),[37] and Edward's sucking of venom from Bella's blood. Edward becomes the ideal (m)Other for the birth of Bella's new self, rerouting her infantile attachments to mother and father, with Carlisle now at hand as physician-father and Alice as seer-sister (or enemy decapitator).

At the end of the film, with her leg in a cast, Bella dances intimately with Edward, her prom date, and offers her neck to him. Although he saved her from death, she wants him to kill her now and turn her into a vampire with his blood (like Mina desires from Dracula in Coppola's film). But Edward, as her (m)Other/mirror, refuses to make her "a monster" like him and kisses her neck instead. Yet Bella's desire to love and live with Edward "forever" only increases throughout the film series. So does the rage of other vampires, beginning with Victoria, James's mate, who watches from a window above them, at the prom dance, like some in the cinema audience, desiring more violence than love onscreen. Spectators may also resent the apparent immortality of the human actors onscreen, inhabiting such characters, in this idealized animal-human-god romance.

The second film, *New Moon*, begins with another mirror-stage scene, highlighting the fear of aging in Bella's erotic and romantic desires. In a beautiful meadow, she sees her grandmother and tries to introduce Edward to her, as he glistens in the sunlight. But she finds she is touching a mirror, seeing her own elderly body with Edward still a teen. And then, she wakes from this dream/nightmare, which inverted their current May–December romance, making her the older lover and giving Bella a vision of their mortal/immortal future.

Though they watch a video of *Romeo and Juliet* in class, and Edward speaks Romeo's lines when called by the teacher, Bella and her beloved are not separated by rival families. Their fathers are friends. Yet entire clans eventually go to war over Bella—increasing her value under the *male gaze*, from Edward's initial fascination with her as opaque in his mind reading, to Jacob's attraction to the white girl returning from his childhood memories, to Aro's amazement when he fails to read her mind, too. He is the leader of a traditional vampire hierarchy in Italy, the Volturi (vultures?), who viciously enforce laws to keep vampires hidden from humans (perhaps as Meyer's parody of the rule-bound Roman Catholic Church). Likewise, Bella, along with the actress playing her, became an object for fans' identification, as ideal ego, mirrored and transformed by the desires of various vampires and at least one werewolf onscreen. But *female* fans displayed even more possessive gazing at the actors in the roles of Edward and Jacob. Those super-natural characters, along with others, became icons of desire and identification—with "Team Jacob," "Team Edward," "Team Alice," and other Cullen fan groups aligning their imaginary kinship on the Web.

In the second film, the Cullens move away from Forks and Edward disappears from Bella's life, not wanting to endanger or tempt her further. Bella screams in her sleep, with a "hole" in her chest—her Lacanian "lack of being" experienced as lack of boyfriend. But then Edward appears to her, like a guardian angel, when she is in danger, cautioning her about what to do to be safe. This seems super-natural, yet it might be that Bella's brain theater projects an alter-ego image of Edward, as good angel to her bad side (and as killer vampire transformed into guardian), in threatening situations, due to her attachment to him as transitional (m)Other/mirror. Like amputees who experience a phantom limb, with the absence of feedback from the missing one, Bella may experience her absent boyfriend as a vampire ghost. This increases her appetite for danger, for the adrenalin "rush" that makes Edward's ghost appear, through Bella's anger at his absence and conflict with his oxytocin desire to keep her safe. Yet this also appeals to thrill-seeking movie viewers, who want to see Bella endangered onscreen, as well as wooed.

Bella's attraction to danger—going on a motorbike ride with a strange man, riding her motorbike too fast after Jacob rebuilds it, and cliff diving like she saw a "cult" of tribal teens do—increases her closeness to Jacob also. Bella first notices his sudden fierceness, toward a human rival, after she, Jacob, and another boy who is interested in her go on a group date to a violent action movie. This suggests what might happen to the teen viewers of *New Moon*, too, but with the caveat that Jacob is more vulnerable to sudden eruptions of testosterone fierceness, as a werewolf. Though Jacob at first seems to despise Sam's "puppies" (the teen werewolves that Sam leads as an adult), because they go shirtless into the woods and jump off a cliff for fun, he eventually joins them. This again might encourage teen viewers toward reckless behavior, alone like Bella or in a pack like Sam's, but the film series also offers insightful tragicomic twists to such immature, limbic, mammalian, "dark play" temptations (Schechner, *Performance Studies*).

Jacob, as a wolf, growls at and tumbles roughly with another teen wolf, Paul, when he threatens to attack Bella because she knows about the pack and she triggered his transformation by slapping him (as a human) in the face. Yet afterward, when they are in human form again, Jacob and Paul shove each other playfully and Paul says, "Sorry," to Bella—while she joins the boys in eating large muffins, baked by Sam's mate, Emily. However, Emily's face is also a tragic warning for Bella and for film fans who want to run with the wolves. She has large scars on her right cheek, from a brief moment of Sam's wolf rage. Earlier, Jacob and Sam's pack, as huge wolves, focused their wild energies to save Bella from imminent death at the hands of the nomadic vampire, Laurent (who was getting revenge for Victoria). Jacob also saved Bella's life during her cliff dive. But as romantic feelings develop between Jacob and Bella, the mortal danger to her becomes a barrier between them, more so than with Edward, especially after Jacob forcibly kisses her in the third film of the series. And yet, she finds Jacob's werewolf (mammalian) warmth comforting, much more than Edward's vampire (reptilian) coldness, and even necessary for her survival, as Edward admits, when the three are camping overnight in a blizzard, before a battle with Victoria and her vicious young vampires at the climax of that film.

Much criticism has accrued with the *Twilight* novels, relating also to the films, as "anti-feminist" (Mukherjea). Meyer's Mormon ideals of chastity before marriage and of men controlling the family and becoming gods in the afterlife are shown with the "passive and dependent Bella" being controlled by immortal vampires (Dietz 111). Bella uses "antiabortion language" when she says giving birth is like breathing air, not a choice but a necessity (McAvan 128). She represents "arrested feminism" (Nicol) and the "effacement" of motherhood (Whitton 133) with father figures around her—Edward and Jacob—taking control of her and Edward's baby, Renesmee, when Bella dies in giving birth. Although Bella is brought back to life as a vampire by Edward, after he delivers the baby from Bella's body by C-section, Jacob "imprints" on Renesmee, forming a werewolf's nurturing/sexual bond with her for life. With everyone, Renesmee can communicate internal thoughts through touch, "invalidating" a unique mother-child bond (135). Thus, some critics find a "backlash" against feminism with the popularity of *Twilight* (Shachar 159), "perpetuating an exclusively heteronormative, patriarchal worldview . . . equating difference with 'evil'" (Donnelly 191). A few critics even see male-centered "rape culture" (Torkelson) and "gynocide" (Averill) in Meyer's novels.

Also, the novels and films, along with their consumer products, such as the Jacob Black Barbie Doll, have been critiqued as racist depictions of Native Americans. Bella exemplifies the postcolonial drive of "*consuming* whiteness" by choosing Edward and the Cullens, as high-class whites, over the Quileutes as stereotyped lower-class darkies (Burke 211). These color-coded tribes of vampires and werewolves may derive from Meyer's Mormon heritage: the fair-skinned Nephites, ancient Hebrews in the New World favored by God, were slaughtered by their cousins, the envious Lamanites, who became the ancestors of today's Native Americans, cursed with dark skin by God, according to *The Book of Mormon* (Burke 215)—although Meyer's Cullens and Quileutes have a peace treaty and Bella increases their collaboration. They also reflect the racist aspects of Stoker's *Dracula* and certain movies deriving from it, such as the South Korean film, *Thirst* (2009), which show "British and American empires as 'civil' and others as 'monstrous' or 'savage'" (Kim and Anatol 196).

In my view, however, such ideological critiques oversimplify and demonize the film series, through left-cortical categories that bear extensive limbic emotional associations: backlash, rape, gynocide, racism, and genocide. As Natalie Wilson argues about the novels, "Bella, via exploring rather than denying her feelings for both Jacob and Edward, questions the tenability of monogamy" in a feminist way (58). Wilson points out that the "global community" of *Twilight* fans with blogs, conferences, and "all sorts of creative work, from fan-fiction to videos to artwork . . . disproves and mitigates the mainstream construction of females as humorless, silenced, and/or hypersexualized objects" (59). Citing Mormon scholar Margaret Toscano, Wilson also says that Meyer, in her novels, "subverts Mormon dogma in two distinct ways—first, she puts love before obedience, and second, she rejects the notion of moral purity maintained by exclusion" (135).

I would add that the film series weaves together, in significant ways, various threads of prior vampire and werewolf images onscreen, reflecting a current struggle that many Americans and others in the world experience with a patriarchal, colonialist heritage, and yet also with the "free love" pressures of the postmodern. Bella is a child of divorce, forced to grow up fast and become the "caretaker" of each of her parents, first as her mother's chaperone and confidant in Phoenix (where her mother is free to date and love younger men) and then as her father's housekeeper in Forks (Benning 94). Especially as a female single child, she suffers from "parentification," a role reversal with each of her divorced parents, where she sacrifices her own needs to care for the logistical and emotional needs of the parent (Tenga 105). Thus, in her teen development, she orients toward older adults through a traditionally feminine identity, against her mother's freedom with younger men and her father's diminished patriarchal status. She exhibits symptoms of a Dependent Personality Disorder (DPD), lacking self-confidence, fixated on her clumsiness, and ashamed of her plain appearance (106). But that changes as Bella finds a greater agency, through new sacrificial choices, with her old yet young boyfriend and his very white, upper-class family, as a strong patriarchy, though it involves dangerous vampires who hunt, kill, and drink the blood of animals with their bare hands and teeth.

The *Twilight* films never show vampires with fangs. But they draw on a long tradition, from folklore to literature, film, and television, which depicts animal drives emerging in undead human figures, or in werewolf shape-shifters, tamed by civilized or tribal orders, yet still bearing wild passions. The Cullens are wealthy elites like Nosferatu and Count Dracula, concerned about territorial rules. Edward bears the self-disgust of Louis. His nemesis, Victoria, makes "newborn" vampire children who are even more vicious than Claudia. Rosalie Cullen shows a fierce resentment at being made a vampire, though she does not get revenge against Carlisle like Louis and Claudia do against Lestat. The Volturi exhibit Lestat's cynical elitist greed, but as rule enforcers and vampire talent collectors, rather than reckless destroyers. Carlisle, living politely and usefully among humans as a doctor, matches Miriam's sense of propriety and her nurturing concern for the (un)dead, while also akin to Sarah's mix of researcher and vampire ruler. Jacob and his tribal brothers grow in strength and speed, like prior teen werewolves onscreen, but they also gain a fully canine form, much larger than their human size, and pack tactics for protecting their kin. Eventually, Bella heals the rivalry between Edward and Jacob, telling them that they *hurt her* when they fight over her. She then brings ancestral rivals, the Cullen clan (with their nomadic vampire friends) and Quileute werewolves, together in an extended, though temporary, kinship, to face the Volturi in battle, in order to save her daughter, Renesmee, from being destroyed as an illicit, animal-human-god form.

Thus, the viewer's perspective is shifted, at key points, toward understanding not just Bella's lower-class aspirations for a better life and family, but also the Native characters' ancestral struggle with colonial invaders. In the third film, Jacob takes Bella to a tribal council where she hears about (and the viewer sees) a white, female, colonial vampire in the past, the "Cold Woman," whose mate was killed by the Quileutes, destroying the tribe in revenge (as Victoria tries to do to Bella and the Cullens). But the Chief's "Third Wife" stabbed herself to distract the Cold Woman (sacrificing her life in the book, but surviving in the film), so that the Chief could stop the vengeful vampire.

Also, Jasper Cullen tells a story (and the film shows a flashback) that complicates the white-skinned versus dark-skinned associations of high-class, colonial Cullens, as civilized vampires, apparently superior to native, savage, Quileute werewolves. Over a century before, Maria, a dark-skinned female vampire and Mexican colonizer, who competed with white colonizers, turned Jasper, as a Civil War soldier, and many other young people into a vampire army to get more "territory." She made Jasper their trainer, manipulating both him and them like puppets, through his love for her—until he found Alice. This parallels Victoria's transformation of Seattle youth into vampires in the present time (of the third film), as her newborn "army," to get vengeance on the Cullens, with Jasper training the Cullens and their werewolf allies to fight that army.

All three females (Cold Woman, Maria, and Victoria) are melodramatic villains, with purely evil motives of colonial greed and/or personal vengeance. Yet their scenarios encourage movie viewers, along with Bella, to experience tragic sympathy and fear for indigenous peoples and children, as vulnerable to territorial conflict and maternal aggression. They also reflect complex, tragicomic alternatives in Bella's association with the Cullens and her desire to become a vampire: to gain personal freedom, pleasure, and power *or* to increase her prefrontal cortex responsibility and restraint networks, even to the point of countering brainstem/limbic, instinctual/emotional drives of survival, with self-sacrifice for the sake of kin, like the Quileute Third Wife.

In the second film, Bella offers her life to the Volturi leader, Aro, to save Edward from being destroyed for showing her too much about vampires; then Aro lets her live on the condition that she become one. At the climax of the third film, following the example of the Third Wife, Bella cuts her arm to distract Victoria, enabling Edward and his werewolf friends to finally destroy her, saving Bella and her family. This might engage Bella's fans with masochistic pleasure/pain, as romantic self-sacrifice, drawing on dopamine/adrenaline thrill networks, as with Bella's earlier motorbike and cliff-jumping temptations. But the many online communities of *Twilight* fans, including the "TwiChurch" meetings of "Cullenism," where fans identify with particular figures in the Cullen clan, may also help to give a wider context of differing perspectives toward more positive, cathartic reflections on the melodramatic temptations and yet tragicomic edges in the film series (and original novels).

Bella learns in the second film from Edward's tragic sacrifice: his attempt to make the Volturi destroy him, because of his depression and self-disgust at the false news of her death. Like Romeo in Juliet's tomb, Edward commits suicide, violating Volturi law by showing his glistening sunlit body in the Volterra public square—though Bella, with Alice's help, stops him just in time, and then saves him again from Aro's capital punishment in his Pantheon-like courtroom (designed by David Brisbin). Here, Bella's fans might realize how Edward's self-sacrifice was futile, stupid, and selfish, even if tragically beautiful, whereas Bella's is genuinely generous, respected even by the villainous Aro. Likewise, as Bella, Edward, and Alice leave the Volturi court, a stream of modern tourists pass by them, going toward it—and then their screams are heard as vampires feast on them offscreen. This might offer a warning to *Twilight* fans that consumers of vampire fantasy, in various media, can become consumed by it, through their brain theater's inner and socially networked, obsessive, even addictive, limbic/brainstem drives.

At the start of the fourth film, Edward confesses that he was a serial killer, as a "newborn" vampire, and flashbacks of this are shown. Then Bella dreams of their wedding, with everyone dressed in white but with the Volturi presiding in black and with red roses turning into a bloody pile of corpses at her feet, including members of her human family. After Bella's actual wedding and then honeymoon trip to a private Brazilian island, where she has violent vampire sex with Edward and gets pregnant, Jacob warns her that the unborn baby in her womb is a "killer." Jacob's werewolf brothers meet and their leader, Sam, decides they must kill Bella to prevent her baby being born. But Jacob and another werewolf, Seth, split off from the pack, forming their own. Bella and Jacob thus exemplify, especially to teen fans, bravery in taking on mature responsibilities—whether with a beast in one's womb or with destructive leadership in one's peer group—even though the neural circuitry in our prefrontal cortex does not fully mature (with myelin insulating the axons) until our mid-twenties.

More courage is required with monstrous problems in Bella's pregnancy, as the fourth film nears its climax. Carlisle, her vampire physician, says that the blood-thirsty fetus is "crushing her" from the inside out and Bella looks gravely anorexic. The Cullens give her a Styrofoam cup with a straw, so she can drink blood from it and lessen her unborn baby's craving. This disgusts the still human Bella, but she complies with the experiment—evoking mixed *rasas* of disgust and courage in sympathetic movie viewers.

Edward communes telepathically with his child in Bella's womb, finding it "happy." As in the earlier sci-fi film, *Alien* (1979), where a baby monster erupts from a man's stomach, these scenes play upon a primal human fear of the animal within. This is felt especially by pregnant women, and yet also by others when brainstem needs and limbic drives erupt, as vertebrate/reptilian and mammalian (or dragon-like) monstrosities of the brain's inner theater and outer actions, against higher-order brain areas and social norms. Bella exemplifies this with a vampire child inside her and with her own power arising from within, as she becomes a super-natural species of human being.

Bella got a taste of vampire power, freedom, and pleasure while riding, like a primate infant, on Edward's back as he speedily climbed up and jumped between pine trees in the first film, telling her: "Hold on tight, spider monkey." Throughout the series, she desires to be a vampire, to be with Edward immortally. Yet she offers her life in giving birth to their child at the climax of the fourth film. (Edward tells Jacob that he must get the baby out of her womb, risking her life, before he can turn her into a vampire.) Then Edward gives her the promised rebirth as a vampire—first using his mouth and then the human technology of a syringe.[38] His alien "poison" is shown flowing through her blood vessels, bringing her body back to life, in a godlike resurrection scene, with her face full, yet whiter, and her eyes shining red, as a demonic "newborn" vampire. This might be taken, along with other Cullen, upper-class, Mormon-inspired powers, as an onscreen warning about Mitt Romney's presidential run in 2011. But many film viewers, female or male, Republican or Democrat, Mormon or not, might also sympathize with, and yet fear Bella's new, tragicomic, good and evil species of wildness.

As a vampire, Bella experiences her own super-primate pleasure to speed-run, leap-fly, and vertically climb, at the start of the fifth film. Also, she arm-wrestles Emmett, finding that her physical strength, as a newborn, is greater than the Cullen elders' and so are her temptations. Edward teaches her to hunt animals in the woods, but she is lured away from him by the blood scent of a human cliff climber. And yet, with prefrontal cortex maturity, she redirects her brainstem hunger toward a mountain lion instead, tackling and killing it for its blood, just as the lion is about to kill a deer, becoming a super-predator in human form (perhaps also rectifying the Dionysian error of the ancient Agave, mentioned earlier).

The Cullens agree that Bella is remarkably "tame" for a newborn vampire. She gets furious at Jacob for "imprinting" on her child and calling her "Nessie" like the Loch Ness monster. But then she reconciles with him—helping her father, too, to accept Jacob as a werewolf and yet an extended family member. She also works with Edward to gather other vampires as allies against the Volturi, who are alarmed that the Cullens may have broken a vampire rule against making a child into a vampire. Thus, the left-cortical (and prefrontal) rule-enforcers again become the villainous antagonists to our right-cortical hybrid heroes, moving toward a melodramatic battle of vampire clans, on a corpus callosum battlefield, slaying to control or fighting for freedom against the other side. The Volturi are on the right side of the screen, relating to the side of the viewer's body controlled by the left cortex, and Cullen clan on the left, corresponding to the right cortex. But Bella, working with Edward and Alice, continues to show a better way to negotiate between such left and right cortical vampires (or werewolves), as outer social networks, just as she did in taming the wildness of her brain's inner theater.

Despite the Volturi's role as villains, tragic sympathy is evoked for them with a medieval flashback scene of "immortal children" as viciously destructive vampires. This led to the law forbidding them, which the elders from Italy fear has been broken with Renesmee, whom they feel they must slay.[39] Carlisle explains to Bella that such cherubic demons were "enchanting" to the nurturing instincts of adult vampires, "to be near them was to love them." But they were "frozen" at a polymorphously perverse stage of human development. "They couldn't be taught or restrained. A single tantrum could destroy an entire village." The film thus encourages tragic sympathy and fear in various directions across the melodramatic battle lines: for and against these ultra-childish vampires (with a bloody-mouthed toddler thrown into a fire in the flashback), for their vampire parents (with a Denali mother vampire pleading with the Volturi not to do that and then being beheaded by them), for a Denali cousin of the Cullens, Irina, who reported to the Volturi her fear that Renesmee was an immortal child (centuries after her mother had been slayed for the crime of creating one), and for the Volturi (though they appear more concerned with increasing their power as rulers than protecting human villages). The tantrum threat of an immortal child with a bloody mouth also reflects a primal attachment beast (with limbic separation/panic and rage circuits) in fictional vampires and in the actual media lures of youthful immortality on various screens today. With babies having billions more neurons in their cute little heads than we do, needing to be pruned, what if they had the bloody appetite and super-natural powers of a vampire? How are the brains of today's children, as well as teens and adults, being rewired by our vampiric mass and social media, especially with the Web's interactive sites and videogames' immortal avatars?

Bella and Edward, with the other Cullens, their vampire allies, and werewolf friends, meet the Volturi on a snowy battlefield and try to explain to Aro that Renesmee is a new hybrid of vampire and human, not a human child turned into a vampire. Renesmee grows faster than a normal human, leap-flies like a vampire, and communicates telepathically like Edward (but in the opposite direction, from herself to others, when touching them). As a monstrous, yet beautiful baby, Renesmee also showed her mother her "first memory," a view from inside Bella's womb, seen earlier in the film. Thus, conflicting emotional perspectives are evoked on the battlefield: sympathy with the Volturi's (and Irina's) fear of monstrous child-vampires, yet sympathy also with the Cullens wanting to protect Renesmee (through their training of Bella to extend her psychic "shield" power), and sympathy with Renesmee, whose hybrid identity and special gift makes her an object of desire, for possession and destruction, as Bella had been. This may lead to a more complex, tragicomic catharsis in film viewers (and refinement of other *rasas*) than the usual thrill ride and emotional relief from romantic, horror, and action-movie melodramas, especially as the fifth film presents a retrospective climax for the entire series.

Speaking to the Volturi on the battlefield, the Cullens prove that no vampire law has been broken. But Aro still feels "uncertainty" about the hybrid child. A left-cortex biased superpower cannot handle such uncertainty (and the movie audience's appetite has been whetted), so a violent battle ensues—giving the film viewer vicarious thrills and mournful moments at various beheadings of enemies and friends, including Aro by Bella and Edward, after Aro beheads Carlisle. The viewer's inner theater *participates* in the battle through mirror neuron signals, brainstem survival instincts, and limbic panic, fear, and rage drives, involving freeze, flight, and fight networks. But this expected, melodramatic climax is then given an ironic, tragicomic twist. The great battle turns out to be a *vision* that Alice gives to Aro of a future path where he himself, as well as certain enemies and allies, will die. This convinces Aro *not* to start the battle over Renesmee—allowing Bella and Edward a negotiated victory. Such an ending exemplifies a better balance of right- and left-cortical networks in

Alice's and Aro's inner theaters, in the movie viewers', and in the outer social drama of opposing clans onscreen.

CORTICAL CONCLUSIONS

Along with his aristocrat to bat and rat (or werewolf) transformations, the screen vampire morphed in the late twentieth and early twenty-first century: from a charming male alien, whose bloodlust reflected the shamanic pagan underside of Judeo-Christianity, to female, homoerotic, teen, and child monsters, either as old-world infections of Americans or as home-grown comical, romantic, and terrorist figures. More sympathetic, yet also more dangerous, they required new types of slayers or slayer teams, as left-cortical limits (with book, cross, and stakes) to their right-cortical living-dead perversities. Like the Jekyll and Hyde werewolves akin to them, the new vampires expressed an inner theater conflict between their own right-cortical wildness and left-cortical restraint, with childish and ferocious passions along with loving and nurturing desires.

Early screen werewolves become disgustingly hairy in moonlight and rabid threats to their friends and lovers, reflecting our mammalian inheritance of emotionally contagious mirror-neuron networks. The original Nosferatu and Dracula disappear in sunlight or mirrors, showing the fragility of our ego identity in the *hall of mirrors* between brain theaters. Their *alienation*, like the werewolves', reveals the biased perspective-taking of other characters, who project in-group and out-group prejudices. Movie viewers might share this in fearing the monster, yet also question it in feeling a kinship with the beast-person. Recent screen werewolves and vampires evoke further sympathetic questions as school sports stars, bungee jumpers, free runners, musicians, a regretful slave owner, stage actors, cliff-diving Indians, super-natural baseball players, and heroic slayers who form packs and families to kill their evil kin. They also reveal anti-heroic twists from their inner theaters—and tragicomic cracks in the mirrors between them—eliciting various mixtures of *rasa* flavors: fear, rage, romantic love and joy, sadness, humor, courage, awe, and disgust, perhaps toward attentive mindfulness in the viewer.

The *Twilight* series shows a culmination of this recent trend, for both vampires and werewolves onscreen. Combining the right-cortical romantic-nurturing conflicts of *An American Werewolf in Paris*, *Blood and Chocolate*, *The Hunger*, and *Interview with the Vampire* with the left-cortical slaying tradition of most werewolf and vampire films, especially the various versions of *Nosferatu*, *Dracula*, and *Buffy*, this series depicts beast-people at various stages of maturity, in extended families and territorial, clan battles. They are still evolving their godlike, animal-virtual powers and, like us today, need better brain and social networks for self-control and common kinship values. Yet *Dracula Untold* in 2014 (dir. Gary Shore), offering the backstory of Count Dracula as a warrior-ruler choosing super-natural evil, as in Coppola's film, also shows that the melodramatic action hero of historical vampire films may return from the dead, again and again, with tragicomic edges.

Bella's difficult choices and disturbing dreams exemplify the pressures on film viewers (especially teens with rapidly changing brains) of remnant animal instincts, erupting mammalian passions, alienating family situations, and social media transformations. Her tragicomic mistakes of suicidal, thrill-seeking behaviors become developmental stages, building toward kinship bonds between humans, vampires, and werewolves, as well as a new form of super-natural hybridity in her child. She sacrifices personal pleasure and independence, yet gains higher levels of neural and social powers. But her early missteps reflect how teens today may get *trapped* by media and social ideals that seem to free them: romantic obsessions, sexual hook-up pressures, rebellious adrenalin (or other) addictions, and destructive peer-group mimicry.

The *Twilight* films, while carrying the Mormon conservatism, white elitism, and consumer idols of Meyer's novels to the screen, offer tragicomic cracks with various refashioned *rasas* in their melodramatic mirrors. Through these cathartic edges, the dangers and insights of our species' wildness might appear, not just projected onto others as monsters, but reflecting our own flaws and hopes—as with the other vampire and werewolf films explored in detail here. The next chapter will investigate this *still evolving wildness* in our rewired circuitry of brainstem instincts, limbic emotions, primate hierarchies, and virtual technologies, as revealed by recent neuroscience findings about body-image projections and right/left-cortical networks. It will also refine the "inner theater" model presented earlier, with new research on our "social brain."

Chapter 5

Ape Egos, Inner-Theater Elements, and Body Swapping

First in Africa and then in diverse environments around the world, humans evolved beyond natural instincts through super-natural ideals, rules, languages, technologies, and revolutionary changes in numerous cultures. This produced great freedom of experimentation, from our internal to external theaters of everyday life, yet also a *new wildness*, with each society developing sacrificial demands to contain or channel it. Vampire and werewolf films exemplify this with left-cortex biased, "chosen" vigilantes (human slayers or moral vampires and werewolves) fighting the undead threat of more right-cortical, infectious monsters, including the "immortal children" of the *Twilight* series. Such righteous genocide and infanticide, against werewolves and vampires, reflects the evolving powers and grave dangers of our brain's inner theater, reshaping its animal instincts into new, super-natural temptations, even when good characters try to "slay" evil.

To explore this further, the current chapter will ponder neuroscience research on the interplay of the right and left hemispheres in our brains, regarding body-image transformations in our ape egos and networking super-egos. How does the interplay of brain areas, corporal communities, and moral contexts relate to child and teen developmental changes? How might the human ape's Self/Other structures be altered through new perspectives with body-swapping experiments in the science lab that relate to our many theatrical media today—and their potential for *rasa*-catharsis? In order to explore these neuro-evolutionary questions, let us first consider the vampiric legacy of our ape ancestors, suggested by the gene-based logic of infanticide, territoriality, and promiscuity in many mammal species today, especially our nearest relatives.

MOTHER NATURE'S THEATER OF CRUELTY

In various mammals, including rodents, lions, bears, monkeys, gorillas, and chimpanzees, a male who finds a female with infants by another male may kill those offspring—so that she will bear his infants sooner and perpetuate his genes (Hrdy, *Mother Nature* 34; Wrangham and Peterson 22, 148–50). But it is not just males that commit infanticide. Female chimps sometimes kill and eat a rival chimp's infant (Hrdy, *Mother Nature* 52). In certain rodent species, the mother kills and eats some of her own pups if they are weak, when resources are scarce, to better the survival odds of others. Or she kills an entire litter, even reabsorbs the fertilized eggs inside her, because a new male enters her territory and she is saving resources for his offspring (89). A mother mouse, lion, or bear may also abandon her litter of offspring if their number is below an acceptable threshold (46).

With a similar rationale, chimpanzees, in competitive male hierarchies, go on border patrols, attacking rival troops and stealing breeding females or beating up and killing a lone chimp at the edge of their territory, sometimes drinking its blood like a vampire (Boehm 345; Wrangham and Peterson 5–17). Yet female chimps also sneak away from the alpha males and mate with such outsiders. A study found that as many as 17% of offspring in a troop were "sired by extra-group males" (Hrdy, *Mother Nature* 85). Bonobos, which have female-dominant hierarchies, usually make love not war when rival groups meet, performing various brief sexual acts across genders and ages to lessen tensions, as they often do within the troop as everyday rites (de Waal, *Our* 89–106). Chimpanzees and bonobos are our closest animal relatives, sharing 98.7% of their genes with ours—and a common ancestor with us four to seven million years ago, with chimpanzees and bonobos becoming separate species about one million years ago (Gibbons).

The vampire logic of our genes might not produce vicious "immortal children" as in the *Twilight* series. But it sometimes produces a supernatural, infectious wildness in human families and societies, akin to but far beyond other apes'. Traumas may repeat across generations, as abuse victims become victimizers or violent vendettas reverberate into clan, ethnic, religious, and international feuds—even to the point of infanticide or genocide. Infanticide also occurs today, like in other animal species, due to limited resources and sex-biased cultures, with baby girls being especially vulnerable in Asia (Hrdy, *Mother Nature* 318–36).

Some scientists see an immortal, apparently "selfish" genetic force using our brains and bodies, and our cultural tools, as vehicles to reproduce a particular code in continually competing generations (Dawkins, *Extended* and *Selfish*). Yet this force of nature has also evolved into many cooperative and compassionate behaviors in our primate relatives. Humans become much more self-aware than other primates through language, technology, and specific cultures, with an *ape ego* that develops super-natural fantasies transcending the given limits of one mortal body. We start with a more *intersubjective* sense of self, but gradually learn to become *independent* individuals (especially in Western culture). We develop as prematurely born primates, through childhood brain pruning in an outer womb of a family theater, into the extended play and social networking of adolescence, toward adult maturity with a desiring ego that balances id drives and superego demands. We continue, though, even as adults, to move beyond our individual brain theaters, preying on or supporting one another, like vampire ghosts and guardian angels—or like voracious, yet nurturing and protective werewolves.

Infanticide has been found in 35 primate species (Hrdy, *Mother Nature* 34). But primate babies have evolved their *cuteness*, through the natural selection pressures of a competitive family environment, to trigger maternal care and paternal protection (390). The human baby's chubbiness, quite different from other primates and surprising to biologists, given its difficult journey through the narrow birth canal, probably evolved to advertise its healthiness to its parents, encouraging years of worthwhile investment (481–83). Human infants favor the mother's *smell*, increasing the intimate bonding between them, and are especially attuned to the *melody* of the mother's voice, responding to that, more than to other voices, even in the womb (411). Babies are also born with a bias toward mimicking facial expressions, focusing instinctively on the patterns of the human face and looking longer at familiar faces, in a theater of mirrors with the primary caretaker as Self-reflecting Other (410–11; Meltzoff).

Studies show that less than an hour after birth, human infants imitate a smiling face, an open mouth, and tongue or lip protrusions (Decety and Chaminade 579). Monkeys, in the first several days of life, will also mirror a human experimenter's open mouth, tongue protrusion, and lip smacking, plus the opening and closing of the eyes or a hand (Ferrari et al.). Neonatal chimpanzees mirror such actions and also tongue clicks in the second week of life (Bard). Thus, *mimicry*, involving visual and vocal cues, probably evolved as an early form of mother–infant communication in our primate ancestors. But we continue to use it throughout our lives, under the surface of our conscious, self-aware, verbal exchanges: through poses, gestures, and facial expressions in shared, interpersonal spaces. These include the cooperative and competitive, conscious and unconscious, mimetic signals of social networking that we do on screens (as with posted selfies and "Skyping") and in the actor–spectator relationships of live theater, movies, TV, and even videogame animation. For we descend from a long line of survivors of infanticide, with the hope of *immortality* still tempting our ape egos, especially when identifying with a certain super-natural Other, mirroring us on the mass-media screen.

SELFIES IN THE SIMIAN MIRROR

Various mirror tests have shown that monkeys do *not* recognize themselves in a mirror, not noticing a change to their appearance (except for one group of rhesus macaques who lived with a mirror in their cage and played with it from infancy). But most great apes—including chimpanzees (after age four), bonobos, and orangutans—notice a dot placed on their faces, soon after becoming accustomed to the mirror. Gorillas do not, except for a few individuals, such as the female "Koko" who learned sign language and thus became more self-aware (Keenan 36–48). Chimps who are raised in isolation also fail the mirror test, suggesting that human "self-recognition," when looking in the mirror, depends on the Other in social networks, which we have internalized (41). Yet why did self-recognition continue to evolve toward the *hyper-theatrical reflections* of the human brain—with a hall of mirrors between us and in numerous stage and screen technologies today?

Human infants develop self-awareness in the mirror much earlier than chimps, usually in their second year, at 15–18 months, but the timing varies depending on attachment, with infants who are less attached to their mothers recognizing themselves earlier (Keenan 67–70). Some emotions, such as fear, emerge prior to mirror recognition. Yet the "self-emotions" of pride, embarrassment, shame, and guilt seem to develop after the child's awareness in the mirror (69–71). Body posture also changes late in the second year, expressing *self and social emotions* in relation to the caretaker's gaze as a mirroring Other. The child raises its arms and eyes with success at a difficult task or slumps and turns its eyes away with failure. Like other animals, human infants begin with a working, *procedural* (anoetic) memory, without self-awareness, and then add an *objective* (noetic) memory, typically at 8 months (73). But a "self-memory system," with the ability to recall personal experiences, to reflect upon current thoughts and emotions, or to predict the future (with autonoetic memory), does not develop until around 18 months, which helps to explain our later amnesia about experiences prior to that stage (73–74).

Such self-awareness has been found *in other apes* around age four, even involving language, as with the sign-language-trained orangutan,

Chantek, who would sign his name starting at that age, while looking in the mirror, and even sign "bad" to himself when alone and ashamed (Keenan 75–76). By the age of eight, Chantek was grooming himself in the mirror and even using an eyelash curler (76). Various orangutans have demonstrated self and Other awareness, or a "basic Theory of Mind," not only when trained to point, like other primates, but in following human gazes (88–89). Also, bonobos show a basic Theory of Mind with their sexual activity, in captivity and in the wild, finding many ways to please the other (86). "Theory of Mind" (or ToM) is the phrase used by scientists for the human ability, going far beyond that of other apes, to form theories about other persons' perspectives.[1] Yet until age three, humans may have a "nonconscious" ToM, like other primates (Timothy Wilson 61).

Chimpanzees, but not monkeys, are able to switch roles in tests of cooperation to get food, showing "their perspective-taking abilities in complex social situations" (Keenan 82). They also deceive others in hierarchical situations, playing "mind games" that suggest a degree of *theatrical* awareness about the desire of the Other, as spectator and co-actor (83). In field tests, where a subordinate chimp knows the location of food and a dominant does not, the dominant chimp will snatch the food away once the subordinate reveals it. But then the subordinate will wander around, as if uninterested, to avoid having it snatched away again, and the dominant will also look away until the subordinate reaches for it. Such "mental state attribution" by a chimp, *as deceptive performer*, was shown, too, with a challenger making an intimidating display toward a rival and then masking his nervous grin by turning his back to the rival and changing the expression on his lips with his fingers. Another subordinate chimp was observed using his hand to hide his erect penis—when trying to lead a female away and confronted by a dominant male (84). Mother chimps have been observed performing the role of "teacher" also, slowing down and exaggerating their use of rocks as a hammer-and-anvil tool to open nuts, with younger chimps watching, even making eye contact with their "students" (85).

Human infants at age three still have trouble modeling their own viewpoint in memory, as well as another person's. In one test, with a box of pencils that had a candy label, 50% of children that age misremembered their own idea of what was in the box, saying they knew it held pencils, as well as imagining that another child would know that fact before opening it, despite the cover showing otherwise (Keenan 95–96). In early infancy (three months of age), other primates perform *better than humans* on tests of recognizing object permanence. And yet, humans continue to build on initial mirror and Theory of Mind skills, developing them far beyond other apes. Scientists have estimated that by 18 months the human infant surpasses a monkey's self-awareness, with the mirror recognition of older apes; by age two it also starts to deceive like apes; and by age three to four it gains the "advanced mental state attribution" of adult apes (96–97).

The infant's brain shifts in its cortical development, with a growth spurt in the right cortex during the first 18 months, then increasing left cortex growth, peaking at age 2 with the development of verbal language (and the start of deceptive skills), and another right cortex growth spurt peaking at age 4 (Cozolino, *Human* 70–71).[2] These right cortex expansions, with initial self-awareness in a mirror and then with advanced mental state attribution, correlate with other research that shows more activity in the adult's right hemisphere with self-recognition tasks. The right prefrontal cortex, right anterior cingulate gyrus (between limbic and prefrontal lobes), pulvinar (within the thalamus at the core of the brain), and right insula (between temporal, parietal, and frontal lobes) are more active in tests of self-recognition, with gradual degrees of computer morphing between the self-face and another's face (Keenan 134, 139, 151). The right frontal cortex is also more active with self-voice recognition and the right prefrontal and temporal lobes with autobiographical memory recall (174, 179, 185). But the *left* parietal lobe is more crucial to knowledge-based memory (192). Patients with right frontal cortex damage show changes in personality and disinhibited behavior bordering on illegality, while also losing a sense of humor and trust (200, 221, 225). The ability to lie and to judge deception has been correlated with right frontal cortex activity (226–27). Also, right hemisphere stimulation can produce an out-of-body experience (203). With Self and Other awareness tests, the right prefrontal cortex and right anterior cingulate are most active (232).

Here we can see the main components of an inner theater of Self-awareness, emerging as the human brain develops in stages akin to, yet far beyond other simian brains. A growth spurt in the right cortex during the first 18 months of life corresponds to the initial development of a Lacanian Imaginary dimension in the "mirror stage," involving mirror recognition and a Theory of the Other's Mind (advanced mental state attribution). Face, voice, personality, proper behavior, trust, humor, body-space awareness, and self/other perspectives begin to develop in specific right cortical-limbic areas, through that early growth spurt and later human stages. But left cortex expansion around age 2, and again around ages 6 and 12, also increases the language, social identity, and Symbolic self/other-deception modes of our egos far beyond other apes.

As parents know, the "terrible twos" can be a time when the child starts to manipulate primary caretakers with temper tantrums and stubborn refusals. "No" becomes a favorite word. But a toddler at that age may also show empathy with another child and comfort it, mimicking other children even if preferring solitary play. A growing independence, with resistance to help from the parent, comes with greater awareness of one's own abilities. Aggression then lessens with verbal development—showing the refinement of left hemisphere and prefrontal cortex networks for language and self-control. At two, the child enjoys helping adults, which increases its sense of a bigger Self, connected with the Other. And yet, as the ape ego grows toward self-reflective, hyper-theatrical, human awareness with new powers of language, along with emotional outbursts, the Little Emperor becomes bossy and demanding, hoarding its toys and wanting things done in ritual ways, with objects put back where they "belong." The rapidly pruned circuitry of the 2–3-year-old brain is absorbing right-cortex images and left-cortex symbols, while cooperating and competing with parents and siblings, to create a social network of personality, as the staging for its inner and outer theatricality.

Having survived infancy with its chubby cuteness, the toddler becomes more of a tyrant and trickster at times, while also luring adults into nurturing it. Thus, the human child, in its first several years, exemplifies the three ways that apes in the wild negotiate survival and sexual drives in their extended family hierarchies: *domination* (with males at the top, except in bonobo troops), *attraction* (especially by estrous females and playful infants), and *deception* (of Self as well as Other, through "Machiavellian Intelligence"). Likewise, but with greater Imaginary and Symbolic, right- and left-cortical complexity, humans extend these modes of theatricality far into adulthood, through cooperative and competitive hierarchies of family, playgroup, school, sports, work, art, and today's social and mass media.

In each of us, this hyper-theatricality involves a switching of ego roles between various brain networks, evolved from our animal ancestors, especially between the "publicly endorsed self-schema" of our left hemisphere and the "suppressed ape . . . with intuitive perceptions and inclinations" in the right hemisphere, threatening or inhibited by the left (Torey 172). How the character of Self (or of many potential selves) develops in *each* brain's inner theater depends on genetics and a multitude of environmental factors, with pleasurable and painful, neuron-pruning experiences at various level of social networking, from family and school to larger cultural groups and the World Wide Web, specific to each person's life. Yet the basic characteristics can be outlined, building on the ancestral stages of our ape ego, as shown by our current simian relatives, performing with each other and in the mirrors we give them.

THE SPLIT-BRAIN THEATER OF THE CHILD'S APE EGO

Let us review some of the research on child development given in Chapter 3, put here in chronological order with added ties to the brain's right and left hemispheres—and with further references to monkey and ape behaviors. I will then add more about ages 4 to 6, leading up to adolescence, to explore the inner and outer theaters of our continuing biocultural evolution. As mentioned earlier, there is an initial growth spurt in the child's right cortex (especially attuned with the mother's right cortex) during the first 18 months, peaking at about age 1, and then a shift toward more growth in the left, which peaks at 2 years (with toddler independence and verbal language), and then back toward the right, peaking, though to a lesser degree, at age 4, and then to the left again, peaking, but less so, at age 6 (Cozolino, *Human* 70–71). There is more asymmetry in the child than in the adult brain, with myelin insulation of the circuitry between hemispheres developing gradually after age 1 (McGilchrist 213).

- Nine-month-old babies reason intuitively about probability, staring longer at the experimenter when she produces red balls from a box that contains mostly white ones (Gopnik 83). This shows a particular network developing between right-cortical intuition and left-cortical logic.
- One-year-olds correctly distinguish one-cause and two-cause items in a cause and effect game (Gopnik 84–85), a characteristic function of the left cortex. They also show right-cortical and limbic aspects of intuitive empathy[3]: "One-year-olds understand the difference between intentional and unintentional acts, and behave in genuinely altruistic ways" (204). They have an "intentionality detector" attuned to the desires and goals of others (prior to an awareness of their own), like nonhuman primates, forming the basis for later aspects of "mindreading" (Whiten 164–65, 168). The right hemisphere plays a key role in the child's emerging self and other awareness, including Theory of Mind, from age 1 to 4 years (McGilchrist 57).
- Regarding limbic and right-cortical functions, infants as young as 14 months will cry along with another child, showing emotional contagion and gestural mimicry (Preston and de Waal 1).
- By the age of 15 months, infants expect fairness in food distribution tests (Schmidt and Sommerville), as found also with capuchin monkeys and chimpanzees (de Waal, *Age* 189–92). At that age, one-third of human infants demonstrate "altruistic sharing" of a favorite toy with an adult when another toy is at hand (Schmidt and Sommerville 5), showing further developments of a prosocial Self in left-cortical

networks.[4] Yet this may also be akin to grooming and food sharing in primates, for delayed benefits in the social hierarchy, as "reciprocal altruism."

- At 18 months, an infant's altruistic behavior continues, as with young chimps, readily helping others to achieve goals (Warneken and Tomasello)—with the shift toward left-cortical growth. While understanding others' goals, however, human infants also begin to find differences between their own and others' desires, fueling self-assertiveness as they approach the "terrible twos" (Gopnik et al. 38). The child develops a sense of time and sequence in relation to oneself, as "the foundation of explicit autobiographical memory," involving the frontal cortex (Siegel, *Developing* 35). This initiates a new stage of defining one's Self in the mirror of others' desires by rejecting their demands and testing the external environment (and caregivers' patience) with the repeated use of "No." During this *double*-in-the-mirror stage, the left-cortical prosocial Self combines with a right-cortical alter ego, as Devil's Advocate. (In Lacanian terms, this involves the child's patriarchal *separation* from the mother, through language and law, as well as primal *alienation*, with the mother's desires going beyond the child as her object of devotion.) Driven by subcortical needs and limbic emotions, the toddler plays with or fights against the desires and demands of the home environment, which may include other siblings as well as adults, to create a distinct sense of Self, building its character and plot in the family theater, through imitation and resistance. The seductive, charming, chubby child becomes, at times, domineering, with temper tantrums and its own demands.
- With the peak of left cortical growth at age 2, involving frontal and limbic, prosocial networks, children start to comfort others who are in pain (Gopnik 39), as found with certain monkey and ape species (de Waal, *Age* and *Primates*). They are also able to use inductive reasoning based on categories (Baumeister 232–33). And yet, at that age humans start to become *tricksters*. They "indulge in pretend play" while starting to realize that the other's viewpoint and beliefs may be incorrect (Whiten 156). This may be akin to, but not yet as developed as adult apes in the wild, which exhibit "true pretense" regarding food or sex when a competitor or superior is watching, as an "implicit theory of mind" (156, 161). With primates, scientists call this form of intelligent deception "Machiavellian."
- At about age two and a half, there are "signs of deceit in children's everyday social life" (Whiten 155), involving right frontal cortex

activation (Keenan 226–27), with a shift from left- to right-cortical growth, from age 2 to 4. If children are put in "the role of the deceiver" in a story-telling experiment with puppets, they will "remove tell-tale cues, and even lead false trails, to mislead a protagonist about the location of some hidden reward" (Whiten 155).

- At age 3, children show a new stage of inner-theater modeling and perspective taking. They have an understanding of causal forces, such as gravity, force, and mass, which chimps never attain (Baumeister 209). They can use the scale model of a room, where a toy is placed, to find the corresponding toy in a real room (DeLoache; Oatley 31–32). They will also make accurate predictions about a character's lack of knowledge about a hidden object—if they are "*actively involved* in the deceiving role" (Whiten 155). But at this age, like a year earlier with the start of "pretend play," children are not as good at using deception for personal rewards as they become later in life. They are likely to deceive by laying false trails even when they could benefit from truthfully informing a partner. They seem to act with "pretense," rather than with the intent to deceive, unlike the functional deception observed in other animals, especially primates with a Theory of Mind.
- Thus, with a growth spurt in the right hemisphere, 3-year-olds show a new combination of right-cortical Imaginary and left-cortical Symbolic functions. They can "create an as-if scenario and invite others to join in with them" (Sodian 398). Yet, they can reason about rules and consider the other's intent when making moral judgments about good or harm (Gopnik 225–26). But they also form in- and out-groups, focusing more empathy toward those like them, even those with the same color T-shirt, seeing them as nicer and better to play with (219). At 3, the child's *ape ego* does not have the "impulse control" of a 4-year-old. But its self-control can be improved, as in experiments with adult chimpanzees, by substituting numbers for the actual candies in a bowl (Carlson et al.), increasing left-cortical abstraction. Through Symbolic distancing, the child or chimp is able to delay an impulsive grasp at the candies and wait to get a larger number instead.
- With a peak in right-cortical growth, involving holistic, contextual perceptions tied to limbic emotions, 4-year-olds can interpret personality traits in a story using puppet animals (Gopnik 98–99). Preschoolers may actually be more conscious of current sensory experiences than older children and adults, with more cholinergic transmitters in their brains and less inhibitory circuits, for greater awareness of the present moment, like adults in mindfulness meditation (Gopnik

> 119–20). Thus, preschoolers have an "attentional lantern," rather than a "spotlight" of consciousness, with less focus on an autobiographical past and future (125, 145–48, 153)—like our ape relatives. Likewise, a chimpanzee trained to understand number order can beat adult humans at holistic (right cortex) recall.[5] The chimp can touch the spots in a checkerboard array on a screen where the numbers 1 through 9 have just been flashed at 210 milliseconds, recalling the randomly generated, two-dimensional pattern like a photo inside the brain, as "eidetic imagery" (Inoue and Matsuzawa). As with a preschooler's attentional lantern of consciousness, the chimp does not have an adult human's more inhibiting, left-cortical, serial spotlight.[6]

By age 4, with its growth spurt of right-cortical, Imaginary functions, the child is moody and throws tantrums at times. It may develop emotional ties to *imaginary playmates*, extending its inner theater outward and pushing on the limits of its social environment. Having an imaginary friend even helps some children from middle school to high school with coping strategies, resiliency, and positive adjustments, though with less of a preference for real-life peers (Taylor et al.).

In their fourth year, children start to find "best friends" with real playmates, too, while enjoying role-playing games and "make-believe" activities in groups. Yet, they may insist on acting independently, get frustrated when unable to do things, or boast and exaggerate the truth. They often appear selfish, not able to take turns, but take pride in accomplishments and seek adult approval. They increasingly use words, instead of physical aggression, when angry—showing the development of Symbolic, left-cortical (and limbic) networks for verbalizing emotions, tied to frontal, motor-cortex controls, along with right-cortical, Imaginary self and other extensions.

This mixture of left-cortical, Symbolic skills with right-cortical emotional ties to Imaginary playmates relates also to reports (mentioned earlier) that 5-year-olds around the world develop a vitalist theory of biological causes, like the Chinese sense of *chi*, with a life force in us and in all living things, which weakens with sickness and death (Gopnik 36). But children also follow or play with rituals of the adult Other, as modeling a super-natural cause and effect. As early as 18 months and increasingly toward age 5, children in both modern and tribal cultures, even that of the remote African Bushman, *over-imitate* the unnecessary, ritually repeated actions of adults in tests with a box that has various mechanisms to reach a prize (Nielsen and Tomaselli). Chimps, however, merely *emulate* the goal of the human demonstrating the ritual actions and skip the unnecessary steps, going directly to the drawer with the prize—when the direct step is visible (Whiten et al.) In these experiments, as with the pattern recall of a random number array, chimps are *better* than humans at quickly solving a puzzle to get a food reward. But humans inherit the predilection to be taught and to "upgrade"—to have the brain's inner theater shaped by the cultural environment[7] and then to change the culture with each new generation, growing, rebelling, and taking control.[8]

Evolving as a distinct species, away from our common ape ancestors, we developed higher degrees of left-cortical, Symbolic, language- and rule-making skills, yet also right-cortical, Imaginary dimensions of hypertheatricality. Today, this means an inherently reflective, Self and Other *metaphysics* (from imaginary friends to a vitalist theory of natural causes) and precise imitation of *ritual* actions in relation to the Other, shown especially at age 4 to 5. Such left- and right-cortical, Symbolic and Imaginary extensions of our ape ancestry can be seen in further developments of the human child, regarding the Real orders of limbic and brainstem networks that involve primal emotions and remnant instincts of survival and sex.

In evolutionary terms, there are both advantages and disadvantages to the extension of Self-recognition and Other-imitation/projection in our brains' neo-simian mirrors. For example, scientists find that people "spend more time in front of a mirror preparing for a date than before heading to work . . . upping the reproductive ante . . ." (Keenan 238). And yet, anorexics spend even more time in the mirror, perceiving a distorted body image through the right frontal cortex, endangering survival and disrupting the menstrual cycle in women, contrary to our fundamental animal instincts (238–39).[9] The Self's image in the mirror, shaped by impossible cultural ideals on various media screens, may become an *undead vampire*, preying on the real body and its reproductive potential. Perhaps this is why screen vampires sometimes disappear in a mirror, reflecting the threat of the "mass media" to consume the consumer. And yet, vampires repeatedly return as icons in the media, especially for young people in recent decades. The stages of child to teen to adult development involve proud, insecure, and critical shifts in the brain's Self-image networks, as reflected by vampires and werewolves with their slayers, as well as other beast-people with their creators and controllers.

FROM CHILD TO HORMONAL TEEN (AND ITS APE DOUBLE IN THE MIRROR)

At age 5 to 6, with another left-cortical growth spurt, children play more cooperatively in groups, sharing toys and suggesting more elaborate and imaginative play ideas. They show affection and caring for others, like young apes, and develop better self-control over impulses and mood swings, with more subservience to caregivers—as their "emotional circuitry comes under stronger prefrontal control" (Goleman, *The Brain* 72). But children at this age may become *anxious* for adult approval, reassurance, praise, and attention, complaining excessively about minor hurts. Self-perceived failure may cause frustration. The child needs *rules* for ethical behavior, with values forming, based on others' views. And it may become increasingly fearful of strange noises, the darkness, or animals—showing new *left*-cortical networks formed by social values and ethical rules, yet also anxiety from alternative, more holistic, *right*-cortical perceptions and projections, relating to limbic passions and deeper animal drives.

At 14 months, human infants show *selective* imitation (or emulation) with a focus on the demonstrator's goal like chimps, but "older and more cognitively mature children (and even adults) . . . [show] more 'blanket' copying" (Whiten et al. 2427). From age 5 to adulthood, humans tend to use a "copy-all, correct later strategy" (2425). In all of us, the left hemisphere develops with a focus on goals and the right with more *holistic, contextual awareness*, involving inner-theater imitations of others' actions and moods through mirror neuron systems and emotional contagion, often subconsciously. From age 3 to 5 and again at age 7 to 11, there is a growth spurt in the *right* cortex (Cozolino, *Human* 71). Especially during such growth spurts, but before and after, too, we absorb *novel* perceptions in the right hemisphere first, which then fit more consciously into or become inhibited by, and yet eventually *alter* the Symbolic rules, abstractions, and verbal meanings in the left (McGilchrist 40–46).

Experiments have found that "helping behavior"—with a child leaving to check on "sounds of distress" in an adjacent room—increases from age 5 to 6, and again to age 7 (from kindergarten to first and second grade), especially when the children are in pairs, but levels off at around age 9 (fourth grade) and decreases at age 11 (sixth grade) to about the same level as at age 5 (Staub 38, 83). This decrease in helping behavior may be due to "children overlearning social rules that prohibited them from interrupting work on their task or entering a strange room" (39). A study also found that Swiss children "developed clear stereotypes of other nations, usually devaluative," by age 9 to 10, related to "blind patriotism" (499).[10] I would add that new peer-group pressures, sometimes against prior-learned "social rules" of proper behavior, with helpful, playful, rebellious, or stereotyping impulses, may increase through *hormonal* changes of the body and brain—toward the teenage years of our human-ape ego and superego.

Puberty starts at age 8 to 13 for girls and a year or two later for boys, continuing well into the teen years. The hypothalamus, near the center of the brain, at the top of the brainstem, sends a gonadotropin-releasing hormone to the pituitary gland at the base of the brain, which then sends other (luteinizing and follicle stimulating) hormones to the ovaries in girls and to the testes in boys. But different parts of the hypothalamus are involved: the ventromedial (lower middle) triggers this in girls, the anterior (front) in *boys*, which involves a "preoptic area" tied to *visual* cues (Panksepp and Biven 249, 255). Estrogen from girls' ovaries leads to breast development, wider hips and thighs, pubic and underarm hair, and eggs with the start of menstruation.

Such changes in girls' bodies become interpreted through the family, school, and peer environment, while interacting with sexual images in the media mirrors of Western culture, which now influence many other developing cultures, producing a worldwide "Lolita Effect." Journalist M. Gigi Durham applies this term to "an adult male fantasy of girls' sexuality" in our mass media, with a rhetoric of empowerment that actually promotes its opposite, "a subjugated and commercialized version of female sexuality" (208–9). According to Durham, this media effect puts girls in a double bind: "be sexy but not sexual. . . . Attract male attention but fend off sexual advances" (219). Films, TV shows, and videogames influence teen "scripts" in male brains as well, associating sex with violence, especially against young women (141–50). But they also show violent acts *by* women as both feminine and masculine, manicured and high kicking, vivacious and vicious.

This was invoked, yet satirized with the vampire-slayer, Buffy: a sexy teen girl who is "chosen" to slay, not just fend off, her oversexed peers—after they become infected by the bloodlust of a medieval male, Lothar, whom she kills with a phallic stake at the climax of the film. This formula continued from the film to the TV series, but with more complex explorations of Buffy's developing desires as a lethal Lolita, along with shifting good and evil identities of her friends and various versions of a "Big Bad" villain each season. In the more recent *Twilight* film series, Bella found a different destiny regarding the Lolita Effect, as chosen object of the male gaze. She embraced it, as Buffy did at various times in her TV series, joining romantically, heroically, and physically with lusting, yet self-restrained vampires and werewolves to battle against the beast-people who were worse.

Boys in puberty transform, somewhat like werewolves in the moonlight but more gradually, gaining a larger penis with sperm ejaculations, a deeper voice, more muscles, and more body and facial hair (as suggested by the *Teen Wolf* films, yet also paralleled with a menstruating girl in *Ginger Snaps*). Testosterone triggers these changes in boys, where it is produced in much higher levels in the testes as well as adrenal glands. But in both males and females, the brain's right hemisphere is more sensitive to testosterone than the left (McGilchrist 33, 39). This may relate to the "wandering eye" of the male gaze, involving the preoptic area of the anterior hypothalamus, plus a more global awareness, sustained attention, and intense alertness with right-cortical testosterone lust and competitiveness.

Strangely, testosterone-fueled aggression is also found in female hyenas, which have an "enlarged, penis-like clitoris primarily for the purposes of sociosexual communication, especially for dominance displays" (Panksepp and Biven 171). In humans, like in female hyenas, several distinct, biologically driven, brainstem-limbic-cortical systems—fighting rage, predatory aggression, and social dominance—may at times merge in the "higher," left and right cortical levels of the brain (172). To some degree, *each* of us mixes sexuality, competitive rage, aggressive desire, and dominance or attraction displays, becoming like were-hyenas in puberty and later life (or the Hyena-Swine character considered in the next chapter). Yet men tend to specialize more in physical aggression and women in psychosocial (172). The lust system also involves distinct brain areas with different hormones in males and females, leading to many social misunderstandings, especially for teens.

Monthly estrus signals in females involve the brain's ventromedial hypothalamus and pituitary hormones triggering the secretion of estrogen for the ripening of "eggs" in the ovaries, and then of progesterone as the egg moves into the fallopian tube (Panksepp and Bivens 256). Estrogen and progesterone control female sexual readiness in most mammals, by producing oxytocin, the trust hormone and neurotransmitter. But in humans, unlike other mammals, "adrenal testosterone" contributes also to females' physical and emotional "receptivity," as if adding some degree of "male sexual urgency" in women (255–57). *Testosterone* plays a much greater role in boys' sexual awakening, though, not only due to its production by the testes, but also because males have larger neuronal fields with testosterone receptors in the anterior hypothalamus and its preoptic area (249–50). This makes men often ready with sexual interests while women are more cyclical and mysterious to men, with concealed ovulation and "psychological estrus cycles," affected by monthly "fluctuations in erotic arousability" (256).

In males, another important hormone (and neuropeptide), *vasopressin*, "promotes sexual ardor, courtship, territorial marking, intermale aggression, and possibly sexual jealousy" (Panksepp and Bivens 251).

It decreases sexual receptivity in females, though it may also increase protective maternal aggression (257–58). "Vasopressin shifts animals toward male-typical attitudes . . . [of being] 'pushy and competitive'" (259). *Oxytocin* encourages "female-typical nurturing attitudes . . . 'to tend and befriend.'" Yet, oxytocin is produced in both males and females through physical intimacy, increasing in the blood and brain during orgasm, or in the mother's birthing and nurturing of a child with loving attachments. Oxytocin also encourages gloating (*Schadenfreude*), along with confidence, trust, parental play with offspring, and the ability to interpret emotion in the other's face (262–63). Its circuits are tied to dopamine *reward-seeking* networks through the subcortical, ventral striatum and ventral tegmental areas of the brain, between the limbic system and midbrain, and with *fearlessness* through ties to the septum, which is near the ventral striatum (Matthew Lieberman 94).

In all of us, but especially in the rapidly changing, werewolf-like, teenage body, testosterone and vasopressin increase the *masculine ape* aspects of the human ego. This helped our ancestors to survive and reproduce, leading to today's inner brain and outer social theaters of passion, courtship, territorial competition, intermale aggression, and sexual jealousy. Oxytocin helped our ancestors by increasing *female sexiness*, as with the "lordosis reflex" seen in other mammals and primates today when the female is ready to mate. She curves her back with her "rump up, and tail deflected to the side," for male mounting (Panksepp and Bivens 265). Oxytocin continued to influence mating gestures in our more recently evolved ancestors, who started connecting in other ways, like bonobos today, which are the only other primates that mate face to face as we do. (They also perform various other sexual acts like us, hetero- and homosexually.) Oxytocin increased the sexual bonding of our animal ancestors, as it does for us today, especially in our hormonally charged teen years. It also aided in mother-child bonding, perhaps with fathers, too. And it continues to increase our partnership confidence, trust, and protectiveness, our parental gloating and yet play with offspring, our reading of each other's emotions in facial expressions (especially through right-cortical networks), and our in-group cooperation.

Yet teens today are exposed to a host of primal, mammalian, ape-ego stimuli, with famous performers (such as Miley Cyrus) showing the lordosis of "twerking" on TV or with explicit pornography available on the Web. Brain scanning by Mario Beauregard and his colleagues confirmed that when male college students (ages 20–42) were shown *erotic film clips*, their sexual arousal involved activity in limbic and subcortical areas: the right amygdala, right anterior temporal pole, and hypothalamus

(Beauregard et al. 166–67). This is because the amygdala evaluates emotional stimuli, the anterior temporal pole imparts emotional color to experiences, and the hypothalamus (tied to the pituitary and adrenal glands) is pivotal to the body's expression of emotion through the endocrine and autonomic, hormonal and nervous systems (Beauregard and O'Leary 132).

But when the students were asked to "suppress" such arousal by becoming "detached observers" of the erotic film and its effect on them, their right lateral prefrontal cortex (rLPFC) and right anterior cingulate cortex (rACC) became engaged, in the advanced primate areas of the brain, while the areas for sexual arousal had lower activity (Beauregard et al. 166–67). Likewise, a study comparing subjects that matched or labeled *emotional* (angry or fearful) faces revealed that *matching* them with one's own face produces amygdala activity, but *labeling* them with words suppresses such limbic emotional contagion, with decreased amygdala and increased right prefrontal cortex activation. In another study, female college students, aged 20–30, were asked to match or suppress feelings of *sadness* in watching "sad film excerpts" (between "neutral" clips). Their brains showed activation in the right orbital frontal cortex (a.k.a. the VMPFC) and right LPFC when "distancing themselves" from the sadness onscreen and becoming "a detached observer," like in the mindfulness training derived from Buddhist meditation (168–69). A similar study with children showed five times more activation in the LPFC, in both the left and right lobes. Thus, *detached observation* of sad film clips "requires more prefrontal work in children than in adults" (186). This finding increases the significance of *teaching* methods, with theater exercises or film watching, that help children to develop such mindful prefrontal networks through new perspective taking (switching roles) and *rasa*-cathartic classroom discussions.

The prefrontal cortex begins to develop "emotional self-regulation" during the first 6 months of life, with "the ability to temporally associate emotional states with specific external stimuli" (Beauregard et al. 174). It eventually becomes one-third of the volume of the entire neocortex (175). But it is the last area of our brains to mature, with its circuits becoming fully myelinated (or "insulated") only in our mid-twenties. So teenagers lack the full strength of their erotic and emotional *catharsis* networks (involving not just suppression, but also reflective clarifying through distanced observation and *rasa*-refinement). At the same time, the PFC circuitry is "rewiring" as teens absorb the social networks of outer theaters that stimulate or regulate such vertebrate, mammalian, and ape systems—with "role models" beyond family ideals (Barrett 194). We go from our mother's womb shaping our brain circuits to the external womb of family and friends, with right and left hemisphere growth spurts in childhood (and developmental differences from other primates) engaging that home theater in various, receptive and defiant ways. And then, we progress toward our teen and adult years with the extended womb of many mirroring theaters and mass/social-media screens, plus werewolf-like, hormonal body changes and vampire-like passions, through specific experiences and choices that prune each brain's circuitry in particular ways. But how does this Other of extending bio-social wombs and symbolic orders *frame* the Self that we each stage inside our brains and perform for others, as a super-natural ape?

INNER NETWORKS OF A SOCIAL SELF

In non-primate animals, oxytocin increases aggression toward strangers, promoting direct care of one's own offspring (Matthew Lieberman 94). Yet in humans, it encourages caregiving not only toward kin and in-group members, but also to strangers, showing how we sometimes become altruistic, with a generosity beyond genetics (95). Such caregiving toward family, friends, and strangers involves higher levels of the human neocortex—with a "mentalizing system" (or Theory of Mind) for imagining oneself *as the other,* beyond other mammals and apes. Yet such animals may also recognize themselves in a mirror, demand fairness, and cooperate with or console their peers (116–17). Various brain mapping experiments show the same areas active with humans *reading* about a character's false belief, instead of just facts, or *viewing* geometric shapes (a video animation of two triangles and a circle) "that could be interpreted socially," or viewing photos taken by others, instead of by oneself. This *mentalizing network of theatrical perspective-taking*[11] includes the dorsomedial or upper-inside prefrontal cortex (DMPFC), the temporal parietal junction (TPJ) between the temporal and parietal lobes, especially on the right side (Saxe), the temporal poles (TP) near the front of the brain, and the precuneus/posterior cingulate cortex (PC/PCC) between the parietal lobe and core limbic areas.

Our advanced mentalizing, perspective-taking system adds a distinctively human layer to the inner simulations of mirror neuron and emotional contagion systems found in other animals, especially primates (Matthew Lieberman 149–50, 178). Teenagers use it, unlike adults, with direct appraisals of Self, thus involving more of the Other's view than we do later in life. But in all of us, the mentalizing system corresponds to the same areas as the "default network," active when the brain is not given a specific task (118–20; Mars et al.). The brain often uses its free association time, as with dreams and shared theatrical media, to "practice processing social information" with its mentalizing circuitry, imagining others' viewpoints. *Empathy* also involves the mentalizing system in addition to mirror neurons and emotion sharing (Matthew Lieberman 156–57). Lieberman explored this by showing subjects in the fMRI brain scanner different kinds of pain pictures, ones with *obvious* suffering, which recruited the mirror neuron system, and ones where *contextual interpretation* was needed, which involved the mentalizing system as well.

Empathy with happy or anxious, as well as painful photos activates the brain's *septal area*, too. Located next to the ventral striatum and key to the brain's reward system, the septal area "has become disproportionately larger across primate evolution, and it has direct connections to the dorsomedial prefrontal cortex" (Matthew Lieberman 158). It shows a connection between personal pleasure and the empathic motivation to help others (157). Studies with rats, mice, and rabbits likewise show that removing the septal area reduces maternal caregiving: mothers no longer make protective nests and provide less milk for their young (159). The septal area also dampens distress signals, such as the "startle response," enabling people to overcome the herd mentality of contagious fear or panic, in order to help someone in distress. "Rats whose septal area has been removed show a much larger startle response. . . ." This area of the limbic system is also "rich in oxytocin receptors," being the highest area for that in some mammals (160). Rodent pups that are separated from their mothers grow up with fewer oxytocin receptors in the septal area and pups that get more parental care have more.

The DMPFC, with its connections to the septal area, is part of the brain's mentalizing (perspective-taking) system. Subjects perform "significantly faster" on mentalizing tasks if that area was active earlier when the brain was at rest, in the "default" mode (Matthew Lieberman 119)—as if *rehearsing* for social and theatrical perspective taking. The medial prefrontal cortex (MPFC, between the dorsal and ventral areas), along with the precuneus (or PC, in the parietal lobe), is more active with "self-knowing," when subjects are asked whether certain adjectives apply to themselves (185). But "self-recognition," in mirror awareness tests, involves more of the lateral or outside prefrontal cortex (LPFC), especially on the right. Thus, the MPFC and DMPFC add a *meta-theatrical network of self-knowing through the Other*, of Self in the Other and Other in the Self, to the LPFC mirror awareness of *self-recognition* that we share with some mammals and primates.

A study on the effectiveness of cognitive behavioral therapy (CBT) for spider *phobia* showed "systematic changes" in the dorsolateral prefrontal cortex (DLPFC), along with the parahippocampal gyrus (involving memory systems)—suggesting ties to Self recognition also (Schwartz et al. 198). This use of "cognitive reappraisal strategies" transforms emotional experience by "diminishing the responses generated by limbic structures" (199). Matthew Lieberman and his colleagues likewise found increased ventrolateral prefrontal cortex (VLPFC) and reduced amygdala activity in spider phobia treatment (220–21).[12] In another study, regarding *aversive* photographs, with subjects mentally *talking* themselves through the cognitive reappraisal of emotions, LPFC activation occurred on the *left* side, tied to language areas for self-recognition.[13] But in the Beauregard study already mentioned earlier, when subjects were asked to *imagine* a detached view, distancing themselves from *erotic* film clips, the *right* LPFC and MPFC were activated (Schwartz et al. 199–200). These studies showed Symbolic and Imaginary, left- and right-cortical aspects of Self in reflective awareness (self-recognition and self-knowing) about the *rasas* of aversive disgust/fear and erotic lust.

The unconscious evaluation of an emotion, determining its initial significance, occurs through the amygdala, at a limbic level (Beauregard et al. 179). But conscious and intentional evaluation of an emotion's "contextual meaning and the adequacy of possible behavioural responses" involves the medial orbitofrontal cortex (MOFC).[14] This is shown, too, in the brain scanning study with aversive photos. When subjects were asked to "reinterpret the picture so that it no longer generated a negative emotional response," the MOFC (or inner part of the VMPFC) became most active when the photos were most negative (179–80). With such *cognitive reappraisal* of aversive photos, the DMPFC was also active (180), suggesting ties to the mentalizing system—and to the theory of *rasa*-catharsis (refining emotional tastes through ironic twists that evoke a reflective distancing), explored in Chapter 1.

In a study of 8 to 10-year-old girls who were asked to suppress their sadness, with distanced observation while watching sad film clips, Beauregard and his colleagues found that the MPFC was activated, along with the LPFC, involving inner self-knowing and outer self-recognition systems, plus the OFC and ACC (176). They point out that the MPFC, which is "heavily interconnected with limbic structures," receives signals about the body and the environment via the OFC (a.k.a. the VMPFC), which also has strong ties to the amygdala and hypothalamus, regulating "visceral responses to stimuli and events" (171, 176). The OFC (above the eyes) is crucial for moral reasoning, or any kind of focus, because it protects "goal-directed behavior from interference" (171)—as a furrowed brow implies.[15] When the OFC is damaged, especially during childhood, there is a "failure to acquire social and moral knowledge" (172). The *right* OFC is crucial for integrating memory, attachment, emotion, bodily representation and regulation, and social cognition (Siegel, *Developing* 40). Thus, in children as well as adults, through its ties to the OFC, limbic, and subcortical areas, the MPFC creates a "metacognitive representation of one's own emotional state," or an "awareness of awareness" of emotion, involving social and moral behavior along with "inferences about the mental states of others" (Beauregard et al. 176). Distanced observation while watching a sad film, as *rasa*-catharsis, involves the MPFC, LPFC, VMPFC, and ACC.

But Matthew Lieberman's studies, comparing teens and adults, found that 13-year-olds use *more* of the DMPFC/PC/TPJ/TP mentalizing system than adults—with more concern for the Other's view when thinking about the Self (193). Questions asking for "direct appraisal" of one's Self (what I think I am) involved the MPFC and "reflected appraisal" of the Self (what I think others think I am) involved the mentalizing system, for teens and adults. But the *teens* showed higher levels of activation in both the Self-knowing and Other-mentalizing areas.

The direct appraisal, MPFC system, sensing an independent truth about one's Self, is especially encouraged in Western cultures (based, in part, on the ancient Greek maxim, "Know thyself"). It seems to be a part of the Self with agency, an *actor* performing various characters in the mirror, in "selfie" photos online, and in different social situations, which then involve LPFC self-recognition and DMPFC reflected appraisal through others' views. But the self-knowing actor may also *mask* "a social construction of the self" behind it, developing in the circuitry of the MPFC during our childhood and teenage years (Matthew Lieberman 194). The MPFC "is not the royal road to personal truth but rather a reflection of various sources by which we learn about ourselves."[16] Thus, the theater of everyday life, with a hall of mirrors between brains, includes inner masks that are like *prisms*, absorbing others' views to create *scripts and images* for self-knowing, self-reflection, and projections of ideals, threats, and desires upon the Other.

As considered earlier here (and in my previous two books), there is a shifting functional cluster of neural pathways that create our temporal sense of Self. It involves a feedback loop between a proto-self in the vertebrate brainstem, with ties to mammalian emotions in the limbic system, and a core self of the ape ego in the neocortex, with its human autobiographical self and super-natural aspirations. But it also involves a sensory-motor feedback loop with the outer environment, especially with the social

Other. Competition and cooperation between neural circuits produce a shifting network that "stages" Self and Other awareness, with working memory at the edges of that stage space and many memory, goal, and conceptual systems shaping it like audiences, stagehands, and directors in a theater (in Baars's model). Yet humans have also evolved hugely complex social groups and vast media networks, especially in recent decades, cooperating and competing for collective awareness on various stages and *screens*, from private physical interactions to public virtual events, which influence the inner theater (and movie theater) of our brains. The inner, self-knowing, MPFC *actor* is staged, too, within consciousness,[17] and shaped by prior experiences in childhood, teen, and adult years (through the *inner audience* of long-term memories and intuitions)—particularly regarding OFC (VMPFC) goal-directed, *moral or rebellious* behaviors. Such personal experiences, along with media models, create *scripts* for a sense of Self, performing reflective, LPFC *characters* in various situations, while absorbing and projecting (or "mentalizing") the Other's view.

Studies show that the MPFC is activated by hypnosis or by advertising messages in "suggestible" individuals (Matthew Lieberman 196–98). A study giving college students persuasive messages about using sunscreen for better health showed that MPFC activation predicted their subsequent behavior better than their own promises about what they planned to do. This points to an inner MPFC Self/Other *actor* who is influenced more by the health message than the outer LPFC (mirror recognized) *character* of Self, when projected toward a later scene in one's life. Lieberman calls this subliminal MPFC activation a "Trojan horse self," covertly bringing into one's brain the others' views (like the ancient Greek trick at Troy). It is revealed with young adults "changing their mental representations of the value of using sunscreen in a way that drives behavior but, at the same time, in a way that they are unaware of" (198). Another study with anti-smoking ads replicated the sunscreen results—with MPFC activation as a better predictor of future actions than "self-reported beliefs and intentions" (199). MPFC activation was also a better predictor than subjects' conscious evaluations of which ad would be most effective in getting them to call an anti-smoking hotline in the future (199–200).

The outer-recognized, mirror-reflected, LPFC *character* of Self is still important to self-control, especially regarding the right ventrolateral prefrontal cortex (rVLPFC), as the experiments described earlier show, with spectators distancing themselves from erotic and emotional films. The VLPFC (a.k.a. the inferior frontal gyrus) is the only part of the PFC that is bigger in the right hemisphere than the left, with that difference emerging in late adolescence, and is the "central hub of the *brain's braking system*"

(Matthew Lieberman 208–9). Studies show that the rVLPFC is especially active in certain types of self-control: (1) overcoming prior beliefs or cognitive impulses, such as loss aversion, when faced with a new logical answer and (2) resisting prior attitudes when considering the perspectives of others (212–15). Thus, the outer-self-recognizing, mirror-reflected LPFC *character* relates to the DMPFC *director* as mentalizing system, to an inner-self-knowing, yet Other-influenced MPFC as *actor*, and to the moral-reasoning, goal-directed OFC/VMPFC as *stage (and production) manager*,[18] with the rVLPFC as *light and sound operator* controlling impulses and prior beliefs through new perspective taking. This also involves temporal-lobe language, movement, attention, and "emotional color," plus memory areas as the retro- and prospective *audience*.

These brain network hubs as inner-theater elements correspond to similar movie-making roles: LPFC character, DMPFC director, MPFC actor, VMPFC *cinematographer*/production-manager, rVLPFC *grip* during shooting, and the temporal-lobe audience that fills in perceptions with meaningful, intuitive associations, like movie viewers stitching the gaps between cinematic cuts and adding to the diegesis beyond the frame. Indeed, related research shows that the dorsolateral prefrontal cortex (DLPFC or upper part of the LPFC) is involved with short-term memory, focused attention, and thus the "temporal chunking mechanism" of the inner theater, *editing* what we experience like *cuts* in a movie (Zacks 219)—or one's working memory as a "stage" in Baars's sense. Also, the DLPFC, as hub of the "rational, analyzing part of the brain," has "no direct connections to the emotional brain," centered in the MPFC (van der Kolk 206). This is akin to the DLPFC *film editor* analyzing what was shot in production, which involved the MPFC actor and others, but changing and integrating it *at a distance*. Such inner theater and filmmaking elements are, of course, merely metaphors for complex brain networks. But they suggest how our neuro-performances are created, extended, and shared in everyday life—with elements akin to the arts of stage and screen, as well as our various mass and social media.

Tests show that it is difficult for any of us to "actively control two things at once," such as resisting food while memorizing a poem, or even to engage in "two forms of self-control in sequence" (Matthew Lieberman 207). So there is often competition between these elements of Self and Other in our inner theater. And yet, advanced brain areas in our foreheads (LPFC, DLPFC, VLPFC, DMPFC, MPFC, and VMPFC) are crucial to a potential "catharsis" of desires, drives, and demands: clarifying our ape ego impulses through meditation or theater, with emotional arousal shifting into detached awareness, at significant moments of perspective taking. Indeed, Lieberman suggests that *mindfulness* training, which enhances VLPFC activity, may help with reappraisal tasks, especially for teenagers, because a study in his lab found a connection between motor self-control and emotion regulation (296–97).

Lieberman makes a distinction between "suppression" and "reappraisal" of emotions, regarding the VLPFC, which helps to refine my theory of *rasa*-catharsis, regarding the theaters of art, everyday life, mass entertainment, and social media. Suppressing a negative emotion involves distracting the brain to avoid feeling and looking distressed, which also interferes with remembering the social interactions of the emotional event, although one's increased heart rate might still affect others (Matthew Lieberman 218). Reappraisal can happen with less intensity, less arousal, and more "mental clarity"—perhaps involving the parasympathetic, quiescent nervous system, instead of sympathetic arousal. Suppression includes VLPFC activity in "hiding an undesirable facial expression" (219). But reappraisal activates the VLPFC *earlier* in the emotional event than suppression, involving a diminished amygdala (fear/rage system) response, with a shift from left VLPFC activity to right (218–19).

As we watch others onstage, onscreen, or in life—and then interact with them vicariously, virtually, or physically—emotional and motor (mirror neuron) contagion occurs, especially through right-hemisphere global awareness. Such absorbing of others' emotions, and the triggering of one's own, can be *suppressed*, with a facial expression masked or a body movement inhibited—through *left* VLPFC activity, maintaining self-control with preordained Symbolic rules and categories, as general functions of the left hemisphere (also suggested by Beauregard's studies). However, with the emotional engagement and yet detached observation of *reappraisal*, an earlier activation in the left VLPFC may alter one's beliefs and expectations. Then, with a shift back toward the *right* VLPFC, more of a complex catharsis may occur, clarifying and changing the emotional and physical reactions performed by an outer, mirror reflected, character of Self. This corresponds to what I theorized in Chapter 1 as Artaudian/Brechtian *rasa*-catharsis (see also Table 7.2).

Lieberman mentions various studies that show the direct effects of *subtle social cues* that change moral performances, by activating right VLPFC self-recognition (230–34). Children aged nine and up were left alone with a bowl of candy after being told they could take one piece. More than half the children took more than one piece, but when a mirror was facing them, only 10% did. Similarly, college freshmen were ten times less likely to cheat on a test in the presence of a mirror (7% vs. 71%). Regarding generosity, adults were three times more likely to give money to another player in an experimental game when they had the minimal social cue of three dots arranged like a face (two above and one below) rather than one dot above and two below. Our ancestors' bio-cultural evolution created a link between self-recognition, self-restraint, and self-normalizing systems in the rVLPFC, so that "our fear of not fitting in socially . . . [would often] override our indulgent self-interests" (234). Lieberman says that we are "built to overcome our own pleasure and to increase our own pain in the service of following society's norms," through the MPFC as a Trojan horse (226–27). But I would argue we also have many contending inner-theater networks and subgroup value systems that often rebel against dominant norms, in two-year-olds, in teenagers, and in many adults who feel collectively oppressed.

Evolution created the premature birth of our big brains, from a womb on two legs instead of four. This extended the mammalian adaptation of emotional connectivity toward a greater pruning of neural circuitry when the child is outside the womb, through the brain's *social* pain and reward systems, which use the same neural pathways as *physical* pain and pleasure (Matthew Lieberman 237, 300). Our primate ancestors' basic reading of minds, through mirror neurons in the lateral fronto-parietal areas, evolved into our advanced mentalizing system about others' thoughts, feelings, and goals—to enhance social connections and avoid the pain of rejection (237). This relates to ape-ego impulses and cultural success in humans, as shown with the effects of school bullying, when social pain leads to "dramatic reductions" in IQ and GRE test performances (278). On the other hand, feelings of belonging can improve GPA scores (279–80). Even imagining an "overall impression" of another person performing an action gives a "social encoding advantage" on memory tests, through the DMPFC's mentalizing system, enhancing one's recall more than just learning about the performance as information (284–85).

There may be conflicts between the various networks of Self and Other in our inner theater: between the LPFC's mirror-reflected, self-restrained *character/operator*, the MPFC's self-knowing *actor*, the DMPFC's Other-mentalizing *director* (as Trojan-horse puppeteer of the Self), the VMPFC's moral-reasoning, context-and-goal-oriented *stage/production manager* (or *cinematographer*), and the temporal lobe's memory-comparing, meaning-finding *audience*. For example, an LPFC network might critically reappraise one's performance, as mirrored by others, inducing autobiographical memories of shame. But MPFC circuits might also insist on pride,[19] while the VMPFC (OFC) triggers guilt[20] and regret,[21] or denies them with a new focus. Neural signals from below these cortical networks compete, too, for influence on the stage/screen spaces of inner awareness and outer actions, like a raucous crowd forming chants in vying waves. From our brainstem and limbic areas—with the vertebrate seeking of food, safety, or sex, the mammalian nurture, panic, play, fear, or rage systems, and the ape-ego impulses of bullying, seduction, or deception in hierarchical roles and territories—contentious signals might be sent at any moment, regarding physical and social stimuli, to the cortical Self/Other networks.

How do the various selves (and others) in each of us come together, as a single mind? Economist John Davis argues for a neuro-evolutionary, complex adaptive systems model to include both the *fission* of "multiple selves," as many neural circuits within one brain, and the *fusion* of those agents into an "individual" self (119–24). Although temporally framed by the "intentional stance" of a "self-concept," based on a "self-narrative," the divided individual continually changes, "re-sculpted" by various social groups and their norms (like teams and rules in game theory), which are other levels of evolving complexity (127–32, 161–62). In this model, an individual's "empathy" extends from kin to non-kin, through "well-developed beliefs" or "theories" of a working self-concept and self-other interactions, involving memories and projections "across multiple sites" (163–64).

Davis does not explore key neural areas for such Theory of Mind and social brain functions. Instead, he outlines a critique of neuroscience, by anthropologist Joseph Dumit, as imposing social norms and averaging out individual differences (Davis 132–36). I acknowledge their critique of fMRI brain mapping, for "isolating" a sample group, "imposing controls," and normalizing the brain-set data of many individuals into "one collective group" (135). Yet I still find such research valuable in showing the functional components of our brains—with the caveat that each brain has distinct neural networks, at a finer level of complexity, developing from particular genes and social interactions, especially in childhood. I hope that my use of theatrical terms (and psychoanalytic theories) helps the reader with the complexities of the neuroscience research that I cite, while also acknowledging individual differences in each person's changing concepts of Self and Other.

Of course, in using theatrical and movie-making terms, I do *not* mean to imply that stage managers or cinematographers use *only* their VMPFC when at work. All brain areas are active in all of us, to some degree, all of the time, unless the brain is damaged. We are not simply "right-brained" or "left-brained," as a common parlance since the 1970s has suggested. Indeed, our inter-hemispheric complexity will be considered in more detail now to "top off" my evo-neuro-psychoanalytic-theatrical-filmmaking model, regarding Self/Other images and narratives. How do films and other theatrical media produce a potential *rasa*-catharsis in the spectator's right and left cortical networks? How does this relate to notions of freedom and progress in the human animal?

MARCHING AHEAD: RIGHT, LEFT, RIGHT

Important roles are played by the brain's right and left hemispheres in its shifting theater of neural networks—interacting with the outer theaters of our continuing bio-cultural evolution, from vertebrate, mammalian, and ape-ego impulses to super-natural, social and mass-media hierarchies, reshaping our remnant instincts. As I outlined earlier in this chapter, growth spurts in the child's brain show surging recombinations of right and left cortical areas, along with deeper limbic passions and subcortical drives, in the changing stages of an inner theater and its family environment. The teenage years involve hormonal surges, too, with radical alterations in male and female bodies, brains, peer groups, and social rites. Through such changes, *right* cortical areas become more active than left with mirror, picture, and voice self-recognition, with autobiographical memory, with deceit, trust, humor, empathy, and guilt, with lust, sadness, and anger (or their control), with bodily/spatial awareness and self/other perspectives. However, *left* cortical areas are crucial to verbal, knowledge-based memory, including the social "schema" of a symbolic Self.

The left hemisphere's sensory and action ties to the right side of the body (and right visual field of both eyes), along with its distinctive areas for language comprehension and production, have led to it being called the "dominant" hemisphere. Ninety-seven percent of the population have typical, asymmetrical distinctions in their left and right cortical functions, including a majority of left handers, as well as nearly all right handers. After experiments with split-brain patients in the 1950s and 1960s, who had their corpus callosum connections between the two hemispheres cut (to reduce epileptic seizures), the left cortex came to be known as the "interpreter," because it focused the alternative associations of the right cortex into verbal categories (Gazzaniga). But a half century of further research has refined that view, with an increased understanding of the roles of both hemispheres in language and the significance of specific right-cortical functions, pointing also to the evolutionary advantages of brain asymmetry in other animals. Such research reveals how we continue to evolve today as *super-natural apes* (through and beyond our natural heritage, toward metaphysical projections and powers), in the personal and cultural shifts of left and right cortical networks, regarding limbic emotions and brainstem instincts. The neuro-agents of our inner theater, cooperating and competing as they unconsciously stage our conscious perceptions and concepts, involve distinctive, often conflicting, left and right functions in their front/back and top/bottom feedback loops.

According to neuroscientist Iain McGilchrist, "In most animal species, intense emotional responses are related to the right hemisphere and inhibited by the left" (26–27). This "lateralization" of brain functions has common elements in nearly all vertebrates, probably from a "common chordate ancestor" (Rogers and Andrew 94). We share many aspects of this with birds, including the right hemisphere's association with "detailed discrimination and topography," while the left in many vertebrate brains, like ours, "is specialised in categorisation of stimuli and fine control of motor response" (McGilchrist 27). This apparently evolved as an adaptive benefit in having one side of the brain devoted to *wariness about predators (or openness to mating)* while the other side can be focused on *hunting for food*—especially in our mammal and primate ancestors who were predators and yet also prey to others. "The right hemisphere yields a broad, vigilant attention, . . . especially of other creatures, who are potential predators or potential mates, foes or friends; and it is involved in bonding in social animals."

With humans, as with "predatory birds and animals, it is the left hemisphere that latches on, through the right eye and the right foot, to the prey" (McGilchrist 26). For toads, this is especially so with familiar prey, but "a novel or unusual choice of prey may activate the right hemisphere," until it becomes familiar, when it again activates the left. Likewise, horses, as herd and prey animals, "perceive new and possibly emotionally arousing stimuli with the left eye," which is tied to the right hemisphere (40). Such animal examples relate to findings in humans that the right hemisphere processes new stimuli, with "*exploratory* attentional movements," which fit into or then alter the left's categorical expectations, which are focused on the "*grasping* of what has already been prioritised" (44). Studies on toads and rats, as well as humans, show that the right brain is the main locus for social interactions (26). So, the exploratory, yet wary instincts of our inner/outer theater, from brainstem drives to limbic emotions to parietal-lobe sensory and frontal-lobe motor action networks, for freezing or fleeing (when there is danger) or for fornication and other social ties, come through the *right* hemisphere (46). Meanwhile, the feeding instinct and other prioritized goals are focused through the *left* hemisphere's familiar categories, which interpret the right's new signals by inhibiting them or changing course—for the animal or human to survive and reproduce.

Most emotions activate the right hemisphere more, which has stronger ties to limbic and subcortical (subthalamic) areas (McGilchrist 33). An exception to this is *anger*, which is "robustly connected with left frontal activation," involving dopamine pleasure-reward circuits, which are stronger in the *left* than the right (61). This indicates why, once the fighting instinct is triggered in personal conflicts, sports, or war, it is difficult for frontal-lobe circuits to contain it and why it often involves good versus evil identifications, through the brain's "binary operator" in the left parietal lobe (Newberg et al. 196).

The right hemisphere is larger (longer, wider, and heavier) than the left, with more overlapping cortical columns for interconnectivity and more white matter for transferring signals across regions, as the right attends to "the global picture," while the left "prioritizes local communication . . . *within* regions" (McGilchrist 33). These divided roles improve survival and reproduction abilities in our animal relatives. In cats, there are shorter reaction times for those with "a lateralised paw preference" (26). Marmosets with more lateralized brains are better able "to forage and remain aware of predators." More lateralized chimps are better at fishing for termites—showing that tool usage by our ape ego is an extension of the left hemisphere's concern with "utility," precision, mechanisms, nonliving things, and objectifying prey (26, 55).

Likewise, there is a strong left hemisphere (and thus right eye) bias for "tool-making in crows, *even where using the right eye makes the task more difficult*" (McGilchrist 27). The bird brain of the pigeon can categorize pictures of everyday natural scenes regarding their content. It uses the left hemisphere for a *local* strategy, "grouping the images according to particular features that must be invariably present." It uses the right for a *global* strategy, "taking account of the thing as a whole and comparing it with an ideal example." This local, objectivizing, "fine control," *mechanistic* bias of the left hemisphere and global, more subjective, "breadth and flexibility of attention," with *new life experience* orientation of the right (involving noradrenaline activation in humans) shows an extension, even in our distant animal kin, from predator/prey specializations to tool use and social interactions (27, 40).

Black-winged stilts peck more successfully with the right eye (left hemisphere), but males more often direct their courtship displays toward the left eye (right hemisphere) of females (McGilchrist 26). Baby chickens and magpies *approach* their parents "or an object on which they have imprinted" with their left eye, using their right hemisphere. Sheep can recognize other sheep or human *faces*, even after a gap of years, primarily using the right hemisphere (61). Rhesus monkeys use more of their right hemisphere for recognizing faces, too, and their left "for the categorical location of a dot within a square" (Hellige 166).

Likewise in humans and other primates, the left side of the face is better at *expressing emotions* and the left visual field is better at *perceiving* them, both involving the *right* hemisphere (Dimberg and Petterson; McGilchrist 61). Human, gorilla, and chimpanzee mothers tend to cradle their infants on the left side, even if they are left handed, increasing the right to right-hemisphere, Imaginary/Real attunement between the human mother and child, in contrast to left-hemisphere, Symbolic detachment. Also, within the limbic system of humans, the left amygdala is active with the conscious, "explicit[ly] *representational* content of the observed emotion, whereas the right amygdala is more closely involved with unconscious emotional processing" (62).

With the evolution of such specialties in humans, the *left* hemisphere became dominant for language production (with Broca's area) and language comprehension (with Wernicke's area), on the front and back edges of the temporal lobe, in the left frontal area and left TPJ. Language-trained chimpanzees also show "left-hemisphere dominance" for processing symbols (Hellige 149). However, in humans the right hemisphere processes language, too, "not in order to manipulate things, but to understand what others mean" (McGilchrist 70). The right is used for *prosody* (tone of voice) and has "a more extensive vocabulary, including long, unusual and non-imagable words" (51, 59). The right is needed for "meaning as a whole and in context," with the moral of a story, with jokes, with nonliteral implicit meanings, with irony, with inferences, and with *metaphor*—which may become difficult or impossible with a right hemisphere stroke (70–71). But clichéd metaphors relate more to the left hemisphere's focus on familiarity (51).

With greater connectivity than the left, and with more insulated (myelinated) wiring to subcortical areas, the right cortex "specialises in nonverbal communication" (McGilchrist 42, 71). Specifically, the right fusiform gyrus (in the lower rear area of the temporal lobe) processes the unconscious interpretation of facial expressions—with "minute facial changes" matched by the observer at 300–400 milliseconds (less than half a second), below the level of conscious awareness. The right hemisphere also attends more to the eyes, while the left focuses on the mouth, so the right is better at *detecting deceit*. People with a damaged left hemisphere, making the right stronger, become better at detecting lies than average observers. This shows that a healthy left cortex can block an unconsciously perceived truth with overly *optimistic*, categorical expectations and confabulations (62, 81, 84). Patients with a right-hemisphere stroke and thus a more dominating left cortex may deny the paralysis of the left limb, some even claiming that it does not belong to them (Ramachandran and Blakeslee 132–41). Experiments with split-brain patients also demonstrate that the staging of conscious experience may involve confabulation, with the left hemisphere's "conscious verbal self" constructing stories to rationalize unconscious perceptions and impulses from the right (Timothy Wilson 97).

The right hemisphere is used for *identifying individual faces* and for valuing their age, sex, and attractiveness (McGilchrist 60). Specifically, the right insula increases in activity when looking at the face of one's romantic partner, compared with an unknown face (168). Right hemisphere activity increases, too, while following another person's gaze, with shared attention, or when looking into someone's eyes. These behavioral aspects of "empathic attention" are disrupted in autism, as is "understanding the mind behind the gaze," and possibly also in schizophrenia.

The right hemisphere is more involved with *primary* emotional expressions (when not inhibited by the left): with smiling, laughter, and tears (McGilchrist 61). But the left becomes active with *social* emotions and with willed or forced emotional expressions (62). The right is key for bonding and empathy, the left for competition, rivalry, and individual self-belief (63). The right temporal lobe deals more with personal and emotional memories, the left with public and factual (54). Even the *mirror neuron system* works differently on each side of the brain. The right hemisphere will "empathise with, identify with, and aim to imitate only what it knows to be another living being, rather than a mechanism" (57). Mirror-neuron areas in the left hemisphere are triggered when the subject sees a real hand *or* a virtual-reality hand grasping an object, sending signals to do the same, but the right TPJ is activated *only with the real hand*. Also, the emotional contagion of *pain mimicry* occurs in the right, specifically in the right anterior cingulate cortex, which is activated both when one experiences pain and when one witnesses someone else in a similar painful experience. Regarding the mentalizing system considered earlier in this chapter, the right inferior parietal and the right lateral prefrontal lobes are crucial for imagining the other's point of view and inhibiting one's own.

Adding to our theater model (summarized in Table 5.1), we might call the left hemisphere a *critic/scripter* (with its abstract, objectifying, category filters, and language processes) and the right a *mime improviser/scene designer*[22] (with its attention to living things and new experiences, including emotional gestures, and with its holistic view regarding social and environmental contexts). Again, this is *not* to say that critics and playwrights fail to use their right hemispheres, or that mimes, improvisers, and scene designers never use their left. But in our brain's inner theater, such general orientations shape, with left and right, Symbolic and Imaginary aspects, the LPFC's mirror-reflected character, the MPFC's Self-knowing and yet Other-influenced actor, the DMPFC's mentalized director, the VMPFC's

Table 5.1

Inner Theater of the Brain (with "elements" added to Baars; Lieberman; McGilchrist)

Theater/Film Element	Cognitive/Affective Function	Neurological Network & Hub
character	outer performing self (awareness in the mirror test)	lateral prefrontal cortex (LPFC)—with left/right LPFC activity for the *cognitive reappraisal* of emotions (left talking about, right reimagining aversive, sad, and erotic photos/films) & with ties to MPFC, DMPFC, VMPFC, anterior cingulate cortex, & limbic areas (Beauregard et al.; Schwartz et al.)
actor	inner knowledge of self (adjectives applied to oneself, reflecting the influence of others)	medial prefrontal cortex (MPFC) and precuneus (PC)
director	sense of other's viewpoint (mentalizing system or Theory of Mind)	dorsomedial prefrontal cortex (DMPFC), temporal parietal junction (TPJ), precuneus (PC), & posterior cingulate cortex (PCC)—plus the limbic septal area with *empathy*
stage/production manager (and cinematographer)	moral evaluation of actions with social and physical reward/pain	ventromedial prefrontal cortex (VMPFC), dorsal anterior cingulate cortex (dACC), & ventral striatum
light/sound operator (and film grip)	control of impulses that come from prior beliefs and attitudes	right ventrolateral prefrontal cortex (rVLPFC), with shift from left to right in *reappraisal*

stage itself (and film editor)	working (short term) memory, rational attention, time chunking, and inhibiting of self-interest	dorso-lateral prefrontal cortex (DLPFC) (Zacks)
audience	memories, intuitions, and meanings, shaping percepts/concepts onstage	temporal lobes & insula
critic/scripter	abstract rules and verbal focus (senses/acts on right side of body)	left hemisphere of neocortex (prefrontal, frontal, sensori-motor, & parietal lobes)
mime improviser and scene designer	holistic, contextual, visual, spatial, tonal openness to the new (senses/acts on left side of the body)	right hemisphere of neocortex (stronger ties to limbic & brainstem areas)
stagehands	emotions and pleasure/pain of survival/reproduction drives (primary emotions: seeking, rage, fear, panic, care, and play)	limbic system (including temporal lobes) & brainstem areas

moral-guidance stage manager (or cinematographer), the temporal-lobe's memory-system audience, and the limbic-brainstem stagehands (plus the DLPFC's editor). These brain areas, as hubs of networks, relate to such roles in the theaters of everyday life and in various stage and screen media between brains today. Also, people with right hemisphere damage, increasing the power of the left, report feeling a "foreignness" of the Self, being disconnected from the world, like "insensible automata, puppets, or mere spectators" (McGilchrist 236)—more like a critic/scripter watching one's own actions.

Table 5.2

Brain Hemisphere Functions

(Based mostly on McGilchrist's research—with Lacanian terms, vampire-film archetypes, and inner theater elements added in bold letters)

Left Hemisphere (Symbolic Patriarchal)	**Right Hemisphere (Imaginary/ Real Mother-Child)**
predator (focused, objectifying, tool using)	*prey or mate (broad awareness, life/death/sex oriented)*
belief, competitiveness, conscious agency	care, cooperation, unconscious socio-environmental influences
abstract/analytical thinking [inhibiting →]	emotional/sensory intuition (with limbic/subcortical ties)
familiar, rule based, orthodox ideas	"anomaly-detector" and Devil's Advocate awareness of new
examining parts in a linear, categorical way	comprehending the whole in a cyclical, contextual way
sequential, cause and effect, literal language	deductive, parallel, paradoxical, and poetic associations
manipulation of known, static, isolated, general	care of individual, evolving, interconnected, incarnate beings
Self-referential (thing/machine oriented)	Other-engaged (toward living world), empathic
Self-certainty, yet toward virtual, unrealistic	responsibility, shame, and guilt, but more realistic
optimistic, yet with projection and anger	melancholic, yet sensitive to tears and alert to change
denotation (with confabulation to repress)	connotation, appreciating ambiguous meanings, ironic humor
affinity to major keys and basic rhythms	minor keys, complex syncopation, and harmonic progression

Table 5.2 (Continued)

Left Hemisphere (Symbolic Patriarchal)	**Right Hemisphere (Imaginary/ Real Mother-Child)**
concerned with social or willed emotions	primary process, bonding, and unconscious emotions
focused attention, grasping *(with right hand)*	sustained attention, exploratory *(left-side with more facial expression)*
looks at other's mouth (detached from body)	*looks at eyes during conversations*
identifies simple, easily categorized shapes	identifies complex, varied figures
produces schematic representations	depth in time/space (Self-image and Theory of Other's Mind)
more dopamine (pleasure) networks	more noradrenaline (excitatory) networks
parasympathetic (quiescent) nervous system	sympathetic (arousal) nervous system ties
schizophrenia, MPD, ASD, anorexia, BPD	depression with anxiety
Slayer's words, crucifix, stake, sunlight	**Vampire/Werewolf's shapes, bloodlust, life-death cycle**
critic/scripter	**mime improviser/scene designer**

In Table 5.2, I have listed many of the left and right hemisphere processes given in McGilchrist's summary. I have also added Lacanian/ feminist terms and characteristics of the "slayer" and vampire/werewolf, considered in the previous chapter, plus the theater/film terms just given. (Of course, the vampire and werewolf are also predators, but often with right more than left hemisphere aspects.) It is important to realize that the networks within and between the left and right hemispheres also involve front to rear (and top to bottom) circuits—and variations between individuals. The left hemisphere not only inhibits many signals coming from the right, across the corpus callosum,[23] but its frontal lobes also modulate other areas toward the back of the brain—as does the right, though in a less pronounced way (McGilchrist 91). The left-hemisphere slayer inhibits the right's vampire/werewolf. But frontal lobes on both sides also filter external sensory stimuli and internal animal drives from the rear of the brain (parietal and occipital lobes) and areas deeper in the brain (limbic and brainstem systems).

According to McGilchrist, "the left hemisphere is the seat of *conscious* self-awareness . . . through the conscious will," involving an "attentional 'spotlight'" (90). He even argues that "the conscious, rational will—grasping and manipulating—may have been responsible for the expansion of the left hemisphere" in our animal ancestors. And yet, the backstage and auditorium of unconscious networks are crucial to what appears onstage in the spotlight of our conscious self-awareness, which then reflects what we experience as "will"—or contending *wills* within the Self.[24] For example, the right inferior parietal lobe, behind the frontal, "plays a crucial role in planning and monitoring the outcomes of one's actions." But the right hemisphere overall, as mime improviser or scene designer (in my model), may also rebel against the "stickiness" of the left's rules and habits (in McGilchrist's terms). "There is a tendency for independence and motivation to be associated with the right hemisphere, and passivity with the left hemisphere." Yet that passivity or stickiness of the left hemisphere, especially involving the modulation of posterior areas by frontal ones, is "a process that resists, but does not negate" (91). It imposes a "necessary distance, or delay, enabling something new to come forward," though that something new happens more on the right side—or when the right disrupts the left.

The two hemispheres do not simply negate or complement each other (McGilchrist 91). Their "incompatibility" creates a "dialectical synthesis" between them. While the right hemisphere prioritizes emotion and bodily sensations, its frontal lobe "brings distance and delay to espousing 'my' position." This allows the otherness of the other person to be perceived, by the inner *mime improviser*, through likeness to me and yet a different perspective: "it enables a broader empathy and the beginnings of altruism." Meanwhile, the *critic/scripter* in the left hemisphere, with "its tendency to abstraction," usually promotes "individual gain" (92). But the left frontal lobe, influenced by the right's shift toward altruism, may offer a sense of "distance" from predatory, grasping, and objectifying animal drives, toward a "peaceful detachment from the material realm and 'emptying out' described by experts in meditation as a mystical experience."

This interplay between right hemisphere sensitivity, yet right frontal delay, and left hemisphere abstraction, yet with a potential, left frontal peacefulness, creating compassionate equanimity overall, may also be involved in theatrical *catharsis* (or *rasa* refinement). Emotions are thus reappraised and yet reengaged through new degrees of perspective taking by the inner mime improviser/scene designer and critic/scripter. While watching a show, through a willing belief (or suspension of disbelief) in its fiction, *cares* develop in the right hemisphere about the characters onstage or onscreen. When various emotions and identifications are shifted into new viewpoints during the course of the performance, through tragicomic twists or distancing effects, the left's previous *belief concepts* might be altered, regarding relationships and responsibilities, with different networks connecting to the right's *holistic perceptions*.[25] Even melodramatic identifications with Buffy or Bella, as the vampire's slayer or girlfriend, could shift into tragicomic catharsis (or the refinement of various opposed *rasas*), through complications in the show and the spectator's personal associations. This might involve a "problem solving" process in the right anterior temporal area and the right prefrontal cortex—with sudden "insight" or pleasurable "aha" signals from the right amygdala (McGilchrist 65).

Generally speaking, positive, prosocial, "approach" emotions involve activation in the left hemisphere and negative, introverted, "avoidance" emotions in the right (Davidson and Begley 39)—with such categories in our ape egos building on the predator/prey asymmetry in the brains of even our distant vertebrate kin. Depression, for example, not only increases right frontal activity, but also reduces right rear (parietal and occipital, body orientation) areas, while involving reduced left frontal (McGilchrist 63–64). This relates to the research mentioned in Chapter 1 on separation panic with rat pups, who call for their mothers for a certain time period,[26] but then *freeze* in a depression-like behavior to avoid predation (Solms and Turnbull 131). Such "learned helplessness" in humans may occur through inconsistent, uncontrollable stressors even in adulthood, which trigger a remnant animal instinct (Sapolsky 218–22).

Likewise, anxiety in the right hemisphere evokes the bottom-up animal circuits of our *fleeing* instinct, beyond left hemisphere controls. "Just as the left can block emotional and visceral input from the right, the right can override conscious processing and emotional well-being in reaction to threat" (Cozolino, *Human* 75). Mania, on the other hand, shows increased left hemisphere activity, with a hyperactive focus on goals (McGilchrist 64). And yet, in comparisons of "normal" individuals, those with more right hemisphere activation, while the brain is at rest, are more susceptible to "negative" emotions. Those with higher left frontal activity, while at rest, are then quicker at minimizing emotional reactions. "Even when we watch movies, greater left frontal activation is correlated with more intense positive affect during the film and a better review afterward" (Cozolino, *Human* 74; from Wheeler et al.). So there may be drawbacks or benefits to specific types of left and right activation in certain situations.

Neuroscientist Richard Davidson discovered that newborn babies tasting sugar water, which makes them smile, have left prefrontal activation and newborns tasting lemon juice, making them pucker, have right prefrontal (Davidson and Begley 36–37). Likewise, 10-month-old babies have greater left prefrontal activity while watching a video of an actress smiling and greater right while watching her crying. With adults, he found higher left prefrontal activation with amusing or uplifting videos, evoking detachment, and higher right prefrontal with frightening or disgusting videos, triggering the danger and immorality monitoring of that hemisphere (40). But he also noticed that while differences between hemispheres in the same person were as much as 30%, differences between individuals reached 3,000% (41). After this, his research explored different "Emotional Styles" between people, with a specific focus on mapping the brains of meditators, from novices to young and old Tibetan monks.

Davidson defines six Emotional Styles as a spectrum of cortical-limbic-subcortical brain networks involving individual variety between people: Resilience, Outlook, Social Intuition, Self-Awareness, Attention, and Sensitivity to Context (Davidson and Begley 5–6). Greater Resilience, for example, relates to more axons (white matter) connecting the prefrontal cortex and limbic amygdala (72). Positive Outlook involves high activity in the ventral striatum, the ventral pallidum (associated with pleasure), and the prefrontal cortex (230).[27] Resilience and Outlook can both be improved with mindfulness training, which increases activity in the left prefrontal cortex (205).

The malleability of Emotional Styles, through meditation and *rasa*-catharsis, reflects the continual competition and cooperation between multiple neural circuits, as agents within the Self. Competition between the hemispheres involves a greater variety of roles for the right, with faster signals traveling from right to left, but a greater power in the left to inhibit the right (McGilchrist 218–19). Split-brain patients, in the first few months after their corpus callosum has been cut, demonstrate the radical ways that the two hemispheres may conflict—when their hands try to do different things. But it is always the left hand (right hemisphere) that rebels, "that pushes the other way, grabs the wheel, chooses the 'wrong' clothes" (219). As McGilchrist puts it, helping to inspire the terms of my inner-theater model: "Once the script has been written and the play half performed by the left hemisphere, an incursion from the right hemisphere is bound to be disruptive. . . ." Thus, the right hemisphere mime improviser (and scene redesigner) can also be a *trickster* to the critic-scripter.

Subcortical connections *below* the corpus callosum provide a trunk for the "tree of consciousness" (McGilchrist 221, borrowed from Panksepp). These are "complexly interconnected, cortico-subcortical loops involving the basal ganglia, deep-lying nuclei in the brain, way below the corpus callosum, which . . . [involve] affective and cognitive functions. These 'loops' underlie subtle, emotionally laden aspects of experience" (225).[28] Thus, consciousness, as proto- to core self, is unified at lower and evolutionarily older levels. But when competing and cooperating neural pathways become more distinctively mammalian, ape, and human, with advanced reflective and mentalizing circuits, "the process becomes *self*-consciousness at the topmost levels . . . [and] the possibility of separation occurs" between the hemispheres' numerous neural branches (220).

Conflicts in the inner theater also interact with the cooperative complexity of our social and technological networks—vastly extending the *inner tree of Self-consciousness*, as the heritage of our *animal evolutionary tree*. Separate hemispheres help us to connect with larger social networks, as is shown, in a contrary way, with psychotics who have "partial agenesis" of the corpus callosum. This compromises the hemisphere separation, with intrusions that can be "creatively fruitful" or greatly disruptive, due to a "premature collapse into unity of elements or processes whose mutual independence needed to be maintained" (McGilchrist 212). Research on schizophrenia and schizotypy shows *"a failure of interhemispheric inhibition,"* with intrusions of the left into right hemisphere and right into left, which isolate the person from society (212–13).

The inhibitory filtering between hemispheres develops in each of us, but only gradually. Recall that myelination of the connecting, corpus callosum fibers "does not even begin until the end of the first year of life, and progresses only slowly thereafter," during childhood and adolescence, as the hemispheres specialize more and more through particular family and social influences (McGilchrist 213). Thus, the predator and prey/mate origins of hemispheric specialization in our vertebrate ancestry involve deeper instincts that persist in us, but are reshaped as we develop toward adulthood. These bloom in our ape egos through (1) the socio-emotional complexity of hierarchical *dominance*, with objectifying anger and territorial defenses, mainly through the left's cortico-limbic circuits, (2) sexual *attraction*, plus nurturing care of offspring, primarily through the right's, and (3) playful *trickery*, especially by underlings manipulating those above—in the interplay of our brain's tree theaters.

A neuro-evolutionary tree-theater model of our brain offers insights about why we accept certain degrees of social *control*, with left-hemisphere orthodoxies and alpha contests, yet fight at times for *freedom*, like a trapped prey animal, through our right hemisphere's remnant survival, mating, and nurturing instincts. In the abortion debate, for example, "pro-lifers" may empathize with the unborn child, feeling trapped in the womb and at risk of death, whereas "pro-choicers" may sympathize more with the mother, being trapped by governmental controls. Soldiers, as another example, might fight for the freedom of others, even those not kin, and lose their lives before producing offspring, while following military orders with feelings of fierce camaraderie. In these examples, the human inner tree theater, through its prior pruning, interacts with other brains and bodies, via in- and out-group conflicts that form bio-cultural feedback networks, often paradoxically combining notions of control and freedom. Today, our interacting, inner tree theaters also extend into cyberspace and other, recently evolved technologies, which transform our bodies and brains, our sense of control and freedom—sometimes in shocking ways.

BACK TO THE BODY (AND ITS TRANSFORMATIONS)

An exhibit in London's Science Museum allows visitors to put their hand in a slot under a rubber hand and then watch the rubber hand being stroked at the same rhythm as they feel their own hand being stroked. In similar lab experiments, people soon feel that the rubber hand, if aligned with their real hand, is an extension of their body, like a phantom double, and that their own hand is closer to it than it actually is, especially the stroked fingers—in a bottom-up and top-down visual, tactile, and coordinated body/brain feedback illusion (Tsakiris and Haggard; Makin et al. 7). Such *identification* with a rubber hand as one's own involves proprioceptive activity in the brain's right temporal and frontal lobes, specifically the right posterior insula and right frontal operculum (Tsakiris et al.). Similar results have been found with monkeys, also involving activation of mirror neurons in the premotor cortex (Armel and Ramachandran). This suggests a connection between seeing an action, sending signals in the brain to perform an action (although inhibited by other signals), and unconsciously shifting one's body image into the body that one sees performing—as in the spectator's somatic identification with an actor onstage or onscreen, while being "touched" by the rhythms of plot and emotion, even without physical touch.

The rubber hand illusion also happens when the subject is *blindfolded* and only feels the touch of his or her finger on the rubber (guided by the experimenter) while being touched in the same way. This creates the illusion of touching one's own hand—in the rubber one—activating the ventral premotor cortex, intraparietal cortex, and cerebellum as sensory and coordination areas (Ehrsson et al.). Furthermore, during the *visual* rubber-hand illusion, measurement of slight changes in sweat level, known as Skin Conductance Response (SCR) or Galvanic Skin Response

(GSR), reveal that when a finger on the rubber hand is bent back, the sympathetic nervous system is automatically aroused, alarmed at a threat to one's own hand (Armel and Ramachandran). With some subjects, the illusion even occurs when a tabletop is used instead of a rubber hand, as long as one's own hand is hidden and the table is tapped and stroked in a similar, synchronous manner. With this table-as-my-hand illusion, pulling an adhesive bandage off the tabletop or hitting it with a hammer alarms the subject, unconsciously raising the SCR/GSR (Armel and Ramachandran; Ramachandran and Blakeslee 58–62).[29] Neurologist V.S. Ramachandran relates this to phantom limb pain in amputees and to the "malleability" of anyone's body image, mapped in the parietal lobe's somato-sensory area, as a "phantom."

Just looking down at one's body, but seeing instead the torso and legs of a manikin, through a head-mounted video display, creates *identification* with the manikin's body as one's own (Petkova and Ehrrson). Again, SCR levels change when the manikin's stomach is sliced with a knife during this illusion (just one minute into the experiment). The phantom body image, projected on the manikin, causes real *alarm* signals in the brain and nervous system (3).

Another experiment showed that the participant's phantom body image could be put into the body of an experimenter sitting opposite, with the head-mounted display giving *that other person's view* of oneself *as one's own* (Petkova and Ehrrson 4). When the participant and experimenter repeatedly shook hands for two minutes, they squeezed in synchrony. But in a control condition without the illusion, they did not (4–5). "Most remarkably, the participants' sensations of the tactile and muscular stimulation elicited by the squeezing of the hands seemed to originate from the experimenter's hand, and not from their own clearly visible hand" (5). Putting a knife to the experimenter's wrist (with a plaster cast over it and the participant's) again produced the SCR alarm response. These experiments reveal how strong the identification can be with characters *onstage or onscreen*, even perhaps with a werewolf, vampire, lab hybrid, or humanlike ape, or perhaps with the virtual figure in a videogame as one's avatar.

Such bodily identification (or swapping) can also be shifted to a hand of a different color, as in an experiment where *white* participants felt the illusion of having a *black* rubber hand (Farmer et al.). SCR again increased with a threat to the black hand. A lessening of implicit (unconscious) antiracial bias was measured after this cross-color rubber hand illusion, with a greater *decrease of racism* in participants who experienced a stronger illusion. Thus, "multisensory experiences can override strict ingroup/outgroup distinctions based on skin colour and point to a key role for sensory processing in social cognition" (1242). This suggests, too, that the multisensory movie-theater experience, involving mirror and intuition neurons, even without physical touch, can alter the spectator's unconscious, in-group/out-group projections—like a white subject's lessening of implicit racism by feeling that a black rubber hand is his or her own.

It may take many movie experiences to shift spectators' long-term expectations. Experimenters gave participants a "vitamin" injection that was actually the stimulant epinephrine, a placebo, or the tranquilizer chlorpromazine, and then showed them a 15-minute comedy film. Those who had a stimulant in their nervous system smiled and laughed more than the others, yet on average did *not consciously* report enjoying the film more than others who got a placebo or tranquilizer. In answering a survey about the film, they rated it with reference to their previous enjoyment of that type of film and its star, rather than the new experience (Timothy Wilson 130–31).

And yet, evolutionary psychologists argue that unconscious projections of racism are *changeable* because they are based on a deeper need, in our animal-to-human ancestry, for "coalitional alliances" (Kurzban et al.). Our hunter-gatherer ancestors, tens of thousands of years ago, did not encounter other humans of different races. But quickly identifying another human as an ally or enemy was crucial for survival. A recent experiment shows that racial identification is easily shifted because it builds on this fundamental mechanism[30]:

> when cues of coalitional affiliation no longer track or correspond to race, subjects markedly reduce the extent to which they categorize others by race, and indeed may cease doing so entirely. Despite a lifetime's experience of race as a predictor of social alliance, less than 4 min[utes] of exposure to an alternate social world was enough to deflate the tendency to categorize by race. (15387)

Movies today, along with other theatrical media, live and virtual, often draw for accessible entertainment on race, class, gender, age, and beauty stereotypes, which can increase enmity between individuals and groups. But movies can also challenge the viewer's interactive inner theater, altering neural networks of coalition and competition between brains. In an "alternate social world," verbal and visual cues may cause *shifts of identification*—akin to the synchronous touch of the rubber hand illusion and the changed perspective of the body swapping experiment. This might induce compassionate affiliation across prior categories, with a playful flexibility in imagining new allies and complex self/other viewpoints.

Shifts in identification have also been measured with faces. An experiment touching a participant's face with a cotton swab while another's of the same gender was touched synchronously on video produced an "enfacement" illusion and its aftermath (more so for younger participants). There was identification with the other's face, perception of similarity to the other's face, and a feeling, though to a lesser extent, that the other's face was attractive and trustworthy (Tajadura-Jiménez et al.). A similar experiment with the synchronous touching of a friend's face and one's own, of the same gender in the same room, using a paintbrush, showed that the enfacement illusion changed how the participant then rated a computerized morphing spectrum between one's face and the other's (Sforza et al.). The point at which the participant saw his or her face shift into the friend's came earlier on the spectrum, showing an increased incorporation of the other's face into one's inner self-image. This effect was mostly unconscious and was greater for participants with stronger empathic traits or for those who found their facial partner more physically attractive.

Likewise, beautiful faces onscreen, especially of familiar *stars*, readily evoke the empathic spectator's *merging* of self with other, through enfacement and vicarious embodiment, whereas less attractive faces provoke aversive reactions and projections. However, radical transformations onscreen, such as a beautiful human turning into a fanged vampire or werewolf, may challenge preconceptions of race (or species), class, gender, and age. This might create new realizations about in-group/out-group affiliations, about our body image as a malleable phantom, and about the character of my-Self staged in others' inner theaters, influencing my own.[31]

IN SUM

The ape ego in humans has achieved tremendous success, populating and altering environments around the globe, even traveling to the moon through a "space race." But its competitive, ideological technologies have also extended the cruelties of Mother Nature, from infanticide to genocide. In comparison with other primates, humans are born very early and very dependent on caretakers, yet they develop much higher orders of theatrical, self-reflective, and new perspective-taking brain systems, shaped by the family and cultural environment. Growth spurts of the right and left hemispheres produce various stages of Self and Other awareness, with early- and later-life experiences pruning particular inner-theater networks and social relationships. The greedy insecurities of the prematurely born, highly theatrical, culturally flexible ape ego in humans, from its initial mirror stage to the "terrible twos" to further permutations in childhood, adolescence, and adulthood, appear also at the group level. With the right hemisphere's orientation, in our animal ancestors, toward global awareness and intuitive tactics, for safety threats, mating possibilities, and novel situations—vying with the left's narrow priorities of food seeking, social rules, and familiar expectations—humans evolved abstract languages and world-changing technologies, yet also repressive orthodoxies and destructive rationalizations.

These super-natural powers and yet tragic flaws of the human ape involve self-deceptive networks within and between brain theaters, with hubs in the prefrontal cortex, tied to deeper limbic emotions and brainstem drives. The LPFC mirror-aware *character* of Self performs outwardly for others while an MPFC *actor* provides an inner sense of Self, imagining others' desires through the DMPFC *director*. Meanwhile, a VMPFC *stage/production manager* or *cinematographer* judges one's actions in a social context (regarding the dACC reward/pain network, considered in the next chapter), with a right VLPFC *light/sound operator* controlling impulses and a DLPFC *film editor* providing time chunks, rational attention, working memories, and the inhibiting of self-interests for the sake of others. The temporal-lobe *audience* of memory traces and the subcortical *stagehands* of emotional drives also support which of these aspects are most active and which become conscious. Performances of Self, on the inner and outer stages of daily life, involve right-hemisphere *mime improviser* and *scene designer* creativity, influencing yet often inhibited by left-hemisphere *critic/scripter* expectations and definitions.

Such inner-theater networks are dynamic, evolving systems in each brain, with specific personality dispositions, memories, and emotional styles. Engaged by the alternate social worlds of movies and other theatrical media, a *rasa*-cathartic "reappraisal" (not just "suppression") of ape-ego beliefs and impulses may occur in the spectator's brain. As shown experimentally, this might involve shifts in brain activity from the left to right VLPFC, as the inner character of Self becomes identified with the Other's desires, conflicts, and transformations onscreen. In related research, the right LPFC is active, along with the right anterior cingulate and right VMPFC, in "detached" observers of erotic and sad films. Experiments with enfacement, body swapping, and the colored rubber-hand illusion also show a potential for physical identifications, changed perspectives, and confirmed or altered prejudices—in viewers' inner theaters when enclosed by the cinematic womb.

Like but far beyond other apes, humans operate in highly complex societies with playful games and serious battles of dominance, attraction, and trickery. We struggle not only for personal or group survival, for genetic progeny and social fame, but also for emotional and symbolic meaning in our life/death awareness and Self/Other theatricality. Movies provide ways to rehearse particular social situations and changes in character, like collective dreams, playing out scenarios of competition and cooperation, of survival threats and erotic desires. Exemplifying this, vampire and werewolf films reflect changing ideas of the animal within "human nature," during various decades of the last 100 years. So do films explored in the next chapter, with lab-created hybrids and partly civilized simians. These popular fantasies show the ongoing evolution of our ape egos, virtual bodies, kinship identities, and scapegoating projections—involving choices of what and how we watch, as chimeras ourselves, reflecting upon the screen.

Chapter 6

Lab Hybrids and Planets of the Apes

Like folkloric horror films about vampires and werewolves, science-fiction movies about animal-human hybrids started early in cinema's history—reflecting the hyper-theatricality of our ape egos in that new medium, as a monstrous potential for devilish passions and angelic yearnings.[1] Recent psychological experiments offer insights about the continued appeal of such films with their various remakes and sequels. Most children have a "natural" (or nature-culture coevolved) tendency to attribute *agency, design, and purpose* to living things, extending from animal to human to god figures, as created and creators (Barrett). Even 3 to 6-month-old babies pay attention in different ways to discs on a video screen that appear to chase one another or just move randomly, showing that they notice agency (36). This also relates to the Theory of Mind tests mentioned in the last chapter, regarding how humans develop views of the other's perspective, well beyond that of our ape relatives.

A majority of 3-year-olds tested in various cultures think that an adult or god would know the contents of a mislabeled or darkened box. But 5-year-olds realize that the adult would have a different perspective, not knowing the real contents like they do, without seeing inside. And yet, at that age, children continue to attribute omniscience to god, depending on their religious context (Barrett 88–90, 99–104, 109–10). They also, between the ages of four and nine, tend to view animals as intelligently designed with a purpose given by their creator (46–49). When compared with children in fundamentalist Christian families, children raised by non-fundamentalists favored such "creationist accounts" less, but still more so than their parents (68). "Developmental psychologists continue to find evidence that the godly properties of superknowledge, superperception, creative power, and immortality are quite intuitive, at least for young children" (80). Thus, the continuum from animal to human, in a child's view and in the intuitions of later life, involves a cosmic theater, extending from the inner performances of one's own brain, with its creative and destructive potential, to adults' power over the animal energies of the child, to the projected superpower of Other beings.[2]

Even without the super-natural powers of a vampire to live beyond the grave, or of a werewolf to transform in the moonlight, other animal-humans express the bestial drives and divine aspirations within us, especially from our early life experiences of a changing body and of big people as role models. From the 1930s to 1970s to 1990s, several movies have been made, based on the H. G. Wells novel, *The Island of Dr. Moreau*, with a cruel scientist surgically or genetically changing animals into half-human hybrids. These films reflect a child's fear, even at home, of a godlike parent/creator exposing and punishing its animal passions—a fear that continues into our adult social lives. Such island movies, like the lab films, *The Fly* and *Splice*, explore the Frankenstein trope of the mad scientist reengineering human life, expressing our continued fear of, yet desire for, godlike biotech powers.[3] The simian societies of the "planet of the apes" films also show our ape egos redefined, more collectively, in a future dream/nightmare of conflicting ethno-species characteristics, cultural ideals, and charismatic leaders. All of these movies play with fantasies involving primal drives that we inherit from our animal ancestors—survival and reproduction, territoriality and pleasure, dominance, attraction, and trickery. Such drives are extended onscreen toward sci-fi threat rehearsals and morality tales of exotic worlds that fold back uncannily onto our own. How do these films reflect our morphing ape egos, our inner movie-theater elements, and our body-swapping identifications with mass media images? How do they also explore specific concerns of the past, which persist today, as our cinematic awareness continues to shape our bio-cultural evolution—with devilish, animal-human characters onscreen and a godlike role for the audience?

PARADOXICAL SOULS

Created just 70 years after the end of slavery in the United States, and soon after the start of "talking" films, *The Island of Lost Souls* (Erle C. Kenton, 1932) adapted an H.G. Wells novel, *The Island of Dr. Moreau* (1896), which had also been made into a French silent film, *Ile d'Epouvante* (*The Island of Terror*, Joe Hamman, 1913). *The Island of Lost Souls* added Depression Era fears of underclass revolution, and yet sympathy for the oppressed, to the novel's concerns about Social Darwinist eugenics (Kirby), colonial callousness, and inhuman scientific experimentation. Although similar in many ways to the popular horror and sci-fi films, *Dr. Jekyll and Mr. Hyde, Dracula,* and *Frankenstein,* which appeared a year earlier, *Island of Lost Souls* was a box office failure. It was so controversial for its suggestions of human bestiality and surgery without anesthetics that it was banned in parts of the United States and in many other countries, including Germany, where the Nazis would soon implement cruel human experiments in their concentration camps (Dinello 367–68). The film was ahead of its audience's awareness as a sci-fi fantasy, revealing the potential horrors of modern civilization and technology, which the holocaust and World War II would soon bring to reality. Yet it also reflected the cruelties of European colonialism, with out-group racial scapegoating, in the previous 400 years.

Sympathy for the film's animal-human monsters is evoked early on. The hero of the film, Edward Parker (Richard Arlen), recently saved from a shipwreck, tries to stop the cruelty of his new captain toward a lowly servant, M'ling (Tetsu Komai), on the ship that saved him, which is transporting dogs and wild animals to a mysterious island. Recently restored to life and human companionship, Parker punches the drunken Captain Davies, getting revenge for M'ling who was punched by him for spilling some food meant for the animals. Parker then notices M'ling's hairy pointed ear and flat nose. Later, after being left on Moreau's island by the vengeful Davies, Parker sees "natives" that are apparently human but with strange heads and bodies covered with even more hair, like the "wolf man" in the film of that name made almost a decade later. Bela Lugosi, a year after his appearance as Dracula, plays the "Sayer of the Law" in this film, with thick hair all over his face transforming him also into a werewolf-like figure.

The title of the film suggests that the island's "natives" are humans with souls who have lost their way. But Parker learns that Dr. Moreau and his assistant, Mr. Montgomery (who saved Parker from the sea), have hyper-evolved them from various species of animals. The title thus raises a question that haunted white colonists ever since Columbus: are the strange-looking natives, discovered (and then redesigned) by Europeans, "human" with God-given "souls"? Played as a dapper, yet sinister dandy by the chubby British actor, Charles Laughton, sporting a goatee and a white suit (like Montgomery and Parker), Moreau explains that he first evolved plants into their future monstrosities and then turned to animals. Assuming a godlike role, he has "stripped away 100,000 years of slow evolution," in order to prove that "all animal life is tending toward the human form." Focusing the *rasas* of awe and disgust in movie viewers, Parker is appalled when he sees Moreau's "vivisection" of a humanoid animal on the operating table, without anesthetic. But Parker also learns that he is the subject of a new experiment, even as a noble American. He is being used in a romantic test of Moreau's "most perfect creation," the Panther Woman called "Lota," who is "the only woman on the island" until Parker's fiancée, Ruth, arrives with another ship's captain, Donahue, to save him.[4]

As Fatimah Tobing Rony points out, some of the animal-human monsters in the film are "coded in racial terms" (168).[5] M'ling appears to be a "dog-man, [yet also] is coded as East Asian—servile and bestial—slightly higher in status than some of Moreau's other beasts who have become his slaves. . . ." Lota (Kathleen Burke), although created from a panther, wears makeup and a costume that make her "a typical 'South Seas' siren" (169). She appears to be fully human with sexy, hairless legs and arms. But when she seduces Parker, he finds that her fingernails can turn into claws. This shocks him even more than the vivisection: "an animal with a woman's emotions," he says, "it's criminal. . . . I could have overlooked those others. . . . Now I'll expose the world to what you really are, Moreau." Thus, through Parker's gaze as heroic lens, the movie may elicit from viewers a cross-species (and cross-racial) compassion, involving an ironic mix of *rasas*: eroticism, disgust, sorrow, fear, anger, and courage. Earlier in the film, too, such *rasas* are evoked when Parker tries to flee with Lota, but they are surrounded by dozens of hairy animal-human hybrids. Lugosi's Sayer of the Law leers lustily into the camera then, putting the movie viewer into Lota's position as object of his bestial male gaze, but also reflecting the viewer's desire for exotic contact. As Rony puts it, "the viewer is made into a monster as well" (169).

If the monstrous "natives" on Moreau's island are "lost souls," were they given that distinctive element of the human spirit by their scientist-creator? He transformed them into humanoid beasts through his *hellish* "House of Pain." And he threatens to put them in there again if they break "the Law" while he keeps them on a *purgatorial* island. Like the slayer to the vampire (in films discussed in Chapter 4), Moreau exemplifies an extreme version of the left-hemisphere critic/scripter, with rationalizing, law-enforcing, language-based functions, while snapping his whip to control his more right-brain oriented, limbic impassioned, and brainstem driven creatures. Yet they seem to have at least partial, paradoxical "souls," through their bio-cultural evolution, expressing an inner-theater friction between the LPFC character, DMPFC director, VMPFC stage/production manager, and MPFC actor of a mirror-aware, Other-influenced, and Law-guided Self—in their conflicts with Moreau.

Unlike Shakespeare's Caliban, who curses his master, Prospero, for teaching him language on a magic island in *The Tempest*, the beast-people of Moreau's island obediently recite his "Law," at least initially. When questioned by Moreau and his whip, Lugosi's hairy-faced Sayer of the Law leads his fellow "natives" in a chant, which restrains their apparent animal drive to devour Parker and Lota. "What is the Law?" Moreau asks. The Sayer replies: "Not to run on all fours, that is the Law. Are we not men?" And the others echo him. He continues to recite the Laws taught by Moreau, which they again repeat: not to eat meat and not to spill blood. Then the Sayer leads the animal-human monsters in a chant of worship and submission toward their creator, Moreau. "His is the hand that makes. . . . His is the hand that heals. . . . His is the House of Pain." Later, Moreau boasts to Parker that giving his creatures language was a "great achievement, articulate speech controlled by the brain." He shows Parker his "less successful" creatures, enslaved to turn a large wheel, to power his further experiments. And he asks Parker, "Do you know what it means to feel like God?"

Moreau, as godlike colonial scientist and tyrant, continues to experiment with his animal-human children. It is not enough to hyper-evolve them from beasts to humans, and then teach them speech and the Law, replacing their animal instincts with cultural, superego demands. He also tests the erotic powers of his "most perfect" Lota, wanting to further evolve her mammalian-ape ego by introducing her to Parker. As he admits later, he plans to get her pregnant via Parker and then exhibit her in London. So he spies on them like the movie viewer. Although engaged to Ruth and calling Lota a "child," Parker succumbs to her simple speech, catlike cuddling, and longing eyes. He embraces and kisses her, and then sees her claws, which disgust him. When Moreau learns of this, he is disgusted, too, by the "stubborn beast flesh creeping back." But he is reassured about her human soul when she, rejected by Parker, shows tears. She thus represents the paradoxical nature of all human beings, with animal drives and brain structures competing against "higher" orders of mind, involving romantic and divine aspirations—as shown with Moreau and his other offspring, more critically, as the film proceeds.

After Parker's fiancée Ruth reaches the island (a departure from the original novel), baser passions emerge in Moreau's creations, reflecting his own. Ruth is shown undressing alone and slipping into bed, evoking the movie viewer's godlike gaze—akin to Moreau's with Parker and Lota earlier. Then a hairy, muscular beast-man, Ouran, not only looks through the bars of her bedroom window, like the voyeuristic viewer or Moreau, but also begins to breaks in, like Frankenstein's monster, causing her to scream. Parker hurries to save her and shoots at the intruding beast-man in the window. Captain Donahue then chases Ouran through the jungle, but gets caught by him and torn apart, in an offscreen *sparagmos* by the animal-human "natives" (like the Dionysian-possessed women in Euripides's play, *The Bacchae*). Ouran brings a bloody shirt to the Sayer of the Law and asks if Moreau is a man and can die, too. The Sayer tells him he broke the Law, but admits that their god is *mortal*.

This film reflects in various ways our inner theater of animal to human to semidivine networks and the hall of *cracked* mirrors it sometimes produces between us. Moreau is mortal (and seems hypocritical to the movie viewer), so his island Law is flawed. His "native" children shift from left-cortical obedience to right-cortical rebellion, evoking limbic rage, group-survival, and fight-for-freedom networks. They improvise a new order of collective performing selves, altering their PFC actor/director, character, and stage manager against Moreau as dominant Other. Like the capuchin monkeys in de Waal's experiment, furious at the experimenter's unfairness (see Chapter 3), their primate egos rebel against Moreau's colonial, cultish science. They charge at him, despite his snapping whip, and carry him back to his lab. "You made us things . . . part men, part beasts," the Sayer leads them in chanting. They corner him, approaching the camera (and movie audience) like determined zombies. Then they put him on the operating table, break the glass cabinet to get his surgical instruments, and alter his godlike position.

When they chase Moreau, the natives hold torches—like the mob chasing Frankenstein's monster in the famous horror film made by James Whale one year earlier. But in this film, Parker flees, too, drawing more sympathy from the audience, akin to them as a vulnerable visitor to the island lab. He escapes with his fiancée back to a civilized world, for a happy ending to the movie. And yet, Moreau's monstrosity, like the natives' final rebellion, casts a shadow on the noble hero and his fiancée as they leave the island. "Don't look back," Montgomery tells them, providing another viewpoint for film watchers to consider, as Moreau's earlier assistant in the experiments who then saved Parker and eventually sided with him against the mad scientist. Can they (Montgomery, Parker, and Ruth) choose *not* to look back? Or will they be haunted by what Moreau showed them of the godlike potential in human technology—and of "all animal life . . . tending toward the human form"?[6]

We might look back with a postmodern critique of this film for stereotyping "natives" as monstrous cannibals attacking the colonist at its climax. We do not know if they literally consumed Moreau, breaking his

Law against eating flesh and making him an even more ironic god or Christ-figure. But this is also suggested earlier in the film, when Captain Donahue asks, at Moreau's colonial dinner table, whether the natives are cannibals, not long before the monstrous Ouran, as a lusty native, visits Ruth's bedroom. On the other hand, the film complicates such simple stereotypes by shifting audience sympathies across the left-cortical binaries of white and dark people, colonists and "natives," civilized and savage, dominant men and seductive, yet helplessly screaming women. Indeed, Lota, as a Panther Woman, not only gains audience sympathy in her attraction to the handsome hero, Parker, she also gets revenge for his fiancée, leaping on Ouran from a tree and clawing him to death during the rebellion, although she dies as well. She thus prefigures many later heroines with claws and other weapons, whose "murderous animal nature" (as Moreau puts it), with righteous acts of vengeance, as slayers of certain monsters, might horrify us, yet also evoke, with tragicomic twists, a cathartic refinement of various emotions in our inner theaters.

In 1932, moviegoers did not know that Josef Mengele and other Nazis would soon outdo Moreau's monstrous island lab in real life, with surgical experiments and other tortures performed on concentration-camp prisoners, both adults and children. But perhaps Charles Laughton in the Moreau role, with a sadistic, greedy arrogance, suggested even then a tragic flaw in the left-cortical hubris of colonial European and Anglo-American cultures.[7] (Laughton later worked in southern California with German playwright Bertolt Brecht, revising his play about Galileo as a response to scientists creating and the United States deploying atom bombs.) The natives' dismemberment of Moreau, and yet Parker's survival beyond the revolution, foreshadows the end of European imperialism after World War II and America taking on some of that mission, in fighting communism around the world, especially in French Indochina. And then, after 45 years of American growth as a world superpower (with Hays Code censorship from 1934 to 1968), Hollywood returned to Wells's novel in the post-Vietnam era, reflecting again upon the temptation to view certain humans as godlike and others as mere animals.

EVOLVING REVOLUTIONS

Genetic engineering experiments at Stanford University in the early 1970s revived popular fears about a Moreau-made world in the future (Kirby 98). Several films were created to capitalize on those fears, including a remake of Wells's novel, which brought back the original title and was directed by Don Taylor in 1977. Obvious differences from the novel and the 1932 film appear in this version, which focuses more on the island's visitor as a sympathetic hero and yet potential animal. This reflects increasing concerns about biotechnology and colonization, after American misdeeds in Vietnam.

At the start, Andrew Braddock (instead of "Prendick" or "Parker") is floating in a small boat at sea, along with a fellow seaman, Charlie, after the wreck of their ship, Lady Vain. (There is a third sailor who dies and they dump him overboard.) While the boat drifts, a musical score stirs the melodramatic tension, which increases when Braddock (Michael York) sees land. Unlike the 1932 film, which had no soundtrack, this one attempts to underscore the emotional participation of the audience and perhaps fails at times—like Moreau trying to conduct the animal passions of his chimeras. This movie cuts the novel's opening, included in the earlier film, about the hero being on a ship, which had rescued him, and fighting the captain there, before being left on the island. Instead, Braddock finds the island directly, more like a colonist himself. He wakes after 17 days adrift, sees land, rows to it, and then carries his unconscious shipmate to the shore. He leaves Charlie near the beach and goes inland to find water, but his companion is soon dragged away by an offscreen force. As the musical tension builds, Braddock sees mysterious figures in the jungle, runs from them, and falls into an animal trap.

With this change in the plot (by screenwriters John Herman Shaner and Al Ramrus) and an added musical score, the 1977 movie focuses its audience on the hero's *subjective* experience of the island, which saves him from death at sea, yet reveals its own dangers. Montgomery (Nigel Davenport) greets the hero when he wakes, restored to health like in the earlier version. But here, he wakes in Moreau's compound and Montgomery is a "mercenary" who refuses to answer questions about the island's creatures. Later, he offers Braddock a drink in his room, from his dripping liquor still, while playing opera music (Mozart's *The Magic Flute*) on a gramophone, but then he loads his revolver—showing several aspects to his colonial personality. Braddock also meets Moreau (Burt Lancaster) who tries to hide his experiments, but eventually informs Braddock of how he became an outcast, as a "man of vision," and was "severely criticized by academicians" for his earlier research, before starting his island lab 11 years ago. He shows Braddock dog, mouse, and human embryos in jars, which start out "virtually the same." And then he asks: "How does a cell become enslaved to a form, a destiny it can never change?" Here, the movie audience is brought into a collegial kinship with Moreau (and with Montgomery later), through Braddock, as colonial explorers not just of the island, but also of the mysteries of Life's forms and fates. But Moreau also asks: "Can we change that destiny?" And Braddock changes this biotech question into an ethical one: "Should we?"

The American actor, Lancaster, is much more fatherly and charming than Laughton's Moreau, complimenting Braddock on his intelligence and including him in his secret research. He also introduces Braddock to his adopted daughter, Maria (Nicaraguan actress Barbara Carrera). Moreau says he rescued her at age eleven from Panama City, where "any man could have her for the price of a dozen eggs," suggesting she was a child prostitute. "She's a very delicate creature," he adds, due to her "bad experiences." There is no suggestion in this film that she was once a panther, but she does have a pet serval (a wild cat) on a leash when Braddock first meets her. She also meets him later while walking her pet on the beach. She hands him the serval, but it leaps out of his arms and escapes into the jungle. They both chase it into the darkness. Braddock loses her, sees M'ling drinking water from a creek like an animal, but then finds Maria's beautiful form once more—thus mixing the *rasas* of fear, eroticism, and disgust for the viewer.

Braddock snoops around Moreau's lab and finds a Bear-Man on the operating table. Moreau then tells him his secret discovery: "a cell particle that controls . . . the shape of life." He injects a serum into the Bear-Man with "a biological code message, a new set of instructions for erasing the natural instincts of the animal." He says surgery is also needed to replace some organs, but the injection already "forces a modification of the body" as the animal-human struggles on the table and Moreau holds him down. With such experiments, says Moreau, he wants to "reach for the control of heredity," in order to "benefit humanity." He foresees "the pain we can ease, the deformities we can avoid."

In this scene, Braddock leads the film viewer as voyeur of such research, stressing left-hemisphere rescripting and *objectifying* of an animal's genetic body. But in the next scene, Maria visits Braddock's room, allowing the spectator a direct erotic gaze, involving the right hemisphere's more *subjective* mime improviser, tied to mirror neurons and limbic passions. She takes off his shirt and he unties the top of her dress. Only their bare shoulders are shown as their bodies touch. Yet more is implied below, evoking a pleasure drive in movie viewers as well. Such "natural instincts" of the human animal might be "erased" by Moreau's serum, but not by this film.

Moreau is also shown, through window slats, looking up from below, but not spying directly on them, as in the 1932 film. This allows movie viewers to enjoy an erotic romance without the perverse reflection of their voyeurism in a watching experimenter. And yet, like Ferdinand with Prospero's daughter, Miranda, in Shakespeare's *The Tempest*, Braddock (played by a British actor) will have to go through various trials, set up by the father-magician-scientist, before he can win Maria as his island prize.

It is unclear whether Moreau sets up the right-cortical mating of Braddock and Maria intentionally, as another experiment, like in the earlier film version (or Shakespeare's *The Tempest*). When Braddock discovers a cave on the island with various beast-people, and the Sayer of the Law views him as being like them, Moreau shows up, fires his gun, and tells them that Braddock is superior to them, like himself. And yet, when he finds Braddock packing to leave the island with Maria, Moreau takes him to his lab and injects him with another serum that reverses his humanity, changing him gradually into a beast. Like the Bear-Man, who faced a bear in a cage as his uncanny mirror, Braddock starts to feel his animal heritage come to the surface, while caged by Moreau. He changes from godlike status, when introduced to the beast-people, to becoming a beast-man himself, losing vocabulary and gaining hair. But he resists the brain and body changes, by reciting his name and childhood memories. Braddock's inner MPFC actor fights against Moreau's altering of his LPFC character and outward appearance—while his VMPFC moral stage manager rebels further at the scientist's obsessive pursuit of knowledge and power.

Moreau's experiments have been failing because his creations revert to bestiality after he makes them almost human. One, a Bull-Man, argues with the Sayer of the Law that animals are better than humans—"strong, proud"—and that the Law is Moreau's, not theirs. The Bull-Man (looking like an ancient Minotaur or the prehistoric Chauvet image) then goes outside the cave, meets a tiger, attacks it, and kills it. Moreau explains to Braddock (before turning him into a beast) that the beast-people retain a memory of the House of Pain when they devolve again into animals. "In the image of our own species, they become vengeful killers." The godlike Moreau thus admits that his theater of cruelty creates a tragic flaw in his animal-human subjects, which he made in his own image, evoking *rasas* of awe, fear, and sympathetic sorrow in the brain theaters of the movie audience.

Ironically, the still-human Braddock kills the Bull-Man, at his request, to prevent him from being punished in the House of Pain for killing the tiger. This triggers a vengeful rebellion by the beast-people. But Moreau faces them down at the gate of his compound and they return to their cave. After that, he catches Braddock trying to escape with Maria and makes him into a new experiment: to explore the "inner battlefield, that war of the cells" between animal and human. This may shift the viewer's sympathies from Moreau as potential victim of the beast-people, behind his fort-like compound, to Braddock's brain and body, as inner territories that the scientist colonizes to create an animal-human border war.

The movie shows this by passing through a skylight window in the laboratory hut and then peering into a mirror over the operating table, offering a view of Braddock, his face and chest becoming hairier, as he wakes and screams. Moreau responds with another injection that quiets him. Montgomery sees what is going on and tells Moreau to let Braddock go. But the doctor insists that to study nature, "one must become as remorseless as nature," showing that the erudite colonial scientist is actually more bestial than his subjects. Montgomery rebels and Moreau kills him with gunshots—while several of the beast-people watch from a hilltop near the compound, like the cinema audience. This tragic twist in their view of Moreau, breaking his own law against killing, leads to the beast-people's melodramatic vengeance and self-destruction. But it might evoke a cathartic refinement of shared feelings in film spectators, in their PFC networks, especially through Braddock's new alignment with Moreau's subhuman, experimental subjects.

As the reversion serum takes effect, Moreau tells Braddock that he is "beginning to think in images, concrete images," suggesting a shift from left to right hemisphere processes. Then Moreau offers Beast-Braddock, while he is in a cage, a smaller cage of *live rats* to eat, after starving him for two days, teasing out his "instincts." This challenges the movie viewer to imagine such a subhuman temptation, involving *rasas* of sympathetic sorrow, fear, awe, and disgust. Braddock's deep mammalian emotions emerge as he tries to hang onto his human identity by recalling his family home in Britain and sledding with his brother, before that brother died; then he breaks into tears.

Mammalian rage erupts from the beast-people when they realize their Creator has violated his Law in killing Montgomery. They pull him from his horse (symbolizing his fall from human mastery over animals) and attack him until he lies on the ground, unconscious and bloody. Then the Sayer of the Law looks at his own bloody hands and says they have broken the Law, too. The others argue that there is "no more Law," but the left-cortical Sayer insists that they need it. The fingernails on his left hand are long and sharp, and covered in blood, but trimmed on his right, suggesting the right and left hemisphere conflict within him.

After Maria frees Braddock from his cage, he tries to use his oppressor's body, ironically, as a shield against the other rebels. He raises Moreau's corpse on a rope above the gate, as if it were an immortal, all-seeing icon, and shouts from a tower in the compound: "Creatures of the Law, He is not dead. He sees you now. Yes, you cannot kill Moreau. . . . Fear him. Obey his Law." But the beast-people soon realize that the corpse is not a god and they burst through the gate of the compound. And yet, without the Law or natural instincts to limit them, their revolution becomes self-destructive. The Sayer of the Law tries to warn them, but the others release Moreau's wild animals from their cages and the beast-people are mauled to death, while the compound goes up in flames. Here, the audience may sympathize with and yet be shocked by the beast-people's rebellion against Moreau's hypocrisy and cruelty, especially when the film first appeared in a post-Vietnam, Civil Rights, and Watergate era (after two assassination attempts on President Ford in 1975). But the film also shows the continued need for cultural evolution, not just genetic success, in nature's risky experiment of a big-brained, theatrically reflective, and radically innovative species.

GENETIC AND CULTURAL DEVILS

In the 1977 film, set at the turn of the century like the novel, brutality emerges in the beast-people when the hypocrisy of Moreau's Law is revealed. Even the delicate Maria must fight, like the animal-human Braddock, when they escape on a boat and one of the beast-people pursues them. She hits the Beast-Person with a wooden oar, helping Braddock to beat him off the boat. But in the 1996 version, set in 2010, such brutality arises at the start of the film with shipwrecked men on their lifeboat fighting to the death, "like beasts," for the final canteen of water—as the *voice-over* of the sole survivor, Edward Douglas, a United Nations diplomat, explains. Two of them, in military fatigues, stab each other with knives and fall into the Java Sea together, spreading blood in the water with a shark below. Then Douglas, played by British actor David Thewlis and dressed in civilian clothes, tells the movie viewer, as his alter ego, "I fought for my life, just as savagely as they did." We see him hit one of the soldiers with a metal oar (like Braddock and Maria fighting the Beast-Person at the end of the 1977 film). But we also see the oar come at us, stressing the violence of bestial survival instincts in the human being. Directed by John Frankenheimer, this remake focuses more on the beast-people as characters, giving insights about their revolution as it turns from righteous rioting to the wild pleasures of mob rule and reckless slaughter—in a film released two years after the Rwandan genocide and a year after the end of the Bosnian civil war and the Oklahoma City bombing in the United States.

Racial coding becomes even more apparent in this latest remake. Although many of the beast-people get full-body animal characteristics, unlike in the other films, the most rebellious and villainous of those living *outside* Moreau's compound, Lo-Mai and Hyena-Swine, are obviously drawn from African prototypes (leopard/cheetah and hyena). The key traitor among his more cultivated "children" living *at home* with him, Azazello, the Dog-Man, has black skin (as does the actor, Temeura Morrison, a New Zealander of Maori and Scotch-Irish descent). This division also suggests the American South with its plantation-era friction between field slaves and house slaves.

In this version, the white American colonist, Montgomery (Val Kilmer), a "neurosurgeon" by training, is a clever, cynical, drug-using, and drug-dispensing *traitor*. He is the first to show Douglas a devilish brutality on Moreau's island. When Montgomery disembarks with supplies from a sailing ship that picked up Douglas, he teases him about the danger of staying with the Javanese sailors onboard, as "real party animals," convincing him to visit the island where he promises a radio for him to contact the outside world. (Douglas was working for the United Nations on a peace mission before his plane went down.) Montgomery says the island was settled first by the Dutch, then the Americans after World War II, and then the Japanese, who built a tourist hotel, which went "belly up." He takes a cage of white rabbits, in a jeep, to a larger cage on the island. While Montgomery moves the rabbits between cages, Douglas reminisces about having such an animal as a pet when he was a boy. So Montgomery allows him to kiss one on the head, to wonder at the world inside its skull, and then breaks its neck, handing the limp, dead bunny to Douglas to hold.

This foreshadows how the devilish Montgomery, having worked with Moreau, a Nobel Prize Winner, for 10 years on his experiments to evolve a better human being, will soon sabotage them, apparently out of jealousy and perverse pleasure. While initially hospitable, the effeminate Montgomery mimics Moreau's voice with a mocking smirk, puts a blue flower in Douglas's pocket, and then locks him in his room, ostensibly for his "own good." But Douglas escapes, glimpses Moreau (Marlon Brando) in his den, listening to classical music, and then finds his lab—with animals in cages, deformed creatures in formaldehyde jars, and several men, including Montgomery, helping a Beast-Woman give birth. Like the movie voyeur, Douglas is tempted to view more than is permitted, even if it might be horrifying.

Douglas flees then and is found by Moreau's daughter, Aissa (Fairuza Balk), whom he had met earlier while she was dancing in a trance. In this version, she is again a Panther Woman, but her sharp teeth and fighting talents only appear later in the film. Here, she leads Douglas to meet a baboon-man, Assassimon, and then through war debris and a village full of beast-people to the tall, blind, and ram-horned Sayer of the Law (Ron Perlman). He holds a staff, looking like an ancient tribal shaman (or the Greek theater character, Tiresias). Soon, Moreau arrives in a shaded carriage on a jeep, waving like the Pope and wearing a white robe, white veiled hat, and white makeup (to protect him from a sun allergy), which makes him appear as a very white imperialist and yet akin to the rabbits seen earlier, some pieces of which were also seen on trees by Douglas en route to the village. Laughton's perversely divine Moreau in the 1932 film and Lancaster's robustly earnest scientist in 1977 change to Brando's clownish benevolent dictator as "Father" here. Yet he demonstrates a calm compassionate ethics, even risking his life by telling Montgomery to give his gun to the "frightened" Douglas, who then aims it at the threatening beast-people and at Moreau. But the godlike scientist shows his power by pressing a device he wears on a necklace, making the beast-people writhe on the ground in pain.

Back at his home, Moreau introduces Douglas to his house-servant "children." All male except for Aissa, they are M'Ling, Azazello, Waggdi, and a tiny man with a bulging brain case, Majai, who later plays a miniature piano on top of Moreau's, accompanying him and dressed like him. Moreau is also kind to the mute Majai when Douglass refuses to shake his tiny red hand, insisting that he do so, because the dwarf only wants "to be polite." At dinner, Moreau praises M'ling at his "beautiful" reading of a Yeats poem ("The Second Coming"). Such scenes evoke an ironic mix of *rasas*: fear, awe, disgust, humor, and perhaps some degree of peace.

But then, as M'ling, Azazello, and Waggdi serve vegetarian dishes to their Father and fellow children, along with Douglas and Montgomery, Moreau explains his discoveries about evil through genetic research. "The devil is that element in human nature that impels us to destroy and debase." Moreau says his work on the island strives "to create some measure of refinement in the human species," by fusing animals with human genes. He has "seen the devil" in his microscope, chained him, and "cut him into pieces," metaphorically speaking. He says that the devil is "just a tiresome collection of genes" and Lucifer is "no more." Moreau's own children represent various stages in "the eradication of destructive elements found in the human psyche." He claims he has almost achieved "perfection" with the creation of "a divine creature that is pure, harmonious, absolutely incapable of malice." But Montgomery (betraying his mentor by ordering rabbit for dinner and with the live rabbits as temptations to the outdoor beast-people, plus further disruptive acts), Azazello (betraying his Father later), and the "divine" Aissa (becoming a beast) will increasingly show that the devil of genetic and cultural destructiveness may rise again, despite Moreau's biotech dismemberment.

When Douglas expresses shock at Moreau's "Satanic" work, the scientist quotes from the Bible to defend himself: "Let he who is without sin, cast the first stone." But by 1996 when this film was released, Moreau's godlike redesigning of natural life had already begun offscreen, Satanic or not. Transgenic mice "fused with human genes" (as Moreau says about his experimental children) were created in 1987. Gene therapy trials started with humans in 1991. Genetically modified tomatoes were available across the United States in 1994. And the first cloned animal, the sheep "Dolly," was born in 1996.

Yet Moreau is not only a scientist-creator, but also a Father-ruler on his island. He shows kindness to his tamer transgenic children at home, while inflicting painful punishments to control the others outside. When he learns from Aissa that one of his more bestial subjects, Lo-Mai, who looks like a leopard or cheetah in humanoid form, broke the Law by killing a rabbit and eating its flesh, Moreau holds a trial in the village. (Similar events occur with a "Leopard Man" in Wells's novel, but the creature flees.) Moreau calls Lo-Mai out of the crowd for this "evil." Lo-Mai charges the podium in a rage and Moreau shocks him with the pain device, forgiving him then as he submits on all fours. But Azazello shoots the submissive Lo-Mai in the head with what looks like a bolt pistol for killing cattle. Moreau is appalled: "Oh my God, what have you done?" Azazello apologizes, hands him the gun, and says, "I thought you wanted me to protect the Law." Yet this act reveals a tragic flaw in the Law: Moreau gave one of the beast-people the technology to rise above it and violate it, unpunished, which triggers a further rebellion later. In this scene, Azazello seems sincerely apologetic, turning toward Montgomery when Moreau asks where he got the gun. But when Lo-Mai's body is cremated, Azazello smiles as a colonized subject who has learned to gather power over others like him.

Hyena-Swine finds Lo-Mai's skull in his cremated ashes and fondles it, like Hamlet with Yorick's in Shakespeare's play. Then Hyena-Swine finds the pain implant in the ashes, attached to a rib bone, which gives him the idea to pull out such a device from his own chest. These acts, of mournful awareness with his friend's skull and of freeing himself from Moreau's control by piercing his own flesh with his finger nails, might cause many viewers to identify with Hyena-Swine, as oppressed victim and yet courageous rebel—through the *rasas* of sorrow, disgust, and awe, involving their intuition and motor-mirror neurons. The film also encourages antipathy toward Montgomery, even more than Azazello, through Douglas's disgust at the doctor's injection of the beast-people with a drug from Moreau, to lessen their regression, plus his own "contribution" of methamphetamine, morphine, and "shrooms," which "keeps 'em mellow . . . [and] keeps 'em coming back for more." Some movie viewers today might even relate this to the pharmaceutical industry, pushing its drug therapies to consumers through television ads and clinical samples. Such "direct to consumer" marketing of therapeutic drugs has "expanded exponentially" since a Federal Drug Administration ruling in 1997, a year after this film was released (Huh et al.).

Further identifications are encouraged when Hyena-Swine visits Moreau's home at night, with several other beast-people. Though they are intruders, Moreau tries soothing their potential savagery by playing Gershwin. Hyena-Swine kneels by the piano and says: "We are not like you. What are we?" Moreau tells them they are his "children," but he does not give them a meaning and purpose for their lives—even though he evolved them beyond other animals, with the human need *to know* what they are.

Pleasure and pain are primal systems in the human brain and body, connecting the physical and social in other mammals as well. An experiment with puppies suffering separation distress from their mother showed that a non-sedating dose of an opiate drug alleviated their social pain and panic emotion, just as it would with physical pain (Matthew Lieberman 50). An area in the limbic system that mammals have but reptiles do not, the dorsal anterior cingulate cortex (dACC) has the highest density of opioid receptors in the brain and is active for both social and physical pain in humans also (51–52). But the dACC, along with the anterior insula, tracks the *distressing* aspect of physical pain, not its sensory component (52). Surgeons have sometimes removed it to treat extreme depression and anxiety, as well as chronic pain (53). Activity in the dACC has been linked to social pain regarding grief at the death of a loved one, a romantic breakup, a negative evaluation, or a disapproving face (64). Another experiment showed that just taking Tylenol lessened the social pain of rejection, as well as physical pain (64–65). Likewise, social and physical rewards trigger activity in the VMPFC (or inner stage/production manager), plus the ventral striatum, especially regarding the feeling of fairness, which therefore "tastes like chocolate" (74–75).

When he takes the bloody implant device out of his chest, Hyena-Swine frees himself from the physical and social pain inflicted by Moreau—as if removing or changing the dACC in his brain. He mourns the death of his friend Lo-Mai, but turns grief into rage at the unfairness of Moreau's Law, finding a reckless, violent pleasure through the pain of his alienation as a beast-person. When Moreau fails to give his chimera a meaning, beyond being an experimental child, Hyena-Swine finds a new purpose for his pain by leading the others in attacking their "Father." They sink their claws and teeth into Moreau and tear him apart. This evokes in movie viewers the ironically mixed *rasas* of sorrow, rage, fear, courage, awe, and disgust.

In real life, Marlon Brando also suffered from an island experiment in making an extended family. In 1966, he purchased the Tahitian atoll of Tetiaroa, which he had first visited while working on the film, *Mutiny on the Bounty*. He built an airstrip there and a vacation home for himself and his third wife, the Polynesian actress, Tarita Teriipaia, whom he had met during the making of that film, and for their children, Simon Teihotu and Tarita Cheyenne. Brando visited his island home often in the 1960s to 1980s. But in 1991 Brando's son by another marriage, Christian, shot and killed Cheyenne's Tahitian boyfriend, Dag Drollet, after she told him about being abused by Dag (according to Christian). In 1995, Cheyenne committed suicide at her mother's home in Tahiti. Thus, by the time he played Moreau, Brando had experienced related tragic twists, involving his own island child and home.[8]

This makes even more poignant a scene that Brando plays with Fairuza Balk as Aissa. Like a teenage daughter, she first comforts her "Father" by putting ice in his metal hat, his "caloric converter," and massaging his neck. At first, he tells her to rub "forcefully," but then it hurts him, and yet he tells her, very kindly, that she has no idea how strong she can be. "You must be very, very careful." She sits on his lap and expresses her fear at how "hideous" she is—with her teeth becoming pointed. He checks her face gently, spreading her cheeks to see her pointed teeth, and also looks at her ears, which she has covered with her hair. He admits there are "changes," but calls them "just chemical imbalances"—something that both teens and their parents in the cinema might relate to. He insists that she is "an absolute angel," beautiful outside and inside, but she wants to look like him. He laughs then and makes her laugh at the idea of how she would really look if she became like him.

This scene mirrors, in a much more positive form, the scene a little later with Hyena-Swine kneeling at the piano, which soon turns into the *sparagmos* and *omophagia* of Moreau as godlike, but flawed Father (akin to the Dionysian rite, described in Euripides's *Bacchae*, of tearing an animal apart and eating it raw). The oxytocin flow in the father-daughter nurturing between Moreau and Aissa, even as she becomes more panther-like, fails to happen with his hyena-swine son—or it happens between that creature and other beast-people as an in-group, against their creator as scapegoat. Moreau has not made them "beautiful," like her, and does not give them an alternative symbolic framework for their deeply mammalian passions, yet burgeoning, higher-order awareness. He fails to frame them as angels, or as anything more than his children and colonial subjects, so they become devils.

For the movie viewer, too, it is easier to identify with Aissa's beautiful face and figure than Hyena-Swine's carnivorous jaws, clawed hands, and twisted walk (on his toes, suggesting pig's feet, though he wears boots). And yet, the performance by Daniel Rigney under the animal mask and costume, especially from his mourning at Lo-Mai's death and removal of the punishment implant to his kneeling beside Moreau, offers many points of sympathetic social and physical pain for the film audience, along with tragic fear at this chimera's fierceness. To the degree that a movie viewer identifies with Hyena-Swine or Aissa or Azazello or Montgomery or Moreau, a *rasa*-catharsis of various limbic emotions might occur in the VMPFC/dACC stage manager, MPFC actor, LPFC character, and DMPFC director. Other hubs of the inner theater's Self-Other networks might be involved, too, through shifts between the left hemisphere's critic/scripter and right's mime improviser/scene designer (especially regarding the Law and the beast-people's revolution)—as the film proceeds toward a catastrophic downfall for each of these characters and for Moreau's island dream of human perfection.

At the cremation of Moreau's body, the Sayer of the Law tells the doctor's *loyal* children (Aissa, M'ling, and Majai) that there is still the Law, that their Father has not left them, and that "His Spirit" watches over them. But Montgomery does not join them in converting the scientist-ruler into a god. Later, he mocks this idea, teasing Douglas by mimicking Moreau's voice on a microphone, while dressed like him, with a padded belly and white makeup, and reading from the Bible. Montgomery also dresses like Moreau in the beast-people's village, sitting in his pope-mobile throne and handing out drugs to them, while again mimicking their creator's voice: "I give my body." When Azazello appears with a gun, Montgomery hugs him, asks about dog instincts (to hunt and run with the pack), and jokes with him about wanting to go to "dog heaven." So Azazello shoots him and then gives his gun to Hyena-Swine. Montgomery suffers a tragicomic death, as sympathetic villain and clown, while the rebellion grows—with plot twists challenging movie viewer's body-swapping identifications.

Aissa becomes more bestial in fleeing the rebel beast-people and turning, when cornered with Douglas, to attack them, like Lota attacking Ouran in the first film version. In this remake, she also dies, hung from a noose by Azazello and his pack. He enjoys this hunt and kill, becoming more doglike as he joins Hyena-Swine's rebellion. But when that alpha male takes full power with automatic weapons, which Azazello helped him to find, Hyena-Swine becomes a "Mad Dog" dictator (as defined in Chapter 3). He creates protean uncertainty among his comrades by suddenly killing Azazello, riddling his body with bullets, as a threat to his rule. All three of these beast-people, while showing positive qualities earlier, as angelic daughter, servant cook, and mournful self-experimenter (by taking the implant out), turn tragically monstrous when they indulge fully in their fighting, hunting, and ruling instincts. They exemplify the creative yet dangerous social powers of the *human* ape and mammalian ego, with sympathetic, genetic, and cultural twists: attractive as catlike angel/beast, tricky as doglike servant/son, and dominating as wild alien/rebel.

Hyena-Swine leads the others in riotous revelry, destroying the dock, lab, and Moreau's home. He tries to force the Sayer to proclaim that he, Hyena-Swine, is now "the Law," but the blind ram-man refuses. So Hyena-Swine presses the pain button, making the other beast-people writhe on the ground, while ordering Douglas to tell them that he is a "god" and to obey him "like they did the Father." Douglas agrees but then tells him that all those who killed the Father and ate his flesh are gods. "So who is the new Father? Who is god number one?" Hyena-Swine then shoots his inner circle of comrade rebels, but is also shot and beaten by others. Wounded and defeated, he walks into the flames of Moreau's burning colonial compound, crying: "Father, why?" This self-immolation by Hyena-Swine, and tragic questioning of his fate, may evoke sympathy again for the hero-villain, despite his destructive, vengeful leadership and wanton cruelty toward others.

The next morning, while standing on the charred dock with the baboon-man, Assassimon, and the mute Majai, the blind Sayer of the Law holds a staff topped with a human mask and expresses his cathartic awareness, changed from what he felt at Moreau's cremation. He finds the courage now to reject Douglas's offer to bring doctors and scientists to help them survive as partial humans. Instead, he will seek a new cultural identity with the remaining beast-people, even as they regress toward animals again. He even suggests that humans might not be at the top of the evolutionary tree. "We have to be what we are, not what the Father tried to make us. To go on two legs, very hard. Perhaps four is better anyway."

Douglas realizes, too, in a *voice-over* to the movie audience, while sailing away from the island, that humans are not totally different from the beast-people who initially disgusted him. With *documentary* footage shown onscreen of people rioting, he says that sometimes his "fellow man" is neither animal nor human, "but an unstable combination of both." This version of Wells's story ends not with the island's melodramatic destruction and Montgomery saying not to look back, nor with the romantic survival of the hero returning to humanness and being saved at sea with his new love. It ends with a tragicomic irony. Cultural revolution, like genetic evolution, risks the emergence of monstrous devils, through the human animal.

CAT-PEOPLE AND FLY-PEOPLE WITH THEIR KEEPERS AND GODS

Godlike yet flawed scientists sometimes appear in horror movies, from Frankenstein and Moreau onward, raising questions not just about human technology, but also about a divine purpose to life, regarding human awareness and creativity. Did God make humans godlike, in His image, unlike other animals? Or are we evolving such powers and changing the animal world in ways beyond our natural instincts and wisdom—with no greater framework or audience to contain us? Are we creating ourselves as monsters, through the power and curse of science, like older forms of magic?

Magically created, animal-human hybrids often originate in exotic places and invade a more familiar realm, reflecting modern fears of the primal Other among and within us, as with Old World werewolves, or the Transylvanian Dracula and Nosferatu. In *Cat People* (1942, dir. Jacques Tourneur), the Serbian-American Irena Dubrovna (Simone Simon) turns into a big black cat when emotionally aroused, as in the tales of witchcraft from her family's Eastern European village—or like Moreau's Panther Woman. The legend here involves native villagers with an "evil" magic who turn themselves into animals—a parallel to the mad scientist who makes hybrids. They "worshipped Satan," Irena says, and fled from the Christian King John (Jovan Nenad) after he "liberated" their village from the Mamluks (or Ottoman Turks). Now, in modern New York City, Irena's personal mimetic rivalry focuses on Alice, who has fallen in love with her husband, Oliver.

Both women visit a psychiatrist, Dr. Judd. He tells Irena, after hypnotizing her to excavate her childhood memories, that the tragic death of her father before she was born and her being teased by others that her mother was a cat-woman may have corroded her "soul." He also explains that there is, "in some people, a psychic need to loose evil upon the world." She later dreams of him as King John with a sword—as Christian or scientific slayer of her black cat magic.

Eventually, the white woman, Irena, turns into a black panther, stalking Alice at a pool and then stalking Alice with Oliver, after he tells Irena that he loves that other woman. Dr. Judd wants to help them commit the jealousy-crazed Irena to an asylum. But instead he kisses her, turning her into a panther. Then he stabs her, a radical alteration of his left-hemisphere's scientific diagnosis and medical ethics, which leads to his death. It also leads to Irena's death, after she returns to her kind in the zoo, where Oliver first met her. During World War II, a beautiful European woman with a devilish animal inside might entice and horrify an American movie viewer identifying with the husband and doctor. But sympathy is evoked here, too, for Alice as a morally troubled friend to Oliver and for Irena as not just a femme fatale, but also a victim of her family lineage and mammalian passions, rekindled in the multicultural laboratory of a modern American city.[9]

In the 1982 remake (dir. Paul Schrader), the setting shifts to New Orleans and the Reagan era with more explicit violence and eroticism, yet again involving folklore magic mixed with modern science, though not with a psychiatrist as keeper or slayer. Irena (Nastassia Kinski) has an older brother, Paul (Malcolm McDowell), a minister, who tells her about their were-panther heritage. He also has a black housekeeper who seems to know his secret. Paul's human to cat transformation occurs through sexual lust and he can only return to human form if he kills as the panther. In this version, Oliver is a zookeeper who leaves his girlfriend Alice when he falls in love with Irena. And Paul wants to have sex with his virgin sister, as the only way to end the curse, like their parents did. So there is a double triangle of mimetic rivalry: Alice in competition with Irena for Oliver's love and Paul in competition with Oliver for Irena's. But Irena rejects Paul, while arousing his lust, thus turning him into a raging panther. He attacks Oliver, but is killed by Alice with a shotgun. Paul then appears to Irena in a dream and shows her their ancestors, black panthers with human souls who "were gods." She mates with Oliver, turns into a panther, leaves him and kills, returning then to human form. And yet, Oliver saves Irena from her murderous heritage and Panther Woman bloodlust. He ties her to the bedposts, making love with her as she requested (to return her to her "kind"), and puts her in his zoo, as a permanent black panther that he feeds and pets through the bars.

Drawing on the idea of Lota in the original *Island* film, as well as the werewolf tradition, both versions of *Cat People* show a beautiful were-panther as *femme fatale* who cannot be tamed and must be killed or caged. The first film, released during World War II, reflects American (and left hemisphere) fears of the dark side in European civilization (and of one's own right hemisphere, limbic, and brainstem areas). But the remake may also reflect, in sly, ironic ways, the fears of white Americans about the "Black Panther Party" in the 1970s, with white characters turning into black panthers, and about racial miscegenation, coded here as incest and cannibal bestiality, but cured by the zookeeper's cage.

More like the Frankenstein and Moreau films, or the various versions of *Dr. Jekyll and Mr. Hyde*, with a "mad" scientist as origin for the animal-human chimera, *The Fly* (1958, dir. Kurt Neumann) has its godlike creator turn the experiment on himself, as in the Christian belief in God the Father becoming the man Jesus. But here, two monsters are produced by the super-natural mixture: a fly with a man's head and a man with a fly's head and arm. A Montreal scientist, Andre Delambre (David Hedison), working in his lab at home, invents a "disintegrator-integrator" that reduces an object or animal to its "billions of atoms." It then transports them across the room "at the speed of light" to another part of the machine and reintegrates them as the same thing again, "like a television signal," as he tells his wife, Helene (Patricia Owens). He boasts that his is the "most important discovery since man sawed off the end of a tree trunk and found the wheel." With it, there will be no need for cars, trains, airplanes, or "spaceships" in the future and no famine in the world. But when he tries it on himself, a fly joins him and his body is reintegrated with a fly head and arm (with a black claw), while the fly flies away with his "white" head and arm on it.

Shocking as this news is to Helene, she tries to help her husband catch the "white-headed" fly and reintegrate as fully human. She enlists their young son, Phillipe, and their maid in the fly-catching goal, without telling them the reason. But after the fly escapes their home through a broken window, the fly-headed Andre destroys his lab device and notes, realizing that his invention is too dangerous for the world to know. Helene completes his wish by crushing his horrid form in a hydraulic press, killing him. She is considered murderous and insane by others until Andre's brother, Francois (Vincent Price), who is also in love with Helene, helps the police inspector to find a spider's web with the human-headed fly trapped in it. The inspector crushes both the spider and scientist-fly with a rock. Thus, a 1950s fear of independent, home-based, Canadian scientists and of television's new powers is expressed and repressed in this romantic comedy horror film—perhaps as a coded, satirical reference to the McCarthyist fear of communist infiltrators, threatening the American way of life, or to the white racist fear of miscegenation (with the mix of white and black in the new Andre and strange fly). But this film also shows a scientist turning himself into a monster and thus realizing his tragic flaw of experimental hubris, before he inflicts it on others beyond his home. His wife gets a happier ending: absolved of the murder with a story invented by the inspector and Francois, she forms a new family with her son and brother-in-law.[10]

Here, the inventor's left-hemisphere, goal-focused *critic/scripter* is initially contrasted with his wife's more bicameral homemaking. He disappears for days at a time, *obsessed* with his scientific work in his basement lab. As the dutiful 1950s wife, she forgives this, though their son suffers the loss of his father. But just when his left-cortical seeking of a great "discovery" starts to pay off, inspired by a right-brain dream of improving the world, the inventor finds that he missed a crucial part of the new global details: a fly in his machine. The scientist's right-cortical *mime improviser/ scene designer* must then engage the help of his wife's more complete brain and body because his left hand (controlled by the right cortex) has turned into a fly's, as has the outer form of his head. He cannot speak but uses a typewriter with his still-human right hand to communicate with her, although his left cortex also loses its inhibitory strength against his increasing depression and despair. After she fails to catch his fly-double, the scientist types to his wife, "BRAIN SAYS STRANGE THINGS NOW . . . FEEL MY WILL GOING" and, as his fly claw wrestles against his typing right hand, "VERY DIFFICULT THINK STRAIGHT."

As with the neurological disorder of Alien Hand Syndrome (Feinberg), the scientist's hand and claw wrestle with each other again after his wife faints at seeing his fly head. He touches her lovingly at first with his right hand, which then struggles to hold back the black fly claw. If people can readily feel that a cross-racial rubber hand is their own, in the experiment mentioned in the last chapter, then perhaps movie viewers might also identify with Andre, to varying degrees, in his cross-species (and black/ white) fate as a tragic hero, not just fear him or laugh at him as a horror monster. Through mirror and intuition neurons, spectators may engage emotionally with the scene onscreen, and then reappraise their feelings and preconceptions—with shifts in their brains' right and left hemisphere networks, especially between the ventrolateral prefrontal lobes.

This 1958 film was so successful that it spawned two sequels, *Return of the Fly* (1959, black and white, dir. Edward Bernds), in which Phillipe repeats his father's experiments, and *Curse of the Fly* (1965, dir. Don Sharp), in which various people are fused together by the teleportation device. Canadian director David Cronenberg remade the initial film as *The Fly*, in 1986, with many altered, intriguing plot points, and Chris Walas added a subpar sequel, *The Fly II* in 1989. In Cronenberg's version,[11] the inventor Seth Brundle (Jeff Goldblum) tries his "telepod" device with a baboon, rather than a cat and guinea pig, and turns the primate inside-out at first. He is not married but in a budding romance with Veronica (Geena Davis), a science journalist covering his work, which encourages the film audience to identify with both of them as active professionals.

At a party, Seth mentions his invention to entice Veronica to his lab. When they arrive, she teases him in return, seductively taking her stocking off when he asks her for something "personal" to teleport. Their subsequent lovemaking inspires him to reprogram his machine and the second teleported baboon comes out unharmed. In this film, the erotic rivalry and passions of the *Cat People* transformations, involving right-cortex, limbic, and brainstem reproduction drives, are combined with the scientist-inventor's obsessive, objectifying, left-cortical hubris. Seth Brundle's rival is Veronica's editor and former lover (with the same initials) Stathis Borans, who wants to rekindle their old relationship. Suspecting Veronica's unfaithfulness with Stathis, Seth rushes to experiment with his own body in the telepod—and mixes it with a fly. Thus, the initial sex with Veronica, which boosted Seth's ape ego, also exposes his tragic flaw.

Seth appears to be unharmed or even improved by the teleportation trip, unlike Andre in the original film. Seth's gymnastic skill, hyperactive thinking, and animal energies increase, so he views his invention as cathartic, as "purifying" his body and brain. "It's going to allow me to realize the personal potential I've been neglecting all these years as I've been obsessively pursuing goal after goal." At the start of the film, just riding in Veronica's car gave Seth motion sickness, which he told her he has suffered since childhood. But his teleporting now alters his neural gut to cortical networks much more powerfully. Veronica notices his personality changing and some stiff hairs growing out of a scar on his back. She tries to warn him that something went wrong. But he rejects her then as a mate, accusing her of only knowing "society's straight line about the flesh, . . . [its] sick, gray fear of the flesh." Seth's counter-cultural view then increases his super-natural hubris. "I'm talking about penetration beyond the veil of the flesh, a deep, penetrating dive into the plasma pool." His de- and re-materializing in the telepod changes Seth's left-hemisphere focus, on scientific goals and social straight lines, toward alternative right-hemisphere orientations, mammalian status passions, and deep vertebrate or invertebrate drives in his animal ancestry.

On the other side of the sexual revolution, from the original movie in the 1950s, this one in the 1980s has the scientist call his mate a "fucking drag" with her social fear of his transformed flesh. Instead of failing to absorb a normal superego in his VMPFC, like Moreau's hybrids, or tearing out a pain implant that was its substitute, Seth liberates himself from traditional social mores by unknowingly implanting animal "plasma" in his body and brain. Yet, as both scientist and subject, he has no one overseeing the dark side of his evolutionary experiment, no divine judge, just a journalist who loves him. He tries to subject her to his invention, too, but she refuses to mate with him *that way*.

This sends Seth on a journey to find a new mate in a redneck bar where he wins a macho arm-wrestling contest, with his super-natural strength, breaking his opponent's arm and stealing his girlfriend. Seth mates with her in his lab-home but she leaves, like Veronica, refusing to go through his machine. He then finds the skin changing on his face, his fingernails coming off, and his computer telling him that he has fused with the fly "at a molecular genetic level." When Veronica arrives and sees Seth wearing gloves and using crutches, his face a mass of scarred flesh, he tells her "the computer got confused" with his teleport. This, along with Seth's initial symptoms, reflects the 1980s horror of AIDS or the so-called "gay cancer" as an aftermath of the American sexual revolution—and a growing fear of computer technology as extending human power, hubris, and lack of wisdom. As Seth dryly jokes: "The computer . . . decided to splice us together. Mated us, me and the fly. We hadn't even been properly introduced." Veronica asks what will happen next and he replies: "I think it's showing itself as a bizarre form of cancer." Then a white digestive fluid spurts from his mouth, increasing the ironic *rasa* mix of love/lust, courage, sorrow, fear, humor, awe, and disgust in the inner-theaters of the movie audience.

The next time Veronica visits, Seth can climb the wall and ceiling. He asks her to videotape him demonstrating how he digests food outside his body by emitting a "corrosive enzyme," like a fly, liquefying it and then sucking it up. Seth retains his wit and wants to share with the world his expansive, right-cortical sense of new possibilities (along with left-cortical optimism) in his "disease with a purpose." Unlike Moreau's creatures, Seth's MPFC *actor* finds, through his body's hybridity, a meaning to his life in relation to his DMPFC mentalizing of others' views. "I'm becoming something that's never existed before. I'm becoming Brundlefly. Don't you think that's worth a Noble Prize or two?" Thus, his LPFC *character* adjusts to a series of mirror-awareness moments as his face degenerates and body parts fall off (ears, fingers, teeth, etc.), which he keeps behind the bathroom mirror, inside his medicine cabinet, as the "Brundle museum of natural history."

Veronica also finds a new potential through her prior mating with Brundle: she is pregnant. But that gives her a nightmare about birthing a huge maggot—a radical altering of her MPFC *actor*-Self through right-cortical *mime improviser and scene designer* twists. When she visits Seth again, he does not listen to her mammalian concerns. Instead, he tells her: "Insects don't have politics. They're very brutal, no compassion. No compromise. . . . I'd like to become the first insect politician." But then he shudders, looks around him, and says he was "an insect who dreamt he was a man and loved it, but now the dream is over and the insect is awake. . . . I'll hurt you if you stay." His insect logic, driven by pure survival and reproduction instincts, threatens to overpower any lingering mammalian compassion or primate ethics. Ironically, this also reflects a persistent political fear in the mid-1980s of Cold War enemies, or cold-blooded serial killers, reverting to nature's *inhumane* dominance and territorial drives—in a godless world.

Another political issue arises, in this sci-fi horror context, with Veronica's desire for an abortion because she fears the Brundlefly offspring inside her body. When she is alone in the abortionist's operating room, Brundlefly breaks in suddenly, through a glass wall, going further than Frankenstein's monster or Ouran in startling her and the sympathetic movie viewer, yet also reflecting the viewer as a voyeur who grasps vicariously at onscreen beauty objects. Seth grabs Veronica and carries her away from his rival, Stathis, who had been helping her to get the abortion. Seth, as Brundlefly, then asks her: "Why did you want to kill Brundle? The baby might be all that's left of the real me." He threatens to control her reproductive body, not just by preventing her abortion, but also by teleporting them together, "you, me, and the baby," fusing them intentionally and creating a new being through his godlike yet devilish technology. "We'll be the ultimate family, a family of three joined together in one body, more human than I am alone"—a perverse version of the Christian Trinity.

But Stathis, an earlier *villain* when he invaded Veronica's home as an ex-lover with a key and tried to control her as her boss at the science magazine, now takes the *hero* role. He brings a shotgun into Seth's lab, in order to save the female victim from the monster. Brundlefly overpowers him at first, but the hero comes back to win, as usually happens in a melodrama. Even with his right foot and left hand melted off by Brundlefly's corrosive enzyme drool, Stathis uses his phallic shotgun to interrupt the hi-tech sex and Trinitarian rebirth. This switching of roles, with the initial hero as monstrous and the menacing ex-lover as feminist friend and ultimate rescuer, shifts audience perspectives, encouraging VMPFC and right/left LPFC changes in moral and emotional appraisals.

Yet there is another, more tragic twist after Stathis shoots at the cable between telepods and frees Veronica from hers. The computer continues its sequence, as programmed by Seth. Brundlefly is fused with the empty telepod, while teleported from his to a third. He emerges from the metal womb[12] as a mix of humanoid insect, broken machine, and thick wiring.

Then the jeopardized woman gets the final heroic moment, but again with a tragic twist, similar to the original film. Veronica picks up the shotgun, yet cannot bear to shoot her beloved—until his claw lifts the gun barrel to aim it at his insect head. Sympathetic viewers might identify with all three characters here, body swapping with them imaginatively, beyond the busted teleporter: with Stathis saving his ex-lover while losing a hand and foot, with Veronica having a potentially alien child inside her while mercy killing her ex-lover, and with Seth dying in his second coming as an animal-Christ figure, exemplifying the perverse potential of a godlike yet unwise, highly inventive brain. He may also foreshadow the increasing fusion of our bodies with various technologies today, while we ignore the damage that this causes to, or invokes through, the "wild life" around us and inside us.

SPLICES OF LIFE IN NATURE AND CULTURE

Mother Nature experiments with us and has done so with our ancestors for millions of years. But through our species, cultures around the world have altered nature's long-evolved outcomes, in radically super-natural ways. The 2009 Canadian film, *Splice* (dir. Vincenzo Natali), explores this continuing conflict between Mother Nature and Father Culture with two scientists, female and male, building a creature through DNA splicing, thus rewriting nature's scripts. Unlike Moreau, Delambre, and Brundle, this inventive pair creates and raises a new species while also in cooperative conflict as a romantic couple. A century after theater experimenters developed naturalist "slice of life" scenes onstage, to study the human animal in its cultural environment, this film adds a third-wave feminist concern about motherhood for career women to the biotech horror onscreen, reflecting gendered elements in our still evolving nature-culture heritage.

Splice illustrates this tension with an initial view from inside the machine-womb as a blob creature, later named "Fred," has its umbilical cord cut, is resuscitated when its heartbeat drops, and is pulled out by scientists Elsa and Clive (Sarah Polley and Adrien Brody). Their newly designed species has DNA taken from various life forms: a bird, horse, fish, kangaroo, lizard, and plant (as shown by the scientists in a PowerPoint presentation). First, they made a female named "Ginger" and then the so-called "Fred." These creatures were vital in developing "medicinal proteins for livestock"—as Clive tells a businesswoman in a meeting. He and Elsa want to splice human DNA into the next version of their designer hybrid, to find cures for human diseases. But the businesswoman tells them that their company will focus instead on "the product stage," isolating the gene in the scientists' creations that makes the "magic protein" that already has medicinal value. So the left-cortical focused scientists shift to right-cortical rebellion in their splicing lab before it is shut down by the profit-minded corporation. Together, they express the experimental creativity of Mother Nature against the money making bias of Father Culture.

Working secretly, Elsa and Clive splice human genes into their chimeric formula just to prove to themselves it can be done. Elsa then wants to implant the embryo into an ovum in their metal and glass womb, as a further bioengineering experiment, despite Clive's moral protestations. He accuses her of "emotional hijacking" and illegality. She replies: "Millions of people are suffering and dying with no hope. We might be sitting on the key to saving them. What are the moral considerations of that?" Thus, these procreative scientists engage the movie viewer's inner theater, especially its VMPFC moral *stage manager*, with left-cortical *critic* and right-cortical *improviser* twists.

Elsa wins Clive over. He presses the button to insert the redesigned sperm into an ovum. Then the romantic couple, sitting in their living room, discuss moving to a bigger home and planning for a child in their future. Yet the feminist geneticist protests: "I don't want to bend my life for some third party that doesn't even exist yet." Ironically, the film shows that they already have such a third party in their lives. The fetal chimera in its machine-womb grows monstrously fast—with its viewpoint also given to the film audience, from inside "BETI" (the Biomechanical Extrautero Thermal Incubator). So Elsa puts her hand in to help it come out. But it latches onto her, endangering her life as many fetuses have done to their mothers through nature's own mechanisms. Clive breaks the machine-womb to free Elsa and then gives her a shot to save her body from reactive convulsions. And yet, their redesigned offspring, a squirming blob with a wild tail, escapes his control, flailing on the floor. So he traps it under a plastic container—like an exposed neocortex with its brainstem as animal tail.

Clive tries to play the role of patriarchal, left-cortical limit to Elsa's right-cortical rebellion and then to the new creature's limbic-brainstem wildness. But Elsa stops Clive from killing their hybrid offspring, dangerous as it seems to be, with her motherly instinct to nurture it. At this point, it has emerged from a barb-tailed, fleshy cocoon as a kangaroo-legged, rabbit-faced, armless creature on the lab floor. Elsa reaches out with her bare right hand to show the baby chimera that she wants to nurture it, although her left hand was already wounded by it. Her oxytocin feelings of *care* overpower her adrenalin survival *fear*. Clive pulls her away from the dangerous chimera, but she insists on gassing it to sleep instead of putting it to death in the lab chamber. Shifting back then to left-cortical scientific mode, Elsa finds that their hybrid creation is aging fast and will die soon of its own accord. She tells Clive they can study "its entire life cycle."

As with many new parents, the wild child becomes troubling to the biochemists with its screams and tantrums. But they learn how to feed the "H50," after accidentally spilling some candy pills that it enjoys. It rapidly grows arms and hands, plus a barbed tail like its original cocoon—developing like a prematurely (or normally) born human, "a fetus outside the womb," as Elsa records in her audio lab notes.[13] She names the female hybrid, now in a dress, "Dren," after it associates the Scrabble letters NERD with those letters on Elsa's T-shirt, which she sees inverted, looking down at her own breasts. Clive and Elsa hide Dren, as inverse nerd, in the storeroom of the lab building, but cannot keep their forbidden offspring completely hidden. In another scene, Dren leaps onto Clive's brother, Gavin, a coworker in the building. But Gavin agrees to keep their secret—even as Dren continues to grow, her eyes shifting from the sides of her head to the front, more like a predator or Dren's left-cortical focused parents.

Meanwhile, the biotech corporation employs Elsa and Clive as performers in a press conference to show their prior project with Fred and Ginger. The company's funding depends upon a good show, with the eyeless blob animals together onstage in a glass case. But the demands of the film's horror genre (and choices by the writers and director) shift the subtle interactive communications between the blobs' nervous systems. Instead of lovely leafy protrusions from their mouths, encircling between them, as shown at the start of the film, barbs protrude and the creatures slash each other to bits, overturning their tank and splashing blood on the front rows of potential investors—and toward the movie audience. Thus, viewers may also feel the violent potential of the more human Dren with her barbed tail—in a subsequent scene of Clive and Elsa making love in the storeroom. Dren watches them through a gauze curtain, seen only by Clive. This reflects, like the press-show bloodbath, the voyeuristic desires of horror movie viewers, involving sadistic drives as well.

The company's concern about the sudden violence of Fred and Ginger, due to Ginger turning into a male, forces Clive and Elsa to move Dren again, to hide her in a rural barn on a farm formerly owned by Elsa's "crazy," dead mother. But then the scientist-parents witness Dren's predatory instincts. She escapes during the move to the barn and they track her through the snowy woods to find her eating the raw flesh of an animal she hunted (like Moreau's rebellious beast-people). Clive also sees where Elsa was traumatized as a child by her mother, living in a small room in the farmhouse with just a mattress on the floor, somewhat like Dren. This sequence of scenes may evoke *rasas* of courage, disgust, fear, sorrow, and awe—at the horrors of our animal ancestry, with predatory instincts persisting and care skills often inadequate, especially when biotechnology objectifies its subjects.

The viewer's emotional perspective also shifts between (1) the scientist-parents' shameful failure with the Fred and Ginger show, (2) the onscreen audience's shock at that bloodbath, (3) the businesswoman's desperation in a meeting afterward, (4) Elsa and Clive's further rebellion, (5) Dren's terror inside a box in their van when moved to the farm, (6) the couple's fear in tracking her through the snowy woods at night, (7) Dren's bloody pleasure in eating wildly but then her obedience in returning to captivity, and (8) Clive's new sense of Elsa's trauma as a child in relation to their romantic and scientific partnership and her empathy with Dren. These vicarious, body-swapping shifts might engage particular elements of the movie viewer's inner theater, such as (1) the MPFC *actor's* fundamental shame, (2) the temporal lobe and insula *audience's* corporal horror, (3) the LPFC *character's* loss of status in the socio-economic mirror, (4) the DMPFC *director's* struggle with others' views as the scientists restage their experiment, (5) the limbic/brainstem *stagehands'* survival terror, (6) the left-cortical *critic/scripter's* seeking to re-contain an escaped hybrid, (7) the right LPFC *technical operator's* acceptance of cultural limits to primal pleasures, and (8) the right-cortical *improviser/designer's* reimagining of roles, interactive contexts, and parenting ties.

The spectator's fear of, yet empathy with Dren is also encouraged with a privileged view of her alone, locked in the barn, like a pet left at home or a child in need of care, while the biochemists are at work in the company's lab. Her pointed tongue, sticking out with delight at water dripping from the roof, suggests the mix within her brain of dangerous animal and cute child. So does her exploration of Elsa's childhood mementos when opening a box with a mirror inside its lid. Like apes in captivity (discussed in the last chapter), Dren not only recognizes herself but plays with her appearance, adding a crown to her head and showing the human "character" of her inner theater's LPFC. She also holds a Barbie doll to her cheek and stares at a faded photo of Elsa as a child with her crazy mother. Later in the film, we learn that Elsa gave Dren some of her own DNA while creating her. So Dren is now discovering her family lineage.

Then Dren hears a cat in the barn. Predatory instincts are triggered again in her left hemisphere, limbic, and brainstem networks. She grabs the cat, runs with it, holds it up, and stares at it. But she also shows her human alienation and loneliness—hugging the cat gently, as if commiserating with the trapped pet.

Dren shows her highly developed intelligence when she uses Scrabble tiles to spell "tedious" and "outside." But she also shows her animal impulsiveness and strength, performing a tantrum for her scientist-parents and wrecking some of the home they are making for her in the barn. She breaks out of it, through a skylight, and when they approach her on the roof, she sprouts wings like a harpy. But Clive says they love her and she returns to his arms, her wings slipping back inside her body. The oxytocin pleasure of human contact becomes more valuable to her than the adrenalin (and noradrenalin) rush of escape and freedom—through her developing DMPFC director and VMPFC stage manager, as the superego networks shaping her plant-animal-humanoid MPFC ego, if her brain is indeed like ours.

Elsa puts makeup and lipstick on Dren, giving her another mirror-stage moment, though with a negative family frame, shaping the hybrid's LPFC awareness of her own appearance. "My mother wouldn't let me wear makeup," Elsa tells Dren. "She said it debased women. But who doesn't want to be debased every once and a while?" Thus, another cultural twist is given to Dren's bestial beauty, reflecting the feminist critique of cinema's male gaze. This may make the movie viewer more aware of his (or her) ghostly voyeurism, or encourage more identification with the females onscreen, in the intimate hall of mirrors between Dren's inner theater of emerging Self-awareness as animal-human hybrid and Elsa's as her godlike creator and yet flawed, human mother.

Researchers have found that children progress from the animal state of *I know* to the human meta-awareness of *I know I know* around age 2 (Lewis 94–95). Even earlier, by 9 months of age, their brains go through a "social-cognitive revolution," from interacting dyadically with people or objects to a "reflective triangle of child, adult, and some outside entity to which they share attention" (Tomasello, "Social" 301–2).[14] This involves "gaze following" (looking where adults look),[15] "social referencing" (using adults as reference points), "imitative learning" (acting on objects the way that adults do), and understanding others as "intentional agents" far beyond other primates—through joint attention with adults, which increases toward age 2 along with emotion regulation (Tomasello, "Social" 302, 310–11; Adamson and Russell 289–91). Even at 6 months, infants show their sense of human action as goal directed, unlike object motion (Poulin-Dubois 266, 275). At around 18 months, they acquire a "naïve psychology" about the people around them having internal experiences, in wanting and requesting certain objects (274). They become *"desire psychologists,"* studying other people as mental agents (275). At age two and a half, children move to the metacognitive level of *I know you know* with an emerging Theory of the Other's Mind and the possibility of deceiving the Other (Lewis 95). Eventually, children become more adult-like, involved in "the interactive and recursive nature of social cognition" and language, with a sense of *I know you know I know*. The child can then perceive itself as an actor with others having different perspectives, including "false beliefs" that are different from what the child knows. Thus, the child's inner theater extends to the outer theatricality of everyday life—taking the position of an actor imagining how the spectator sees the character.

Splice explores such developments in Dren as plant-animal-human hybrid. It shows the external womb of a cultural environment shaping her super-natural progress from *I know* as animal to *I know I know* as child, to *I know you know* as captive teen, to *I know you know I know* as devilish monster. Thus, her scientist-parents violently contain her, lovingly nurture her, rashly transport her, and harshly reprimand or desperately fight her. Clive is initially more fearful of the creature, wanting to kill it or contain it. Elsa is more nurturing, despite being bitten by it inside the mechanical womb. But she also disciplines the hybrid child, yelling at Dren for running away and later taking the cat away from her in the barn, without giving her sympathy or a substitute, transitional object. She says they cannot risk the germs the cat might give Dren. "It could make you sick. You can't always get what you want. That's a part of growing up, too." Yet this also reflects how Elsa's "crazy" mother may have infected her parenting impulses, making her repeat the alienating acts she suffered as a child.

Dren shows her gaze following, social referencing, imitative learning, and intentionality understanding as she develops from a child to adolescent with growing sexual desires: from an initial "imprinting" on her mother (as Elsa calls it) to drawing pictures of her father (which Elsa finds) to copying Elsa's lovemaking with Clive, turning her mother into a rival. Dren's early lab traumas are also shown, not just her parents' alienation of her as a grotesque newborn that they trap and almost kill, but also her father's near drowning of her later as a child. In that scene, she suffers from a fever in the lab storeroom, so Elsa and Clive put her in a metal sink with cold water to bring her body temperature down. Dren starts choking in panic. Against Elsa's protests, Clive forces Dren underwater (warily holding her barbed tail). This triggers her fish inheritance, her ability to breathe water as well as air, which Clive pretends to have expected, when Elsa later questions him. The scene evokes in viewers both the frantic uncertainty of nurturers with a sick child and the horrific anxiety of one parent suspecting evil in the other. But we also see Dren following her parents' gazes and using them as reference points during this traumatic episode—from comforting care to apparent violence—setting up her later mysterious mix of good and evil attachments and intentions. (In a parallel scene near the end of the film, the demonic Dren almost drowns Clive. He, too, revives, yet dies in another way.)

After Dren's death and resurrection as a child, about 40 minutes into the film, the plot jumps to an adolescent Dren, with eyes now in front of her head, being given a Barbie doll by Elsa, from her own childhood keepsake box. A shift in Dren's identity comes here—and in the viewer's potential identification with her—as a bare-headed hybrid, now with a more human face, though still with unusually wide-set eyes, holding the blonde doll. The gaze of the movie viewer follows Dren's gaze from Elsa as social reference to the doll as imitative model for learning humanness (even if "debased" as an object of the male gaze). Dren also shows that she is starting, as a budding "desire psychologist," to explore the cultural dimensions of her creators' intentions, and their perverse limits, with her sketches of Clive as father and then her dancing with him as a man in the barn—while Elsa is absent—about an hour into the film.

Clive puts on an old jazz record, sets the needle, and invites Dren to dance. But as he "leads" and she twirls joyfully, he feels something is wrong and suddenly leaves. Dren looks confused and rejected, yet starts to sense a vexed mixture of emotions at the splice of nature and culture—in her and her creator/father/friend. Clive then accuses Elsa of putting her own DNA in Dren, suggesting she wanted more *control* of the child as an experiment, instead of having a natural child with him. Evoking the left-cortical *critic* and VMPFC *stage manager* of movie viewers' inner theaters, he accuses Elsa of unethical behavior when she did not consider her "family history" in giving her genes to Dren, ironically suggesting his prior doubts, too, about having a natural child with her because of her mother's craziness.

Elsa tries to make amends with Dren by bringing the cat back to her but Dren kills it suddenly, with her barbed tail, and then threatens Elsa. So the scientist-mother knocks her out, ties her to an operating table, and (like Dr. Moreau in his House of Pain) operates on her, cutting off the barbed tip of her tail. The movie spectator is given Dren's viewpoint of Elsa as cold-hearted scientist-surgeon, documenting her work with a microphone, and then a view of the naked Dren from above, her arms out, bound at the wrists, with her legs turned to one side, like Christ on the cross. But this female Christ figure, with occasional, identification-inducing moments, also becomes more and more monstrous—due to her super-natural genetics, her cultural shaping by the fearful scientists, and the genre demands of the horror film audience.

During the surgery scene, Elsa tells the microphone (and movie viewer) that the H50 experiment shows "dangerous psychological developments," which may be caused by "a disproportionate species identification." With her patriarchal, scientific, left-cortical *critic/scripter* inhibiting her right-cortical maternal feelings, Elsa removes a necklace from Dren, cuts off her dress, and wipes the lipstick off her face, which she had put on her earlier in the film. "Cosmetically human affectations should be eliminated wherever possible." Elsa also cuts off the "stinger" at the tip of Dren's tail. The creator's objectifying of Dren here, as a dangerous "H50" that requires surgical modification, and the creature's scream along with a close-up of her tail being sliced, may shift audience sympathy away from the practical, professional experimenter (or mad scientist) toward the monster. Clive is likewise shocked when he arrives at the barn and sees what Elsa has done. But Elsa takes the cut-off tail-tip to the lab, as a fresher sample than the flesh from Fred and Ginger, to find the protein that her corporation needs as a marketable product, to survive and thrive. She thus aspires to a higher ethical and cultural goal: "I'm going to solve this thing. I'm going to make things right." Yet Elsa's utilitarian cruelty shapes Dren's further monstrosity—through Clive's compassion for their hybrid offspring and then his perverse lust for her naked suffering female body, which might be evoked in movie viewers as well.

While Elsa is working in the lab for the corporation, Clive takes a drink at home and watches Dren swim naked in her tank, through his computer screen and the video cameras he installed in the barn. He reaches to touch the screen, like a Web voyeur, and is startled when Dren reaches toward him with her three-fingered hand, as if seeing her Peeping Tom—and the film spectator. Then Clive sleeps on the couch, perhaps dreaming, and calls out for "Elsa." He visits the barn, calling again for "Elsa," but Dren spreads her wings behind him, as if giving him angel/devil/bird hybridity. She also gives him a firm hug and kiss on the lips, her wings retracting into her naked body. At first, he tells her, "No," but soon he is on the floor with her, lowering his pants—the sex drive of inner limbic/brainstem *stagehands* overpowering his VMPFC moral *stage manager* and rVLPFC *operator* controls. As they make love with Dren on top, her wings spread wide again, showing shared hybridity. But when Clive is on top, the barb reemerges from Dren's shortened tail, revealing that her sexual drive still bears a deadly sting. She goes from the animal *I know* to a more human *I know you know I know*, but also to *I know I can deceive you*—through your desire for me. This was also shown earlier when Elsa brought back the cat as a loving offering and Dren killed it with her barbed tail.

Elsa interrupts the coitus of Dren and her scientist-father by entering the barn. Later, Clive tries to explain to Elsa what happened: "We crossed a line. Things got confused . . . about right and wrong." At first, Elsa blames Clive alone. But gradually she admits that they had both tied up Dren, kept her from society, and "maimed" her—a tragically cathartic realization. Despite their "love" for her, they shaped Dren's monstrosity, physically and culturally. Depending on the movie viewer's personal sympathies and judgments in prior scenes, involving body-swapping identifications, the moral hybridity here matches the morphing mix of animal and human in Dren, through Elsa's nurturing yet cruel treatment of her and Clive's coupling with her, showing culture as well as nature being respliced.

But then the scientists alleviate their shame by shifting toward *cold aggression*, with left-cortical, *critic/scripter* narrowness and predatory objectivity, inhibiting right-cortical holistic sensations of a tragic paradox and the VMPFC stage manager's confusion. They decide, as Clive puts it, that "the experiment is over [and] responsibilities end" to their hybrid creature. They go to the barn to kill it. But they find that Dren is already dying from inner causes, so their sympathies (and perhaps the movie viewer's) shift again toward right-cortical compassion. Dren dies naturally and they bury her at night near the barn. The film might end here on a tragic note, with further cathartic realizations for the characters and viewers—about the continued "responsibilities" of scientists and the culture at large toward lab creations and test animals, especially those with human attributes, as neuroscientists are discovering in mammals and primates. But the popular horror genre demands more melodramatic spectacle and monstrous reincarnations.

Perhaps representing the irresponsibility of many human cultures toward the earth and its toxic potential for revenge, Dren rises from her grave, zombie and vampire-like, with wings, talons, and a male body and face.[16] As a Christ/devil-figure or phallic offspring of Mother Earth, but also like a bird of prey, Dren swoops down and takes, in turn, the scientists' male boss (who came to the barn after learning about their extended experiment) and Clive's brother (who came with him), killing each of them. Recent research by archeologists shows that our primate ancestors may have sometimes been killed in a similar way by large birds of prey, as are some primates today, attacked from the sky, giving us a legacy of that horror ("Ohio").

The male Dren-devil draws Clive into a dark pool. He struggles to return to Elsa at the bank—nearly unconscious, yet alive, like the young female Dren whom he almost drowned in an earlier scene. But then the male Dren emerges from the pool, like a creature from the black lagoon, attacks Elsa, strips her with his barbed tail, and rapes her (perhaps in revenge for her circumcision/castration of Dren earlier). Clive and Elsa manage to kill it, but Clive is stabbed in the heart by Dren's tail and dies. Sometime later, the pregnant Elsa meets with the businesswoman and signs a contract, turning over the product of her womb and lab work, a multiply hybrid creature, for marketable products. This frames Elsa as heroic survivor, yet with her inner DMPFC *director* converted from truth-seeking scientist toward the corporation's profit motives. Thus, her temporal lobe and insula networks may be altered, too, as her inner *audience* of memories, intuitions, and meanings—in relation to the outer movie audience reimagining her in earlier parts of the film, through their own inner audiences.

INVERTED THEATERS OF THE APES

Splice depicts Dren's evolution from plant-animal creature to mammalian child to humanoid adolescent to super-natural monster with various brain, sex, and bodily changes. This evolution is not just genetic, but also cultural. It is radically altered by an external human womb, both technological and familial, in the lab and barn, involving primate social dynamics considered earlier in this book: kinship, territory, attraction, trickery, and domination. The *Planet of the Apes* film series shows these dimensions of ape egos, superegos, and ids more collectively, in future societies where simians walk upright, wear clothes, ride horses, shoot guns, and live in cave-like buildings, but have premodern technologies. Or, in the two recent versions, it shows them closer to our time, evolving in that direction, within and against human culture.

In the first film (1968, dir. Franklin J. Schaffner, with a screenplay by Michael Wilson and Rod Serling, based on the novel by Pierre Boulle),[17] the apes enslave and experiment with humans like the scientists do with their animal-human hybrids in the Moreau, Fly, and *Splice* films. Of the three astronauts who survive a crash landing on a distant planet, which turns out to be earth 2,000 years in the future, one (Landon) is eventually shown with a scar on his head from a neurosurgical experiment. Another (Dodge), who dies when the gorilla soldiers capture the astronauts along with native humans, is later shown stuffed, as a black man with false white eyes, put on display in a museum, like we do with animals in taxidermy scenes of "natural history."

The third astronaut (Taylor, played by Charlton Heston) cannot speak initially, after being captured by the apes, because he was shot in the throat. He is kept in a cage like the native humans who also do not speak. But he is seen as different and called "Bright Eyes" by the female chimpanzee psychologist, Zira (Kim Hunter). Treated as a savage by the gorilla workers, yet studied respectfully by Zira and her chimp fiancé, the archeologist Cornelius (Roddy McDowal), Taylor tries to communicate by writing, gradually convincing Zira and Cornelius of his ape-like intelligence, especially when he regains his voice. Fortunately, they speak English exactly as he does from 2,000 years before—even though apes today do not have the vocal chords that allow for human speech (Hu) and our modern English is very different from its medieval roots a thousand years ago. But the social dogmas and religious beliefs of the chimps' superiors, the orangutan rulers in "Ape City," and of the gorilla workers and soldiers, continue to suppress Taylor's name, speech, and individuality.[18]

This sci-fi film projects the revolutionary changes in American racial awareness and class structures of the 1960s into a parable of the distant future.[19] Taylor, the white man, finds himself treated like an enslaved animal along with other, nonspeaking, more primitive-looking, mostly white humans, who were already on the planet. The film's horrors express white cultural guilt about the colonial treatment of conquered peoples, especially indigenous Americans and black Africans, by inverting the color scheme: (1) the black-haired gorillas capture humans as slaves; (2) the black-haired but lighter-faced chimps study humans in cages, museums, and archeological sites; and (3) the orange-haired orangutans judge humans as having no rights. Racial identifiers are flipped and cultured animals shown as wilder in their cruelty than their "wild" human subjects.[20]

Early in the film, when the three astronauts first see the native white humans, wearing animal skins and eating crops in a field, Taylor says they will soon "be running this planet." Yet the ruling orangutans and gorillas view such humans as deserving domination: annoying as pests eating their crops, but useful as slave labor. The gorillas also find sport in hunting them on horseback, shooting at them, and capturing them with nets and poles. The gorillas even take a photo of themselves, standing proudly over some of their dead human victims, like wild game hunters or picnickers at a lynching. Thus, the astronauts, as adventurous colonizers, whose clothes were stolen by the other humans when they swam in a pond, get stereotyped with the colonized slaves, as sub-ape animals, when captured along with them.

Feminist issues arise, too, with Zira's defiance of a patriarchal tribunal of orang scientists, judging Taylor as a primitive animal. This also involves her orang mentor, Dr. Zaius (Maurice Evans), who is in favor of "experimental brain surgery" on humans and their eventual extermination for the betterment of "simian survival." Zira argues instead for Taylor's rights as an intelligent, speaking ape. She gradually persuades Cornelius, but the tribunal rejects her legal argument.[21] So she and her fiancé, along with her young nephew, help Taylor to escape from Ape City. But to balance this depiction of an intelligent, stubborn, rebellious female, especially for the traditional male gaze, movie viewers are offered one of the native white humans, a sexily clad, silent "Nova" (Linda Harrison), as a companion and potential mate for Taylor.

Zira uses Nova to give Taylor a blood transfusion when treating his throat wound. The chimp scientist, somewhat bonobo-like here, orders that Nova be put in the cage with Taylor, telling him with a smile that the female is a "present." Although Nova does not speak and Taylor treats her condescendingly, Nova shows that she remembers the blood connection they had, surprising Zira. Nova is not just eye candy for the male cinematic gaze, but also shows intelligence, beyond the apes' prejudice. She becomes Zira's substitute in her physical closeness to Taylor, which the engaged chimp scientist seems to desire, through her rebellious right-cortical *improviser*. Zira (like ape egos in the movie audience) uses the captive woman as a romantic surrogate due to her socially correct left-cortical *critic/scripter* and DMFPC *stage manager*. And yet, Zira's *I know you know* studies of Taylor and his *I know you know I know* reactions betray the potential misrecognitions, too, between genders and individuals, or across species, not to mention the contending animal-human networks of just one person's inner theater.

Such a 1960s allegory of the outer-space colonizer enslaved and the feminist professional in conflict with herself allows movie viewers to identify with Taylor as ironic hero, fighting for his rights, or with Zira as active ally fighting for scientific truth and legal fairness, while restraining her wayward desires. Heston's body, like Harrison's, becomes a potential object of desire for the movie viewer at many points, shown shirtless or fully naked (from behind), or at other erotic moments when covered with the torn animal-skin blanket that the apes give him. For example, after failing to escape from the apes the first time, without Nova, Taylor is put in a cage with her, but held back by a gleeful gorilla with a phallic firehose shooting water at him,[22] while other gorillas hold the wet, squirming, human female. And yet, after the gorillas leave, Taylor gives Nova a cynical critique of the 1960s era and its sexual revolution, saying that he left earth because there was too much "lovemaking with no love." He also wonders aloud whether Nova can ever understand him on his new planet. Toward the end of the film, however, Taylor shows more hope, encouraging Zira's young nephew, Lucius, to continue rebelling against the older generation of apes.

Taylor also kisses Zira "goodbye" on her chimp lips, before leaving with Nova on a horse, in their escape at the film's end. She lets him, after saying he is "so damned ugly," perhaps for the benefit of Cornelius and Nova, as they watch. The physicality of the Taylor-Zira kiss, despite the rubber chimp-mask, may evoke "enfacement" and embodiment signals in the movie viewer (identifying with either or both the human and ape), through mirror and intuition neurons, as with the rubber hand and body swapping experiments discussed in the last chapter. Then there is the sensuality of Taylor and Nova riding away together on the beach, wearing little, with her hugging his back on the same horse, again suggesting a romantic happy ending, for audience embodiment, until the final sci-fi shock.

A new camera angle shows spikes of corroded metal, framing a distant Taylor and Nova. The cinema audience gradually sees that its point of view is actually that of the Statue of Liberty, mostly buried in sand. Then we get Taylor's view of it and see him get off his horse, kneel, and beat the sand—realizing that he is on earth with humans who have lost their civil liberties and cultural superiority through misrule. This tragic warning at the end of a romantic, sci-fi action movie reflects its Cold War, nuclear threat context, yet also its meta-historical questions about the destiny of our animal-human, bio-cultural evolution—with or without God or Lady Liberty or movie viewers watching.

As Louise Allen has argued, the Taylor–Zira–Nova erotic triangle, with Cornelius and Dr. Zaius in the patriarchal background, may appeal to a postmodern viewer today as "transgressive," even beyond its original context (8). Throughout the film there is a "species drag" of human actors in ape costumes, exposing the structures of our current stereotypes, in the performance projections of gender, animality, and humanity (7–9). This climaxes with the Taylor–Zira kiss, which appears stronger than earlier Zira–Cornelius pecks and thus "perverts traditional romantic narratives" (6).

But the film also subverts melodramatic movie expectations with Dr. Zaius as not just a villainous ape, cruelly endorsing neurosurgical experiments and the ultimate extermination of humans. He becomes an increasingly sympathetic figure with an inner tragic conflict. As Minister of Science, Zaius sits on the three-orang Tribunal that judges Zira and Cornelius guilty of "heresy" for proposing that Taylor is the "missing link" in primitive human to cultured ape evolution. Privately, Zaius interrogates Taylor for the "truth" of where he comes from, threatening him with "emasculation" and brain surgery, which would result in a "living death," like he saw with his fellow astronaut, Landon. Zaius rejects Taylor's answer about traveling through space from a distant planet because that contradicts what most apes believe through their "sacred scrolls," like Zira's psychological and Cornelius's archeological theories. But Zaius also wants greater knowledge about other potential "mutants" like Taylor and fears they have formed a "tribe" in a jungle beyond the "Forbidden Zone" that threatens ape culture. This may evoke viewers' sympathy for Zaius as he seeks truth and fears outsiders, even while they sympathize with Taylor, mixing the *rasas* of courage, fear, anger, and awe across species and melodramatic embodiments.

After the chimps help Taylor and Nova to escape captivity, Zaius finds them on the beach near a cave, which is Cornelius's dig site. The chimp archeologist shows his orang superior the artifacts he found there, including a doll that speaks, which "prove" that humans had an earlier culture, like apes in their time. It is then that Zaius reveals his deeper knowledge and fear, through his belief in the sacred scrolls along with science: human culture is a self-destructive aberration that may reemerge as a danger to simian survival. Or, as Cornelius reads, at Zaius's urging, from the twenty-ninth Scroll of their Lawgiver, "Beware the beast, man, for he is the devil's pawn. Alone among God's primates, he kills for sport or lust or greed. . . . Let him not breed in great numbers, for he will make a desert of his home and yours." Zaius admits to Taylor that he always feared him as a man, "a warlike creature who gives battle to everything around him, even himself," and whose ancestors destroyed a "paradise," turning it into the Forbidden Zone. Yet the minister calls his soldiers back from chasing Taylor and Nova, letting the astronaut discover "his destiny" and potentially revive a more advanced human species with Nova as his Eve.

Although Taylor and his mate escape from the orang minister of science/religion and his gorilla soldiers, the astronaut's discovery that he is actually on a future earth exposes a conflict in the hero's inner theater that matches his opponent's. Both Taylor and Zaius are tragically troubled by a potential in the ape ego, id, and superego of becoming too humanlike and therefore super-naturally wild. Kinship bonds, territorial drives, attractive lusts, deceptive greediness, and power politics are limited in the nonhuman ape species of our world by balanced instincts that evolved in millions of years of natural negotiations between genes, changing environments, and small societies. But we humans broke through such limits, as a Pandora's Box species, with an ever-growing godlike knowledge of good and evil, and with super-natural technologies that far outreach our reflective wisdom, in addictive and destructive ways. The apes in this first film of the series show stable kinship classes (gorilla soldiers, chimp scientists, and orang rulers), territorial respect (such as not entering the Forbidden Zone except with permission as Cornelius has done as an archeologist), sexual restraint (as Zira shows in her loyalty to Cornelius), standards of truth (with the Tribunal following Ape Law even if it denies rights to humans), and generational rivalry without civil war (although Zaius blows up Cornelius's dig site to repress the human remains). These elements of ape-human ego, id, and superego continue to develop in the film's several sequels, set on the planet earth in the future and current time.

The nuclear threat allegory becomes the main focus for the second movie, *Beneath the Planet of the Apes* (1970, Ted Post). The ape ego of gorilla General Ursus joins forces with Dr. Zauis's scientific superego—to invade the Forbidden Zone, expanding their territory for food and truth. Meanwhile, Brent (James Franciscus), a human astronaut who somehow followed Taylor's spaceship with his own and crash landed on the future earth, seeks to find Taylor with the help of Nova, after also meeting Zira and Cornelius. But Brent and Nova, like Zaius and Ursus with their gorilla army, discover an advanced society of mutant humans under the earth's surface, who use telepathy to project cinematic illusions, bodily pain, and violent impulses into their enemies' brains. The telepaths also worship a nuclear missile with Christ-like symbols on it, the Greek letters alpha and omega, in an underground cathedral that seems to be St. Patrick's in New York because the ruins of other such sites are found by Brent and Nova, including the city's subway and Radio City Music Hall. "Holy Bomb of Peace," the telepaths call the missile, singing hymns to it and taking off their human-face masks in order to reveal the "inmost self" of radiation disease to their "god."

Thus, screenwriters Paul Dehn and Mort Abrahams invert the ape masks, introduced in the first film, to show the detached, pseudo-religious savagery of an advanced human culture, which extends the cinematic power of our inner theater's mirror and intuition neurons toward the direct manipulation of others' brains—toward fear, rage, and mass suicide. One of the telepaths, explaining that they are people of "peace" and do not kill their enemies, puts Brent in a cage with Taylor, forcing them telepathically (perhaps through DMPFC director and limbic stagehand networks) to fight to the death, like a movie viewer desiring such scenes. To stop the ape army, the telepaths project horrific scenes of fire, crucified gorillas, and a huge statue of their ape Lawgiver bleeding from its face, as virtual 3D images on the earth's surface. But the apes, led by Zaius, disregard such illusory scare tactics and overrun the underground telepaths' territory. One of the bomb-worshipping (Mutually Assured Destruction or MAD) peaceniks then commits suicide in the tub, like an ancient Roman. Another starts the arming of their nuclear weapon, which the gorillas aid by pulling down the missile icon. After escaping the telepath's virtual power and mourning the death of Nova from a gorilla gunshot, Taylor cynically completes their apocalyptic wish, blowing up the planet earth.

Thus, *Beneath the Planet of the Apes* extends Vietnam-era, Cold War doubts about the sanity of today's humanity into its depictions of future apes and humans, with pre- and postmodern cultures. Tribal ties turn into territorial battles between the apes with guns and the telepaths with advanced mental technologies. Attractive ideals of a sexual and peaceful revolution appear with Nova again dressed as a cave-woman fantasy for the male gaze, riding on a horse with Brent, and the young chimps with protest signs trying to stop the gorilla army's march. But these ideals turn into the horrors of underground New York ruins, of the gun-happy gorillas' persistent drive for dominance, and of the telepaths' cinematic projection tricks, religious self-deception, and nuclear disease. In a film released just 8 years after the Cuban missile crisis, the plot twist of going beneath the earth's surface, showing future telepaths' having rituals within and virtual projections onto (but also outside) the cave walls, suggests a tie to *prehistoric cave art* as the start of something monstrous in the ape-human id, ego, and superego, which might climax in a real-life *nuclear holocaust*. We evolved beyond other apes to create super-natural art forms, technologies, and ideologies. Such developments in recent times almost led to a nuclear war in 1962, through Capitalist-Communist territorial conflicts, and did lead to napalm bombing in World War II, the Korean War, and the Vietnam War, escalating from tens to hundreds of thousands of tons.

Humans are, as Dr. Zaius fears, monstrously creative and destructive. But the film ends with a further tragicomic twist, after the telepaths' nuclear warhead explodes, destroying earth and turning the movie screen bright white. A voice-over puts the movie audience into the role of observing gods in outer space, telling us that "a green and insignificant planet" of a medium-sized star among the billions in the universe "is now dead." This encourages *rasa* engagement and then cathartic distancing of the human-ape's inner theater, with right-cortical sympathies, involving limbic id passions, altering left-cortical preconceptions of ego and superego, in this reappraising perspective. Perhaps we still have time to transform into wiser apes and telepaths, unlike the creatures onscreen—through our evolving kinship ties, territorial disputes, beauty ideals, Machiavellian deceptions, dominance drives, and hyper-theatrical rites.

The apes refer to their god as the Creator, in relation to their Lawgiver and Sacred Scrolls. But the telepaths extend such Judeo-Christian ironies, singing hymns in cathedral pews, with "Amens" given and a minister leading them, while worshipping nuclear explosive power.

> The heavens declare the glory of the Bomb and the firmament showeth His handiwork. . . . And His light unto the end of the world. He descendeth from the outermost part of heaven and there is nothing hid from the heat thereof. There is neither speech nor language, yet His voice is heard among them. Praise Him, praise Him, my strength and my redeemer. Glory be to the Bomb and to the Holy Fallout as it was in the beginning, is now, and ever shall be, world without end. Amen.

The leader of the communal rite wears a missile-shaped medallion on his chest, instead of a crucifix. Yet he refers to the Bomb as if it were Jesus Christ, connecting heaven and earth through the phallic missile and its sacrificial altar: "Almighty and Everlasting Bomb, who came down among us to make heaven under earth, lighten our darkness. Oh, instrument of God, grant us thy peace." Then each person unmasks, revealing "the truth that abides in us . . . unto that Maker" and showing grotesque, irradiated faces to the film viewer. But the Big Bang at the end offers a supernatural viewpoint, with a *voice-over* (of whom?) explaining the death of the planet that ironically caps the telepaths' prayer. The "world without end" is somewhere beyond earth, and yet also the end of their world. This reframes the film, putting the movie audience in the divine role of creator, lawgiver, and alpha revolutionary power, remaking the world through the brain's inner theater, as "the truth that abides in us."

The third film in the series, *Escape from the Planet of the Apes* (1971) was directed by Paul Taylor, but again scripted, like the second, by Paul Dehn. The first two films suggested an animal rights issue by flipping the experimental table on humans, making Taylor, Landon, and Brent the subjects of ape control and experimentation. The third develops that issue further, combining it with a question about human destiny and a providential God, again shifting the role of the movie audience.

Cornelius and Zira escape the future earth's destruction, along with a chimp "genius," Dr. Milo. He repairs Taylor's spaceship and pilots them back in time to the current audience era, though the chimps see the nuclear explosion on the future earth while leaving it. Milo is then strangled to death by a gorilla when the three chimps are kept in a cage next to it in a zoo in 1970s Los Angeles. But after humans realize the astronaut apes can speak intelligently, they become celebrities with a luxurious hotel room and fashionable clothes. Cornelius wants to keep some secrets but Zira leaks them, rebelling against him, with her desire to tell the truth, but also by getting drunk. Then the chimps are treated more like enemies, in a CIA interrogation room. A Kissinger-like advisor with a German accent, Dr. Otto Hasslein, warns the president that if the pregnant Zira is allowed to give birth, or Cornelius and Zira are allowed to procreate again, the future earth will end in the nuclear explosion that they saw when they left it.

Ironically, Hasslein's explanation to a TV audience about multiple universes in the future, branching out from different potential human actions now, belies the error in his overly anxious warnings. He misses or represses the fact that Cornelius and Zira's traveling back in time could not be the cause of apes acquiring language and a more advanced culture in the future, leading to war and nuclear destruction, because the future they came from could not have included that past branch, prior to their trip. (Indeed, Cornelius gives an alternative, evolutionary explanation, shown in the next film of the series.) And yet, Dr. Hasslein encourages the TV and movie audiences to imagine a godlike perspective—of an artist creating a painting of himself painting himself painting, and so on, like a hall of mirrors showing "infinite regression." This is his way of rationalizing the killing of Cornelius, Zira, and their baby (even against the president's order) to assume his own godlike role in stopping the future destruction of the earth.

When initially talking with the president, Dr. Hasslein expresses uncertainty about changing the future by killing the chimps. "Which future has God, if there is a God, chosen for man's destiny? If I urge the destruction of these apes, am I defying God's will or obeying it? Am I His enemy or His instrument?" But after Zira and Cornelius escape Hasslein's custody with their baby, he piles up the problems of the world—pollution, population control, and nuclear weapons—to convince himself and his colleagues to take action against the chimps. "We think we've got all the time in the world? How much time has the earth got?" Hasslein draws on his inner theater's deep *survival* circuits and limbic fears as *stagehands* to move his brain's higher areas, and those in others, to take immediate action against a threat that is 2,000 years in the future. At the film's end, he kills the speaking chimps, but dies also in the sacrifice for the future. And yet, the baby chimp survives, shown in a cage (and played by a real chimp), apparently saying, "Mama," not knowing that he will never meet her again.

Through such tragicomic twists, this film, like earlier ones in the series, provokes its audience to question the Judeo-Christian foundation for modern Western science: a belief that humans have "dominion" over the earth and its animals (Genesis 1: 26). Under the influence of sodium pentothal as a truth serum, Zira tells her human interrogators the details of how she vivisected humans in the future, for the benefit of the apes (like Dr. Moreau with his hybrids). This *disgusts* even her zoo veterinarian friend, Dr. Lewis Dixon—and perhaps the movie viewer—into rethinking what humans do with lab animals.[23] Cornelius also reports a common gorilla and orangutan belief, as given by Dr. Zaius earlier in the series, that the Creator made apes "in His own image." This highlights the human *hubris*, in the 1970s and our time, of justifying the use of animals in scientific research with the exceptionalism of our species, whether given by God or Mother Nature.

The series shifts the audience's perspective from godlike, trans-historical, and outer-spatial views, in the first two films, toward questioning such aspirations with the villain of the third, Dr. Hasslein. It evokes sympathy for the intelligent, witty, loyal chimps, Cornelius and Zira, who are "pacifists," they say, like all chimps in the future. Yet, the third film reveals Cornelius's attempts to hide the truth from humans and Zira's memory of surgical experiments with live humans in the future. The fourth film, *Conquest of the Planet of the Apes* (1972, dir. J. Lee Thompson and again written by Paul Dehn), focuses on the rise of their offspring, Caesar (played by Roddy McDowall, the same actor who portrayed his father), as a Messiah-like communal or communist leader, in the violent rebellion of apes against their fascist human masters, in a future 1991.[24]

At the start, sympathy is evoked for Caesar, the baby chimp seen at the end of the previous film. He is now an adult being led on a chain, with a metal collar, by his foster father, the animal-loving circus owner, Armando (Ricardo Montalban), who seems to be showing him the rest of the human world for the first time, warning him not to speak. Armando shows Caesar a memorial to the pet cats that died 8 years ago, along with dogs, in 1983, and explains that a virus killed them all, leading to apes becoming pets and then slaves—as Cornelius had told from his future knowledge in the prior film. The viewer and Caesar witness various acts of callousness and cruelty by human owners, trainers, and guards, sometimes dressed in black uniforms, looking like Nazi SS officers. This politicizes Caesar, who yells from the crowd, "human bastards," at the officers beating a fellow ape for misbehavior. Armando helps Caesar to flee but then turns himself in, bluffing that he yelled during the incident and does not know where his circus chimp is. Yet, when interrogated under a special lamp, Armando cannot continue lying and then jumps out a window to his death—sacrificing his life for the chimp.

Such an opening act of the film draws melodramatic identifications of viewers with the chimp hero, human ally, and ape victims against fascist villains, in the binary operator of the brain's left parietal lobe (as part of the left-cortical *critic/scripter*), involving also the moral VMPFC *stage manager* and emotional limbic *stagehands*. But this melodramatic framework is complicated by *tragicomic edges* in various scenes. When Caesar hides in a cage of newly imported orangutans, the non-chimps crowd around him with territorial aggression, making him hiss to defend himself against what initially seemed to be his fellow victims and potential allies against the humans. Yet when Caesar is led through the training facility, empathy is again evoked from him—and potentially from the movie viewer (through limbic distress circuits of emotional contagion)—for other apes suffering harsh treatment by humans with clubs and flame throwers. They are being "conditioned" to behave and serve, to wash their hands and pour water into a glass, which Caesar performs exceptionally well. He is purchased then at an ape slave auction by the local Governor (Don Murray), who becomes suspicious that Caesar might be the offspring of Zira and Cornelius, previously switched with another baby chimp that died when they were killed. (All the adult apes are played by masked humans, so the viewer can pretend they look similar to real apes even when walking somewhat upright and speaking.)

Governor Breck, a white man dressed in black, develops into the main villain of the film. But his black aide, MacDonald (Hari Rhodes), initially aligned with the dominant humans against the apes, becomes a secret *ally* to the intelligent, black-haired, chimp slave. After Armando's suicide raises suspicions, MacDonald helps Caesar to flee. But the chimp is caught and tortured with electrodes attached to his head (perhaps reminding viewers of the brain surgery performed by chimps in the future, according to this film series, or perhaps of brain mapping today). And yet, MacDonald helps Caesar to survive and escape again, despite the governor's sudden command to execute him. With Latino and black actors cast as Caesar's only human allies, race issues are clearly emphasized in this 1970s film. Armando and MacDonald are Caesar's noble allies in his melodramatic fight against a white villain's cruel regime. However, both sympathy for and fear of the apes may grow in viewers' inner theaters as Caesar leads them toward revolutionary violence against their oppressors—drawing on tragic associations with Vietnam, Kent State, and Watergate, yet also communist threats, the Black Panthers, Black Muslims,[25] and other signs of civil war at home and abroad, then or more recently.

Here, Caesar becomes more central as a tragicomic hero than Zira or Cornelius in the earlier three films. He is shown in alienated mourning, peering into the dark night with tears falling from his eyes, after the death of his foster father Armando, due in part to his own error of shouting at the brutal humans, perhaps also revealing his earlier suffering and alienation after the death of his chimp parents. Yet one of the "primitive apes," a female chimp that Caesar was apparently bred with by his human captors earlier in the film (like Zira started to do with Nova and Taylor), touches him sympathetically then, connecting the inner theaters of their ape egos, as well as ids, without spoken words. Likewise, Caesar's vengeful and yet justice-seeking gaze meets the eyes of other apes, in various locations, inspiring their personal rebellions. These feelings (evoking audience *rasas* of sorrow, anger, and courage) become collective, too, as Caesar gathers the apes in wordless covert meetings, with various stolen knives, forming a terrorist or freedom-fighting organization across species. Caesar also leads them, without their speaking, in protest marches against the state's security officers. When many of them, just carrying knives and clubs, are gunned down by those authorities, tragic memories of the Kent State shooting by National Guardsmen of unarmed student protestors and bystanders, just two years before this film's release, might be evoked in viewers' inner theaters. And yet, as Caesar's ape militants then overwhelm the human officers, beating and stabbing them to death, the viewers' visceral sympathies may again shift toward the new victims, through mirror neuron and contagious pain networks, evoking body-swapping identifications and *rasa* reappraisals.

Such visual, tragicomic twists become verbal when the apes capture MacDonald and Governor Breck. In their collective frenzy, the apes (many in red uniforms, suggesting communist revolutionaries) are about to bludgeon them, but Caesar makes them pause. He questions Breck about why humans turned apes from pets into slaves. The Governor explains that humans feared the savagery of apes as apes themselves. "Man was born of the ape and there's still an ape curled up inside every man, the beast that must be whipped into submission, the savage that has to be shackled in chains. You are the beast, Caesar. You taint us. You poison our guts. When we hate you, we're hating the dark side of ourselves."

Governor Breck thus reveals that his oppressive regime was not simply evil. Like many in the film audience, who enjoy a godlike dominance in watching the film's mayhem, Breck and his colleagues *projected* a tragic flaw from their inner theater and in-group, a fear of their own savage evil, onto other types of apes as melodramatic scapegoats. The governor's tone incenses Caesar to club him with his rifle, as a scapegoat also. But the chimp leader pauses, as his higher brain networks process Breck's insight, inhibiting primal revenge impulses. Not wanting to indulge in savagery himself, Caesar tells the other apes to take the fallen oppressor away.

MacDonald tries to aid Caesar's VMPFC superego as *stage manager* of the revolution's violence. As the other apes threaten to club Breck, MacDonald says to Caesar, "This is not how it was supposed to be. . . . Violence prolongs hate. Hate prolongs violence. By what right are you spilling blood?" Caesar exults in his current victory, predicting that "apes on five continents" will do likewise and his "people" will rule the earth after humans destroy their own cities in nuclear war. But then he sees the apes with rifle butts raised over Breck. He also sees the female chimp who befriended him, now gazing at him with a "No" to such violence (as one of the first primitive apes to speak). So he tells the apes to put down their weapons. "And we who are not human can afford to be humane. Destiny is the will of God and if it is man's destiny to be dominated, it is God's will that he be dominated with compassion and understanding." Caesar tries to be a better leader or a better god than the humans who had dominion over his fellow animals—maybe like the film's creators and audience.

The final film of the original series, *Battle for the Planet of the Apes* (1973, dir. J. Lee Thompson, story by Paul Dehn, screenplay by John and Joyce Corrington)[26] begins with the orangutan Lawgiver (played by the famous filmmaker John Huston) telling ape and human children in 2670 about the strife between their species more than six centuries before, which the movie viewer sees in scenes from the prior films.

> In the beginning, God created beast and man, so that both might live in friendship and share dominion over a world at peace. But in the fullness of time, evil men betrayed God's trust and, in disobedience to His holy word, waged bloody wars, not only against their own kind, but against the apes whom they reduced to slavery. Then God in His wrath sent the world a savior, miraculously born of two apes, who descended on the earth from its own future.

The film then shows Caesar, as "savior" in that prior time, but through conflicts with gorilla General Aldo and with "mutant" humans who have survived nuclear warfare. They live in the underground remains of their destroyed city, but emerge like a procession of ants to make war against the apes—led by the evil Governor Kolp who wants to exterminate the apes out of boredom underground and greed for new territory.

A few good humans still live and talk with the apes, some as teachers, others as servants. Caesar gets advice from another MacDonald, brother of the one in the previous film, and from a wise orangutan named Virgil. But the gorilla Aldo vies with Caesar for alpha leadership of the apes, disputing whether they should learn from or just control the humans as servants. Caesar then leads the apes in repelling the mutant humans who attack their village with motorbikes, jeeps, a yellow school bus, mortars, and machine guns. Yet Aldo and his soldiers take the violence further, massacring the fleeing mutants. He also orders the death of the local, friendly humans, which he had corralled (like Japanese Americans during World War II), as an evil species. Caesar forbids this, but the gorillas band together to assassinate him—until Virgil declares that Aldo previously killed Caesar's young son, breaking a sacred law: "Ape does not kill ape."

The apes then support Caesar's revenge on Aldo, whom he overpowers while they are both on the branch of a tree, killing him in revenge. But Caesar prevents further vengeance against his human friends, building the basis to renew their shared "dominion" in the Lawgiver's story. Thus, the original series ends not only with melodramatic, interspecies battles, between gorillas and chimps/orangutans, or between apes and "evil men," but also with a note of tragicomic hope that they might balance their rivalries in the future—if sci-fi desires of the godlike filmmakers and mass audience allow.

RESURRECTING THE SPACE RACE OF HUMANOID APES

After the original film series, 14 episodes of a live action *Planet of the Apes* television series were broadcast on CBS in 1974 and 13 episodes of an animated TV series on NBC in 1975. But it took another quarter century for a remake of the initial film[27] with a very different plot but still with prosthetic makeup on the actors as humanoid apes, along with higher-tech visual effects. Created during the late 1990s, in an era of Y2K, end of the millennium, apocalyptic anxieties, the new *Planet of the Apes*, directed by Tim Burton (with a screenplay by William Broyles Jr., Lawrence Konner, and Mark Rosenthal), was released in July 2001, just a couple months before 9/11. It presents a society of talking apes and enslaved humans on a planet called "Ashlar" in 5021, where the film's hero, astronaut Leo Davidson (Mark Wahlberg), finds himself after passing through a time warp, like Taylor on a future earth in the original film.

But here, the primitive humans talk and the apes move in more ape-like ways, hunched and swaying when on two feet, running on all fours, and swinging or leaping with super-natural strength.[28] Male gorillas and chimps ride horses and fight in armor and helmets, but without guns, like ancient or medieval warriors, more primitive than in the original series. Several female chimps have hair styles (on their heads) and one is shown as the trophy wife of an older pudgy orangutan. A comical, orangutan slave trader, Limbo (Paul Giamatti), extends the egoistic Self-awareness of the real-life orang Chantek (mentioned in the previous chapter) by perfuming himself with rose petals while hanging upside-down and looking in the mirror. The wealthy apes also party together with snobbish attitudes and their children mock the humans, more as a collective class with various ape characteristics than as the species-organized classes in the original. The new *Apes* film explores "species guilt" more than racial guilt, with Leo's mention of apes' survival, on earth in his time, only in zoos and labs, like the one on his space station (O'Hehir 13).[29]

Initially, too, there are differences from the original film. Leo arrives on the apes' planet not with fellow astronauts but alone, after disobeying his commander's order at the start of the film. Stealing a space "pod," a small travel craft, from their space station in 2029, Leo follows a chimp-astronaut friend, Pericles (played by an actual chimp), who was sent into an electromagnetic storm in a similar solo vehicle. Thus, Leo's crash landing on a strange planet three millennia later, after following Pericles through a time warp, echoes the beginnings of the "space race" in the 1960s, when various animals, including chimps, were sent on experimental vehicles prior to humans, as expendable sacrifices in the conquest of space.[30] It also reflects the 1990s, with a public fear of millennial change, based on the Christian calendar.

Pericles (named perhaps after the ancient Greek democratic leader or Shakespeare's character) appears again late in the film, touching down on Ashlar in his space vehicle. He is viewed by the apes, at least at first, as their returning god, Semos. Prior to that, Leo finds the ruins of his space station on the planet and learns that it crashed there, too, but three millennia earlier. Its other ape workers, led by the genetically enhanced Semos, overthrew the humans, starting a new culture on the planet. They evolved into the apes and humans that Leo found there. The crashed space station eventually became a holy site to the apes, a temple of their god, Semos. Pericles' sudden reappearance makes him a *deus ex machina* in a double sense: a super-natural figure who changes the drama's ending, actually stepping out of a machine, yet also a resurrecting, godlike, revolutionary leader, akin to Caesar in the earlier films.

Here, movie viewers' identifications might shift, through such tragicomic twists, from sympathizing with the alienated ape ego of Leo, and fearing (yet enjoying) his initial rebellion on the space station and in various survival dangers on the strange planet, to reflecting on human-animal relations in science and technology today—or within our own brains and bodies. Which leaders of the past and persons in our immediate lives do we project as godlike in our inner theaters and in other, collective media? How is this due to a particular cultural shaping of our super-natural, human-ape departure from preprogramed instincts, with an alienated need for meaning in life? And which alien or idolized projections relate to the territorial race to keep specific spaces around us as our own, or that of our kin, against other groups and races?

Leo, as alien visitor on Ashlar, gets a female chimp ally with hints of cross-species sexual attractions, plus a primitive human female, as his love interest triangle like Taylor's. This time, however, the primitive human speaks and the chimp female is not a scientist, but an animal and human rights activist named Ari (Helen Bonham Carter). She buys Leo and the human female, Daena (Estella Warren), from the slave trader, Limbo. Ari then helps Leo to free his fellow humans (including a little girl, the caged pet of a young female chimp) and escape from the home of her father, a wealthy senator. Leo's main opponent among the apes is the sinister General Thade (Tim Roth), a power-hungry chimp and descendent of Semos, who functions mostly as a melodramatic villain here, more like Ursus and Aldo than Zaius as antagonists in the earlier films.[31] For example, during a dinner party of upper-class apes, held by Ari's father, General Thade pries open the mouth of Leo, a house slave at the table, to see whether humans have a "soul" inside, as a brutal joke about Ari's argument that they do. Thade (who seems to have a humped back, like Shakespeare's Richard III) is also courting Ari, despite their political differences, as an alignment with her powerful father. But Ari rejects him and he eventually brands her hand with a hot iron, like the humans are branded, as a rebel conspirator with them.

And yet, the viewer is offered a sympathetic scene of the villain Thade with his dying father (played by Charlton Heston, like the ghost of Taylor returning in ape form), who warns him that humans are dangerous, due to their prior civilization with its guns. Thade also has a gorilla ally, Colonel Attar, who leads the marching troops and fights fiercely in the final battle. He kills Ari's friend, Krull, a former general whose career was ruined by Thade. Although a gorilla warrior, like Attar, Krull is more sympathetic for the movie viewer because he sides with the humans in loyalty to Ari, helping them to escape her house and destroying Leo's gun to avoid further bloodshed.

However, after Krull's death in the final battle, tragic sympathy is evoked also for Colonel Attar, a dedicated worshiper of Semos, when the gorilla realizes that Thade is not the noble leader he thought he was, in a divine lineage, but an egoistic trickster. "Everything I have believed is a lie. You and your family have betrayed us." Even the villainous Thade is given a tragicomic twist at the end. Trapped by Leo in the command module of the ruined space station, Thade shoots a gun (brought by Pericles to the planet) many times at Attar and then at Ari, after each declines to free him. The bullets ricochet off the thick glass door, eventually forcing the power-mad chimp to hide from his own violence under the command console.

In another tragicomic twist at the end of the film, "borrowed from the 1963 satirical novel by Pierre Boulle," which inspired the original series (O'Hehir 12), Leo returns to earth in Pericles' solo pod, through a fortuitous, space-storm time-warp, leaving behind his female human and chimpanzee friends on Ashlar. When Leo crash lands on what looks like the steps of the Lincoln Memorial in Washington, he finds a statue there of General Thade[32] and a modern ape society, suggesting surreal questions about one of America's greatest political gods. Was it apelike of Lincoln to promote a brutal civil war to hold the United States together, using the antislavery issue as a political tool? Or might such a war have produced, in an alternate universe, a different ape ego as alpha, eventually worshiped like Lincoln in a Greco-Roman temple?

LAB-GUILT AND APE-MESSIAH REVIVALS

The 2001 remake was profitable at the box office, but not as valuable to many critics as the original film.[33] And yet, its ending might be seen as foreshadowing the shift in America's millennial fears toward a certain dark-skinned race and Abrahamic religion, with the terrorist attacks on capitalist and governmental monuments, in New York and Washington, later in the same year. It also led, after a decade of the "War on Terror," to the making of a seventh version in 2011. However, *Rise of the Planet of the Apes* (dir. Rupert Wyatt and scripted by Rick Jaffa and Amanda Silver) focuses more on lab guilt, human aging, ape-ego alienation, socially networked rebellion, and viral contagion. It also uses advanced computer graphics and motion-capture technology to create realistic-looking yet virtual apes performed by humans.[34]

Based somewhat on the fourth film, and on actual experiments in teaching apes human language, this version starts with a baby chimp, named "Caesar," whose mother becomes violent while on doses of an experimental, virus-based drug to increase intelligence, ALZ-112, a potential cure for Alzheimer's disease. After the mother (named "Bright Eyes" like Taylor in Zira's lab) goes wild at a board meeting and is shot to protect the public, the black biotech boss, Steven Jacobs (David Oyelowo), decides to "put down" all the chimps that the corporation owns. But a scientist working in the lab, Will Rodman (James Franco) takes the baby home, rather than letting it be killed. Will lives with an Alzheimer's victim, his father

Charles (John Lithgow), who names the chimp, "Caesar," influenced by Shakespeare's play.[35] (It is not clear whether he also saw the original film series.) Thus, the 2011 version adds a lab-hybrid-brought-home motif to the *Planet of the Apes* precedents, akin to Dren being saved by her scientist parents after the blob creatures go berserk before the public in *Splice*.

The computer graphic imagery (CGI) in this film may increase spectator identification with the chimp, Caesar, and some of the apes he meets later. Instead of watching human actors walking upright in ape masks, as in the original series, or given super-natural running and leaping abilities, as in the 2001 version, viewers see animated yet highly realistic apes with subtle facial expressions and detailed gestures, communicating with the humans and each other nonverbally. Also, most of the chimps here, like in earlier films, have white sclerae in their eyes (which real apes and other mammals do not), making them more humanoid with their gazes, despite the otherwise realistic depiction of their bodies. The film begins with a godlike view of trees and then of wild chimps in Africa, shown in close-up with dark brown sclerae. Suddenly, they are chased by black humans. One chimp is caught and caged. It is driven away in a truck from the rest of its troop. The close-up of its agonized cry zooms into one of its eyes, with a brown sclera, golden iris, and black pupil—and then pulls back showing a brighter yellow iris "with flecks of green" (as Will tells the board meeting) and a white sclera. Thus, the transition to the next scene identifies the movie viewer's gaze with the chimp "Bright Eyes" in a lab cage, who then completes a puzzle in record time, astonishing her handlers. The audience is led from a wild chimp's alienation to a captive chimp's intelligence, from the browns to the whites of the eyes, even though Bright Eyes goes suddenly wild and attacks human until she is shot and killed.

Sympathy is also evoked for the orphaned Caesar (Andy Serkis within the CGI) as a frisky youngster wearing clothes but swinging on things in and around Will's home, and then as an adolescent let loose in the redwoods near San Francisco—with the movie viewer given his treetop perspective, involving *shared* primate spatial circuitry in the parietal lobes and cerebellum. As a baby, youngster, and adolescent, Caesar becomes an oxytocin-inducing object of care for Will, his father, and his new girlfriend, an Indian-American primatologist named Caroline (Freida Pinto). With the ALZ-112 brain boost, inherited through his dead mother's blood when he was in her womb, Caesar soon learns American Sign Language (ASL) and eventually uses it to ask his foster father existential questions about whether he is a pet (when he sees a dog on a collar and leash like him), who his father is, what he is, and what happened to his mother. These expressions of an ape's inner theater, with various stagings of self and other, encourage the movie viewer's cross-species, body-swapping identifications with him.

So does the poignant triangulation of Charles's awakening awareness and Caesar's, through the viral drug given them illegally by Will—in an all-male trinity of (grand)father, young chimp, and scientific progenitor.[36] Will's research product, though leading to the death of Caesar's mother and eleven other chimps, resurrects Charles, nurturing him to a new life, while also making Caesar more human. Women (Charles's nurse and Will's girlfriend) play only minor roles as maternal caregivers. But then a tragic flaw is revealed in the patriarchal hubris of Will's biotech solution. His father slips back toward subhuman dementia as his diseased brain builds resistance to the viral drug. Caesar seems the first to notice this, while seated at the dinner table with Will and his father. The chimp sees that Charles is using the wrong end of the fork to cut his food and gently turns it around in the old man's hand.[37] Thus, while Caesar's human acculturation increases, through family life and the viral drug, Charles's declines. This also reflects an existential anxiety about the aging process that many movie viewers face, for themselves and family members, dramatically accelerated here.

Various aspects of Caesar's development as a human-raised ape relate to an actual experiment in doing this, conducted in the 1960s–70s by psychotherapist Maurice Temerlin, his wife Jane, and their son Steve, with their "chimpanzee daughter" Lucy.[38] Temerlin and his family first adopted a male chimp as an "exotic pet" (to go with a large collection of birds), but he died in a "freak accident," choked while playing with his "security blanket" (2–3). Maurice and Jane, though "grief stricken," immediately adopted a newborn female chimp, taking her away from her captive mother and making her their "child" (3–4). Lucy lived with them for 12 years (as Steve grew from preteen to adulthood), learning over 100 words of ASL.[39] But then she became too physically strong to live in their home, near the University of Oklahoma. She was sent to an ape preserve in Gambia where she gradually became accustomed to living with other rehabilitated chimps, and then was released with them into the wild (through the dedicated work of Janis Carter). But Lucy died there a couple years later.

Caesar's high-energy swinging throughout Will's house and in the redwood forest reflects what Temerlin describes with Lucy being "far more active" than a human infant (8). Like Caesar in the film, Lucy quickly learned to hold her own baby bottle to drink (11). By 15 months of age, she was sitting at the table and using silverware, after watching other family members do so, without being taught (15). Like Caesar with Charles in the fork-turning scene, Lucy would comfort her "mother" Jane when she was in distress, by putting her arm around her, grooming her, or kissing her (165). The chimp was also very protective of Jane. When Maurice argued with his wife, Lucy would attempt to groom her or distract him, by getting him to play, or then bark and threaten him (177). With strangers she became more aggressive, as when children played roughly with Steve, or adults were visiting, because their status was lower to her than that of her family members (176).

Yet once, at the age of nine, with genitals swollen in full estrus, Lucy almost bit her "father" Maurice. He was being affectionate with Jane and they ignored Lucy, so she grabbed a phone and started dialing to annoy them. Maurice took the phone away and cursed her. Then Lucy charged him, screaming, and put her mouth on his arm, but stopped before breaking the skin. Her "scream of anger changed to a scream of terror . . . terrified of her own impulse to bite" (190). Temerlin understood this as an instinctual taboo against attacking a family member, with *shame* expressed afterward, perhaps akin to the incest taboo observed with her and by Jane Goodall with wild chimps (132–35). But Lucy would bite others to protect her family, toys, or territory, as when a housekeeper moved furniture she was sitting on, or simply to set a limit to aggression and establish "dominance" when humans did not (44–45, 63–64).[40]

Likewise, Caesar's identification with Charles, as his grandfather, takes a tragic turn when dementia makes the old man get into an argument with a neighbor, Douglas. Caesar, as full-grown chimp, attacks the neighbor to protect Charles. (Earlier in the film, Douglas threatened the three-year-old Caesar with a baseball bat when he entered his yard.) Caesar leaps on Douglas, beats him with his fists, chases him, leaps on him again, and bites his finger. But the chimp's face then shows *regret and shame*, as he returns to Charles's arms, with other neighbors surrounding them and staring in horror. This may evoke fear and pity (plus other *rasas*) in movie viewers who identify with Caesar's tragic flaw of impulsive, protective rage and with his subsequent alienation from his home community.

As a danger to his human neighbors, Caesar is placed, by court order, in a primate shelter. The viewer's sympathy is again evoked with the chimp's agonized facial expressions and gestures when separated from Will and his girlfriend by a thick glass window in a door. After they leave, Caesar is tricked by a cruel keeper, Dodge (son of the shelter's manager), into going from a large play room, through a wire-mesh tunnel, to a small cage, where he is trapped, while other apes around him start screaming and banging on their cages. Ironically, it is Dodge who then calls the shelter "a madhouse," repeating what Taylor called the lab where he was held captive by apes in the original film. Yet here, traditional assumptions are challenged not with a human hero but with a chimp's perspective as prisoner of cages, apes, and human handlers (like the start of the third film)—inverting the Judeo-Christian view of a God-given right to dominion over all animals.

Repeating the apes' punishment of Taylor for misbehavior, Dodge, as sadistic villain in this film, turns a fire-hose on Caesar, punishing the rebellious chimp for throwing food at him. And yet, in a subsequent scene, movie viewers experience more of the apes' own theater of cruelty, involving domination and scapegoating, when they are released into the common play area where Caesar first experienced the shelter—seen by Will and Caroline as a positive, stimulating environment when they left him there. With a large troop surrounding him, Caesar is attacked by the alpha chimp, Rocket, who tears the newcomer's shirt off, chases him on artificial tree branches, and makes him fall to the ground. As with body-swapping identifications evoked in earlier scenes of the infant Caesar moving dynamically through Will's home, or climbing and swinging through the redwoods at various stages of his maturing form, viewers here might merge their motoric mirror neuron and emotional contagion circuitry with the animated ape's onscreen. But in this scene, they would also experience his tragic fall from the branches, in their inner theaters, extending the virtual uncanniness of the film—with Caesar struggling to stand then, yet falling again, made unconscious by a dart from Dodge's gun.

When Caesar wakes in his cage, he finds a new friend, a circus orangutan, Maurice, who uses ASL gestures to communicate through the cage walls, warning him, according to the subtitles: "Careful. Humans no like smart Ape." Using his intelligence to gain another ally without ASL, Caesar steals a pocketknife from a human visitor, turns it into a key that can unlock his cage at night, and also unlocks the cage of a large gorilla named "Buck." (Lucy was also proficient with stealing and using a key to escape her home enclosure [Temerlin 94].) Caesar then shows Rocket his new ally, Buck, and the former alpha submits with a bow and extended palm.

Movie viewers, having first sympathized with Caesar's loss of his mother as a baby, and then of his human family as a youth, see him mature now from an ape wearing human clothes and alienated from others in the shelter to its new leader—through *body language*. Unlike the earlier *Planet of the Apes* films with "civilized" talking apes, the viewer here becomes immersed in a "primitive" ape culture, with male chimps vying for status, sometimes brutally. Yet Caesar gains the trust of his fellow inmates, against the domineering alpha, through cleverness, kindness, and camaraderie, as they join with him then to escape from their common enemy, the unethical and sadistic human keepers.

Meanwhile, in the human realm, Will fights with his biotech boss, Jacobs, about whether to restart ape trials with an improved version of the viral drug, ALZ-113, in aerosol form, and then how fast to do so. The casting here of a black actor with an African accent as the Gen-Sys CEO evokes not just a "post-racial" worldview (Baum 116), but also the postcolonial role of African rulers and businessmen with difficult decisions about the use or abuse of natural resources, such as primates and other animals. Jacobs is the executive who quickly ordered the killing of a dozen apes at the start of the film, after Bright Eyes went wild. Yet now, after being told how ALZ-112 not only helped Will's father with his disease but increased his intelligence, Jacobs orders Will to resume the trials with the new drug. As he later says, when Will wants to proceed more cautiously, "You make history; I make money." Jacobs also threatens Will with blackmail, knowing that the scientist gave the drug illegally to his father.

However, Will has been influenced by Caroline, who cautioned him that some things are not meant to be controlled or changed, and by Charles, who refused to try the new drug (ALZ-113) and accepted death instead. Will chooses to be wary now about the risk in seeking godlike biomedical power, even as Caesar and his ape followers "rise" in their collective intelligence. Although Will bribes the shelter's manager to release Caesar, the ape leader refuses to go home with his former father. He escapes on his own, steals the viral intelligence spray from Will's home, returns to the shelter, and administers it to the other apes—causing a revolution there while evoking audience *rasas* of courage and awe.

The two heroes, human and ape, make different decisions in their parallel battles. Will quits his job at Gen-Sys, rather than continuing to counsel his boss about the risky drug trials. He also ignores the accidental spread of the airborne viral drug to a chimp handler, which eventually leads to a deadly epidemic in humans by the end of the film. Caesar, on the other hand, shows concern for his fellow apes in a human way, teaching Rocket to share a communal bag of cookies with the others. He tells Maurice that together they can become stronger, demonstrating this with a single stick, easily broken, and then a group of sticks, which are harder to break. This might evoke, for some viewers, the dangers of "fascism" (from the Italian word *fasces* for "bundle," an early Fascist symbol). Caesar then escapes from the shelter and visits Will's home, voyeuristically viewing Will and Caroline asleep together in bed, like the monstrous Ouran visiting Ruth in the first Moreau film. But Caesar does not dwell in family rivalries or perverse pleasures. Instead, he steals the new drug from Will's refrigerator, as a benevolent thief, rebel, and ruler, returning to the shelter to liberate and hyper-evolve his peers.

Viewers are encouraged to sympathize with Caesar, the super-natural chimp, even as he grows more monstrous and fearsome: from the lab where his mother went wild to his home-life mischief and danger to neighbors, to his climbing to the top of the redwood forest (with a view then of the human city), to his caged socialization with other apes, and his rise to alpha power. At home, as a full-grown chimp, Caesar was shown staring at himself in the mirror, focusing on a "birthmark" (as Will called it) on his right shoulder. A similar moment of outer-self awareness (reflecting the LPFC *character* and DMPFC *director* in human viewers) occurs when Will offers to take his adopted son home from the primate shelter. Caesar looks at the other apes in the cages around him and into the eyes of the orangutan Maurice. Then he closes his cage door, identifying more at that point with the nonhuman apes, even across species, than with his family.[41] His alienation in losing his mother (like Caesar in the original film series), in being raised with humans, and in being trapped with hostile apes in the shelter, shifts now to a sense of *purpose* with his distinctive birthmark and monstrous intelligence (reflecting also the viewers' MPFC *actor* and VMPFC *stage manager*). Belonging to his new tribe, with a role to play in leading them toward freedom, means more to Caesar than belonging with a human family.

Strangely, or perhaps to minimize the strangeness of animated apes as lead characters, this film does not show their genitals. It also has Caesar wearing pants in the shelter, even after his shirt is torn off, until he returns there to share the intelligence drug. Female apes are not featured, unlike in previous *Planet of the Apes* films. Even the human females in Will's home life, his girlfriend and his father's nurse, are minor characters and few are shown at Gen-Sys. At the primate shelter, women are not employed, though two are brought there at night by Dodge and his friend, as a way of teasing their dates. But the women soon become scared of the apes, when the men taunt them, and insist on leaving. So this film does not explore gender dynamics like some of its predecessors. Nor are racial issues represented by ape species as distinct classes. Instead, it focuses audience sympathies and fears on the existential struggle of Caesar, as he negotiates survival—if not sexual—situations, involving group status and territoriality, through attraction, domination, and trickery.

Like Lucy and Chantek in their real-life human worlds, and various primates observed in the wild, Caesar tricks his superiors in the shelter. First, he grabs Dodge's male friend by the shirt collar, through his cage, threatening him with violence, but actually stealing his pocketknife.

Next, he fashions a key by tying it onto a stick, unlocking his cage and Buck's. Then, after Buck enjoys some free time in the playroom, Caesar lures Rocket there and makes him submit—involving the animal values of attraction and domination, status and territoriality, freedom and belonging. Later, he refuses to go home with Will, but then goes there himself and steals the ALZ-113. He sprays it onto his fellow apes, without offering them a choice, and they inhale it while sleeping. The next morning, their eyes are green like his and they have white sclerae.

Caesar then lures Dodge into the playroom, after the other apes have returned to their cages. This causes Dodge to call another keeper, Rodney, who had been watching the original *Planet of the Apes* on TV, to go to the observation platform with a tranquilizer gun as his backup. Dodge uses a Taser rod on Caesar but the chimp then grabs his arm. Ironically, the human villain repeats Taylor's heroic phrase from the original film: "Take your stinking paw off me, you damn, dirty ape." In reply, Caesar speaks for the first time, "No," and overpowers the human handler, knocking him unconscious. Caesar then stops the other apes from beating Rodney to death (after he failed to use the tranquilizer gun and left the observation platform). Yet when Dodge revives and approaches the apes with the Taser rod turned on, Caesar shoots him with the water hose, electrocuting him. The surprise on his face shows that Caesar killed Dodge by mistake, but also through the *hubris* of his newfound power. Ironically, Dodge dies by a similar error of hubris with his prior technologies of phallic control (the Taser rod and water hose). Thus, the scene gives a *tragicomic* twist to the death of this villain and to the reaction of his killer, our ape hero, shocked to see the deadly mix of electricity and water.

Some viewers might start to sympathize with the humans as Caesar's super-natural, bright-eyed apes become an alien (or native) horde of invaders, pouring over the hills, from the shelter to the city—in the god's eye view given to the film audience. Caesar's wild tribe then breaks through the glass walls of the Gen-Sys laboratory, smashing equipment and freeing the test apes. Even Jacobs, Will's main antagonist, might become a focus for movie viewers' left-cortical, *critic/scripter* identifications, as he enters the Gen-Sys building and finds no human colleagues, but then stands in an atrium with broken glass and overturned chairs. The apes look down on him from the balconies above, unseen by him at first, yet viewed by the movie audience as a potentially vengeful chorus of right-cortical, *scene-redesigning* passions.

On the other hand, the apes (mostly chimps) might also remind viewers of World Trade Center victims on 9/11 as they leap through the glass of the Gen-Sys building several floors above the ground. Yet these primates land safely, super-naturally, and continue their vengeful rampage. Jacobs escapes from the building, too, and gets into a California Highway Patrol helicopter to rise above the fray, through that super-natural technology, wanting to "kill the leader," as he tells the officers with him. We watch from the ground but also from above, along with Jacobs and the CHP officers in the helicopter, as Buck (the gorilla) bends the fence of the chimp enclosure at the zoo and releases his comrades there. The chimps, as right-cortical *improvisers*, also pull out some of the metal fence posts as spears to carry with them.

They climb up the outside of a building like King Kong, but then appear on the rooftop, like a primal tribe with spears. They crash through office cubicles and chase pedestrians in the street. Maurice and Buck hurl a sewer "manhole" lid and an uprooted parking meter at police cars. The apes also terrorize innocent people in cars stuck in a traffic jam on the Golden Gate Bridge. Such spectacles of primate terrorism provide many points for personal associations in viewers' inner theaters—shifting potential sympathies toward the human victims, with the VMPFC *stage/production manager's* balancing of order and freedom.

The usual melodramatic formula of action movies, with heroes facing more powerful villains, yet managing to win in the end through courage, earlier realizations, and the power of goodness over evil, becomes twisted in *tragicomic* ways during this final battle. The apes, as former sympathetic victims in the Gen-Sys lab and primate shelter prison, led by the heroic Caesar, turn from freedom fighters for animal rights into vengeful terrorists against San Francisco citizens. But as police start to get the upper hand with a blockade on the bridge and charge the apes on horseback, beating them with clubs, viewers' sympathies may *shift again*, stirred by the movie's soundtrack, toward seeing the apes as oppressed victims and heroic rebels.

When a gorilla, presumably Buck, pulls a cop off his horse, dominates him, and raises huge fists to pummel him, Caesar creates the potential for *tragicomic*, VMPFC awareness, with a Brechtian distancing effect. Like the female chimp to Caesar in the fourth film, this Caesar stops the silverback's violence with a vocalized, humanlike "No!" Despite viewers' intuition and motor-mirror neurons being engaged with the apes' pain in being beaten and Buck's courage against the cop, Caesar sets a *limit* to the gorilla's vengeful rage. Buck, teeth bared, just roars at the cop's face.

This film's territorial space race, between city and forest, shows the apes climbing the suspension wires of the Golden Gate bridge while a helicopter cop, encouraged by Jacobs, shoots one off—again reminiscent of the various *King Kong* movies. Yet, Mother Nature aids the apes, with a fog that provides cover for their counterattack at the bridge blockade. Will, the scientist-parent, also tries to mediate, running there to find Caesar. Viewers' sympathies are thus focused on the human scientist, his family-raised chimp, and the chimp's right-cortical rebellion against left-cortical authorities. With Caesar leading the charge on horseback (like a Western hero), the apes attack the CHP and SWAT officers from above and below, as well as on the bridge, overwhelming them, dragging and tossing their bodies, and then celebrating their victory over the empty patrol cars while the cops flee. And yet, Will shouts Caesar's name, perhaps as a superego, VMPFC reminder to him, like his own "No" to Buck earlier, not only about vengeance but also about *hubris* in the bright-eyed apes' apparent victory.

A techno-angel of death appears then, rising over the edge of the bridge, as a helicopter cop sprays bullets at Will and the apes, who take cover behind the abandoned vehicles. Caesar heroically throws a chain at the chopper. Buck goes further, leaping from bridge to helicopter, despite being shot many times. He takes it down King-Kong-like, falling with it to the bridge explosively. Here, the viewer might identify with Buck's super-natural heroics, and his tragic warrior's fall, yet also sympathize with the humans in the helicopter—given their filmic point of view, as the ape leaps at them. But then the camera shifts to outer, overhead and lower angles, offering angelic or godlike views of the crash onto the bridge, as if the movie audience were watching from another, invulnerable helicopter. On the bridge, Caesar pulls Buck from the wreckage and watches his friend die, touching him gently and closing his eyes. He thus shows an advanced human-like awareness of death—again shifting the viewer's perspective, with the face of the virtual ape as a window or mirror between inner theaters.

This leads to a more melodramatic moment of vengeance, unlike Caesar's earlier "No" to Buck (or the prior Caesar's decision with Governor Breck). Here, the ape messiah finds Jacobs, bloody but alive, in the wrecked heli-copter as it teeters at the edge of the bridge. Jacobs pleads for Caesar's help to survive. The viewer knows, even if Caesar does not, that Jacobs ordered his death at the start of the film, along with the deaths of the other lab apes, after his mother went wild under the influence of the drug. Jacobs was thus Caesar's main antagonist throughout the film, though ironically he helped him and the other apes increase their rebellious intel-ligence by pushing for experimentation with ALZ-113. Perhaps Jacobs's name also recalls the biblical Jacob who stole the birthright of his first-born, twin brother Esau. Yet here, the apes steal an Eden-like inheritance back from the humans: a super-natural fruit in their bright-eyed, higher-order knowledge of the potential for good and evil.

As the ape alpha and savior, Caesar could show tragicomic kindness toward his enemy by taking Jacobs's hand and pulling him from his personal fall into the bay below. But instead, he nods to Koba, a bonobo who suffered through multiple experiments at Gen-Sys and prior labs. The grey-eyed Koba gets melodramatic revenge by pushing the chopper off the edge of the bridge, punishing the evil black businessman while showing his own cruel grin.

Taking a police car, Will follows the chimps to the forest, where they swing through the trees. After an angry chimp leaps on Will, knocking him to the ground, Caesar arrives, sending the vengeful chimp away and *this time* extending his hand to help the human survive. Caesar also convinces Will, by saying, "Caesar is home," that he and the other apes have found a better primate shelter and super-natural lab in their new redwoods territory. Sharing Will's smile at this, the audience may cheer the apes' melodramatic victory over corporate bosses, cruel keepers, and city cops. But there is a further *tragic* twist during the film's final credits, setting up the sequel. Ironically, the intelligence-enhancing drug, as the good/evil fruit of biotech progress, which sacrificed the lab apes yet also led to Caesar's knowledge revolution and return to Eden, becomes a deadly virus in its human subjects. It spreads quickly as a worldwide plague through airline travel, shown in branching lines on a global map—as if repeating the tragic fall of our ancestors, after their temptation of godlike knowledge at the biblical tree. This, along with the film's earlier tragic twists, may remind viewers of their own *hubris* in assuming Judeo-Christian or scientific dominion over other apes and the rest of nature, without enough care and wisdom.

Uncannily, the sequel, *Dawn of the Planet of the Apes* (dir. Matt Reeves with the same screenwriters as the prior film, plus Mark Bomback), was released in summer 2014, just a few months before the Ebola outbreaks in West Africa, Europe, and the United States, with a virus carried by humans through airline travel. The sequel begins 10 years after the prior film, in 2026, with a small number of human survivors of the plague living in the ruins of San Francisco and with Caesar's ape colony thriving in the forest across the bridge from them, without their need for technological power. The humans want to fix a hydroelectric dam, as their power supply, but it happens to be in the apes' territory. Thus, the space race of dominion, attraction, and trickery continues in this latest installment of the series, but the planet is our own, with apes as neighbors, not masters, after being freed as lab and zoo specimens, not enslaved workers.

The film initially depicts the apes as tribal natives hunting deer in the woods—having evolved to the stage of humans tens of thousands of years ago, using wooden spears. While the rest of the troop moves on, a bear suddenly pins Caesar's son, Blue Eyes, threatening his life. But his father, with the help of the experimentally scarred bonobo, Koba, returns to save him. Thus, an initial kinship is shown between Caesar, his son, and the more vengeful Koba from the previous film, with further *tragic twists* developing in this sequel, within the *melodramatic thriller* formula.

The apes have also built a village in the woods where the orangutan Maurice teaches the young chimps with images and words (such as "APE NOT KILL APE") on the rock wall behind him—reflecting the prehistoric cave art of humans, as well as earlier films in the series. That village context appears when Caesar (with a John Wayne swagger) returns home to be with his wife while she is giving birth. Apparently, these apes have evolved a monogamous mating system, unlike the various species of wild apes and almost all mammals today (Connor).

Most of this film is about male alpha and territorial rivalries, within and between the ape and human groups, focusing on a leader who wants to build trust between the tribal species (Caesar for the apes and Malcolm for the humans) and a more bellicose leader (Koba for the apes and Dreyfus for the humans). Conflicts are also caused by the human Carver, who panics when found in the apes' territory by Blue Eyes and Ash (Rocket's son). Carver shoots and wounds Ash. But consilience is created by Caesar, recalling the kindness of his human family, lost to him now. He allows the current humans to come into ape territory to work on the hydroelectric generator. Malcolm, his wife Ellie, and their son Alexander form friendships while working with the apes. Ellie also helps Caesar and his sick wife with antibiotics. But Carver threatens their newborn baby with his weapon. And then Koba (with a humpback like Richard III or Thade) gets revenge for Ash being wounded—and for his own maltreatment as a lab test subject years before—by beating Carver to death. Eventually though, Koba kills Ash when he disobeys him, out of sympathy for the humans, evoking an ironic twist of *rasas* in the movie viewer.

Through shifting sympathies, fears, identifications, and projections, the movie audience is positioned as godlike judges to this cinematic experiment with evolving fictional apes. First, the apes' home is shown—with the humans as intruders. Then Caesar and Koba, using sign language and a few verbal words, debate whether to fight the humans or make peace with them. We, the movie viewers, are brought inside ape politics, involving male chimps, the alpha Caesar and loyal Rocket, and their rebellious sons, plus Caesar's advisor, the orangutan Maurice, and the bonobo Koba. All are naked (with white sclerae) but their genitals and sexual activities are never shown.

We then watch as Caesar, with red and white face paint like a native warrior and with white stripes on his chest like a skeleton, leads numerous apes on horseback and on foot, also climbing and swinging through the urban jungle, to confront their enemies in a show of force. Most of the apes are chimpanzees, but a few are gorillas. All, except Maurice, are black. Camera angles show, and give the point of view (POV), of a mostly white human group, crowded behind the walls of their San Francisco fortress, threatened by these American-Indian-like, dark-colored, terrorist invaders. Caesar tells the humans: "Apes do not want war, but will fight if we must." Later, Malcolm reminds him: "War is not what you want. There must be another way." And yet, the mass audience's appetite for violence pushes this film toward bigger and bigger battles between apes and humans—and within their groups.

Dreyfus opens an armory to provide automatic weapons to the humans to defend their urban citadel. When Caesar is accused by Koba of favoring humans over apes, reasonable debate turns to *rage* and the alpha chimp beats up his bonobo rival. But later, Koba tricks two of the humans at the armory, by laughing, tumbling, and drinking with them—then stealing one of their guns and killing them. For the movie viewer, this mixes the *rasas* of humor, fear, and sorrow, with ironic twists as the ape mimics a circus stereotype of the silly chimp, but turns that against the more foolish humans, gaining dominance through trickery and weaponry.

Koba also uses the gun, while hidden, to shoot Caesar and then blames the humans for that assassination (and the burning of the ape village which he actually caused). He takes command after Caesar falls. But the former alpha, apparently killed by his friend like Julius Caesar (*et tu*, Koba?), returns from death's door, aided by his son and their human friends, Malcolm and Ellie, who have medicine. This focuses viewers' sympathies into a melodramatic mix of *rasas*: awe, sorrow, courage, and vengeful rage—aligned with specific ape and human characters, as good guys, using righteous violence to stop or slay the bad guys.

Caesar's son, Blue Eyes, after defecting to Koba's side, sees the evil emerge in the new alpha, especially after he kills Ash. So when Blue Eyes has a chance to kill the defenseless Malcolm, he restrains himself and follows Malcolm to where Caesar is healing in his former family home in San Francisco—evoking viewers' memories of the previous film. Yet in this film, Caesar becomes more of a melodramatic action-hero messiah, getting revenge on Koba, killing him as "not ape," and thus winning the battle within his tribe. Tragically, however, he is not able to save his tribe from further war with the humans. Malcolm informs him that Dreyfus, before he died, contacted another group of survivors, which are on their way to fight (setting up the sequel for 2017). This may also remind viewers of apocalyptic endings in the original series. But the audience is offered a choice then, too, through ironic *rasa* twists and cathartic distancing, not to go that way in real life.

CONCLUSIONS

There are many human and ape relationship movies not considered here, such as the *King Kong* series (1933, 1976, and 2005) and its various sequels, including *Mighty Joe Young* (1949 and 1998). There are other beast-people movies, such as *The Relic* (1997, scripted by two writers who penned the recent *Apes* reboot), with its human-fungus-animal "Kathoga." Or the *Percy Jackson* films (2010 and 2013) with a satyr helper and centaur teacher for the human hero. Or *The Wolverine* (2013) with a "Viper Woman" as antagonist to a were-wolverine as superhero (who has steel claws coming out of his knuckles), in an American and Japanese, medical research and robotics context, or the Wolverine's appearance, too, in other *X-Men* movies, or *Men and Chicken* (2015), a Danish absurdist film with five brothers who find that they were made from various animal elements on their father's farm. *Life of Pi* (2012) deserves even more recognition with its complex, tragicomic twists of survival, competition, and friendship, between an Indian boy and several zoo animals on a life raft, including a tiger—showing various types of animals within human beings.

In this chapter, however, I have focused on lab-created monsters, with the *Island*, *Fly*, and *Splice* films, and on sex-driven transformations and bio-cultural chimeras with the *Cat People* and *Apes* movies (akin to the werewolf and vampire examples in Chapter 4). A scientist who hyper-evolves animals into partly human "children," or transports himself into a human-fly hybrid, or a scientist couple who become parents to an angelic/demonic offspring, suggests horrific warnings about people playing God. But it also invokes the inner theaters of movie viewers, as collective Other to the characters onscreen, to develop new networks balancing the egoistic and libidinal potential of an MPFC *actor*, LPFC awareness of *character*, temporal-lobe *audience* of salient meanings, and limbic and brainstem *stagehands* (reflected together in the hybrid beast-person)—with DMPFC *director*, VMPFC *stage manager*, and rVLPFC *operator* as superego (shown in the beast-person, too, but more so by the scientist-parent).

Likewise, films about lusty were-panthers or about nonhuman apes as scientists, politicians, warriors, and religious leaders, with alternative cultures and super-natural powers, engage the viewers' left-cortical *critic/scripter* and right-cortical *mime improviser/scene designer* at different points in the mythic, sci-fi transformations. The initial *Apes* films reflected Cold War fears of nuclear annihilation and Civil Rights hopes about evolving new perspectives for our tribal passions, along with ideological revolutions of race, class, and gender. The 2001 movie continued such concerns into a new millennium with postmodern wit. The more recent *Apes* films combined these issues, from earlier in the series, with the scientist-god problem from lab-monster movies: first in a male-parenting, tragic aging, and big pharma context, and then in a post-apocalyptic, terrorist-clan conflict. And yet, by presenting computer-animated simians with human eyes, the last two films also reflect the mass audience's immersion today in virtual, social-networking realms, with temptations of melodramatic good-versus-evil violence and illusions of videogame immortality—involving our *brain's* motion capture (mirror neuron) and body imaging (or swapping) *technology*.[42] The final chapter will reconsider the bio-cultural evolution of that inner theater, as expressed by the films discussed earlier, with comparisons to prehistoric cave art, especially regarding our animal-human instincts, social-ordering morals, and prodigious playfulness.

Chapter 7

Morals of the Tale Still in Play

We began this journey through inner and outer theaters—of the brain, prehistoric caves, and modern screen monsters—with the question of human wildness, of what "drives" us to transform our environment, ourselves, and each other in hypertheatrical, good and evil ways, far beyond the evolved instincts of other animals, in many cultural varieties. Such differences may be consolidating now, through global travel and shared media spaces. Yet current technologies also expand our species' transformations, as we become godlike in our screen imaginings, though often without the wisdom or consensus to manage the real-life changes that we make. This final chapter will summarize what came before it and then consider how our species' super-natural wildness and yet moral (or *rasa*-cathartic) awareness might be evolving today, through inner and outer theaters, inherited from the past.

Popular media manifestations of vampires, werewolves, lab hybrids, and simian civilizations may seem to be frivolous entertainments. But such films reflect our still evolving animal-to-human awareness, in the particular decades when these movies were made and in the brains of current viewers. So does Herzog's recent documentary of the Chauvet cave, with its various animal figures, including a bull-head, lion-head, human-vagina-with-legs chimera on one stalactite. So, too, do images of animals, abstract signs, hand prints, and animal-human hybrids in many other caves, painted with charcoal or ochre and etched on the walls of *selected theatrical spaces*, by brains that were structured like ours, tens of thousands of years ago. The exploration of such cave and film details in Chapters 2, 4, and 6 provides a bridge from prehistoric spaces and images to current movie, home, and handheld theaters, to glimpse what our "inner theaters" may share, then and now.

You might believe in a cosmic theater of gods or one God framing our bio-cultural evolution. Or you might see that process as existing without a metaphysical spectator—evolving toward higher levels of *consciousness*, through each human being and perhaps through life on other planets, or just evolving here as a survival and reproduction trick. But through any of these lenses, a common "human nature" can be glimpsed in the super-natural figures on cave walls and movie screens, developing its wild impulses and moral restraints in diverse cultural ways, through and beyond the instincts of our animal relatives. Our hyper-theatrical consciousness, playing with remnant instincts to form new moral worlds and creative environments, on earth and onscreen, is our greatest gift and burden. It makes survival for most of us much easier than for wild animals, yet it makes some of us *wilder* in our social conflicts, emotional impulses, and metaphysical needs for purpose and meaning beyond just surviving. Thus, we reimagine lost friends and pleasures, foresee our own aging and demise, but also hope, fight, and sacrifice ourselves for more that might come from our lives, like the beast-people and their slayers or creators onscreen.

ODD AND EVEN

Chapters 1, 3, and 5 combined neuroscience research and psychoanalytic theories with movie-theatrical paradigms to create a model of inner performance elements, regarding the artifacts on cave walls and cinema screens, explored in the even-numbered chapters that followed. Chapter 1 considered the "staging of consciousness," according to cognitive psychologist Bernard Baars, along with theater metaphors (such as "puppeteer") and other terms from neuroscientist Antonio Damasio, plus related research. It refined Baars's reference to Paul MacLean's evolutionary model of the brain's animal levels: vertebrate/fish brainstem (a.k.a. "reptilian") areas with basic drives, a mammalian limbic system for emotions, primate fronto-parietal lobes for Self/Other and spatial orientations, and advanced human prefrontal cortices with distinctive right and left, Imaginary and Symbolic functions (adding Lacan's terms). Such levels relate to Damasio's definitions of a proto-self (assessing internal bodily states) and a core self (interacting with objects), which we share with other vertebrates, especially through mammalian emotions and social systems, plus an autobiographical Self with higher orders of hyper-theatricality, which is uniquely human.

The staging of consciousness in our brains involves an inner theater of animal legacies, with the bottom-up and top-down feedback loops of survival and reproduction values, Deep Goal and Conceptual Contexts, immediate intentions and expectations, and executive controls, as the "director" and "operators" of a largely unconscious self. Baars relates his model to Gerald Edelman's theory of "Neural Darwinism" within the brain, interacting with the outer theaters of "experiential selection" (Edelman, *Second* 27–30). Baars's model also includes "audience" members lobbying (or Web users "liking") through automatic, implicit, verbal, declarative, and autobiographical memories. Such neural networks cooperate and compete for which perceptions and ideas appear as "players" in the "spotlight" of the mind and on its "stage" of working memory. I also mentioned Politser's actor-critic-audience model for neuro-economic pleasure, learning, and liking, as evaluated by the brain.

There are cinematic elements, too, of the brain's inner theater (as Kosslyn and Zacks point out). The mind's eye zooms in and out, rotates, or pans across imagined or remembered objects and scenes. There are "flashbacks" of memory associations, variable frames, and temporal and spatial chunks in the brain's "filling in" of perceptions, involving past and future possibilities. There is a backdating and rearranging of the present continuum like the "cuts" in film editing. And there is an internal monolog in our brains like a film "voice-over," involving Self and Other, left- and right-cortical networks. When we go to the movies, or watch films on personal devices, we are sharing in various kinds of dreamlike, memory-like, and reality-like performance spaces. As with dreams (in Revonsuo's theory), we use movies to "rehearse" our animal drives, emotions, and actions, as well as specific fantasies and realities—from remnant instincts and daily experiences to long-term cultural patterns, with contending social rules and bonds. Between brains, inner director/operators and audience members signal (mostly unconsciously) through mirror-neuron mimetics, emotional contagion, and Imaginary and Symbolic networks. Each person's circuits are thus *pruned* in unique ways, especially through childhood experiences,[1] with influences from family members, peer groups, school, work, and media environments.

Nine neuro-aesthetic "laws," as proposed by V. S. Ramachandran, were considered in Chapter 1, along with the nine *rasas* (refined emotional flavors) of ancient Indian theater theory.[2] Through these ideas and Aristotle's ancient notion of the purifying of emotions in reading or watching Greek tragedy, I developed a Western and Eastern flavored, neuro-aesthetic theory of "*rasa*-catharsis," also involving the modern aims of Bertolt Brecht and Antonin Artaud. Next, I considered the primary and social emotions defined by current neuroscience, including panic, fear, rage, lust, seeking, care, sadness, and joyful play in other mammals, as researched by Jaak

Panksepp. Antonio Damasio gives a similar list of primal emotions: joy, surprise, disgust, sadness, fear, and rage. To this, he adds a list of social emotions: sympathy, pride, indignation, shame, envy, guilt, and admiration. Jonathan Haidt offers "families" of emotions to point to universal moral foundations. There are parallels, too, between the animal-to-human emotions in Panksepp's, Damasio's, or Haidt's list and the aesthetic feelings involved in ancient Greek and Indian theater traditions (Table 1.1).

Regarding such developments, from primal to complex, animal to human and moral emotions, I considered prehistoric cultural changes, from episodic to mimetic, mythic, and theoretic stages (according to Merlin Donald). This also relates to right and left cortical, Eastern and Western orientations in humans today. There is increased stress in socialized self-control from primates to humans, with herd mentalities and alpha competitions, especially through our dreamlike media and "consumer" demands, far beyond basic food and water, as reflected by the voracious *beast-people onscreen*. Yet I also described the potential for *rasa*-catharsis, as a cognitive remastering and reappraisal of primal and social feelings, with tragicomic twists at certain moments in the film examples of Chapters 4 and 6.

Chapter 2 explored prehistoric cave art as showing an early development of human Self and Other awareness, from the brain's inner theater to its shared performances. Whether in narrow passageways, possibly as painful tests of shamanic initiates, or in large chambers with different acoustic effects for the collective audience, the cave art shows *projections* from Paleolithic brains. Perhaps these were hallucinations seen by the artists on the rock wall, and then recorded there, which also became for others, in the flickering firelight, *moving pictures*. Even for us today, the animal images, especially when combined with human features, handprints, and abstract signs, reveal a growing awareness of human nature as evolving culturally, then and now, super-naturally akin to, yet moving beyond other animals.

Chapter 3 related the animal-human "staging of consciousness" model from Chapter 1, plus its prehistoric representations in Chapter 2, to the "hall of mirrors" between brains: the multiple reflections and projections of Self and Other, of individual and group identities, revealed by cognitive experiments and neural mappings. This involves the human infant's mirror recognition of Self, in the first two years of life, which appears only in a few, highly intelligent, and very social mammal species, such as elephants, dolphins, and apes. These animals also have Von Economo spindle cells or "intuitive social decision making" neurons, but humans have them in much larger numbers (Gazzaniga, *Who's* 39–40).

Prior to the full awareness, at about 18 months, of being distinct from, while also mirrored by the primary caretaker, the human infant lives in a self-other *continuum* more like other primates, such as baboons. According to psychoanalytic theory, this Real order of natural being with the (m)Other is lost as the child becomes shaped by specific Imaginary and Symbolic experiences. Yet it persists as a "lack of being" and "want to be," with particular fantasies and desired objects of Oedipal sacrifice and maternal reunion.

The movie screen, as window and mirror, may encourage pleasurable fantasies of fetishistic, voyeuristic, or sadomasochistic reunion, through alluring fictional objects—activating the spectator's inner theater while the body is in dreamlike immobility within a darkened space. *Melodramatic* violence with good heroes and innocent victims fighting, fleeing, getting revenge against, or "slaying" evil villains may consolidate stereotypes, along with erotic objectifications, affecting the neural circuits, social networks, and real-life acts of audience members beyond the screen. This may involve the danger of "cathartic backfire" with aggression increased and stereotypes reconfirmed.

And yet, even within this general formula of revenge melodrama, of courageous heroes and evil villains, some films may have more complex effects, especially in the genres of romantic horror, monster comedy, and sci-fi. Viewers' critical awareness of their own primal emotions sometimes increases during vampire, werewolf, lab-hybrid, or simian-civilization movies, through a paradoxical mix of desire and fear, attraction and repulsion. *Tragicomic* twists in such entertainments might evoke a *rasa*-catharsis of various feelings, with shifts between Artaudian immersion and Brechtian distance. This cognitive reappraisal of one's bottom-up, impulsive, animalistic emotions, through a greater top-down awareness, also elicits a questioning of binary, objectified stereotypes—of certain groups as primal others, as more animal than human—in the hall of mirrors between brains during the film and in everyday life.

Human *group* identifications begin around age three, with in- and out-group *projections*, involving more empathy toward those "like" us. Understanding that others have different perspectives starts at age three to four, and then perspective taking about others' false beliefs at age four to six. Parents and teachers, along with peer groups, influence the child's performances and neural development with their expectations—as indicated with lab rats performing differently when handled by humans who assume they are smart or not.

This gives key evidence for the significance of the teacher's role in what I call "classroom catharsis." When teaching any subject, but especially theater, film, or media studies, a teacher may increase the students' awareness of their inner-theater drives and emotions—their mindfulness regarding how the material affected them and what it reflects in our society—through what is selected, which questions are asked, and which exercises are given to "play" with it. The teacher might also reappraise her own cognitive and emotional assumptions, including unconscious expectations for her students, regarding gender, race, class, and other attributes.

Tests with rats and monkeys show that emotional contagion, empathic helpfulness, and self-sacrifice are part of our mammalian heritage. Fairness expectations have been found with wolves, coyotes, dogs, ravens, and monkeys—and in 15-month-old humans. Vengeance has also been found with chimpanzees. Yet reward and pleasure networks in the human brain are activated with altruistic anger and the righteous punishment of others—suggesting how mammalian empathy and primate fairness may become *vindictive scapegoating* in our species, especially with contagious group identifications and binary projections (of "us versus them" or "us versus that one"). This can be *rationalized* by our objectifying left cortex, like the shift from social norms of generosity (with more right-cortical caring) to market norms of fair trade.

Levels of *cooperation* appear in our animal relatives: (1) emotional contagion and motor mimicry, in bird flocks, mammal herds, and human mobs, (2) the coordination of shared goals in wolf/dog packs and ape troops, (3) active consolation for an injured colleague in apes, and (4) perspective taking, targeted helping, intentional imitation, and various kinds of altruism in apes (with targeted helping also in dolphins, whales, and elephants). These levels might be taken as a natural foundation for ethics in humans. But there is a dark side to animal-human group cooperation. It increases with outside threats, sometimes projected onto others (as in genocidal scapegoating), and it involves *competitive* drives within as well as between groups (as with the genetic logic of infanticide in various mammal species). Regarding the brain's inner theater, this also includes left-cortical Self distinctions competing with right-cortical self-other relations.

But the "hall of mirrors" effect, between our hyper-theatrical brains, evoking uncertainties and dangers, has positive as well as negative aspects, reflecting our super-natural heritage of youthful creativity. Human *neoteny* extends far into adulthood: our youthful appearance as apes with rounded heads and flat faces, our promiscuous sexuality (more like bonobos than chimps), our creative flexibility from art to technology, and our playful behaviors, even with organized games, sports, and dramatic media. This also relates to a double edged, reflective mirroring with romantic, communal *bonding*, which involves the neurotransmitter oxytocin, plus an opioid rush of pain reduction, bringing pleasure in togetherness, yet also *mimetic rivalry*—as we mature from primal attachments to new mates and group identifications.

Personal degrees of communal engagement—or of ego defensiveness against the mirroring influences of others—stem from an individual's "attachment style," which may be passed unconsciously from mother to child. Yet *group egos* can also be defensive, with positive in-group bonding but also negative out-group projections, often (hypocritically) of odious, yet repressed traits within one's own group. This involves *overgeneralizing* similarities within the group against the dissimilarities of others, producing "empathy gaps" and stereotypes (Van Boven and Loewenstein; Ames). The likelihood of such projections in everyday life makes even more meaningful how movies and plays may rehearse *effortful perspective taking* by the audience, especially with tragicomic, *rasa*-cathartic twists onscreen or onstage.

"Machiavellian intelligence" is another aspect of the super-natural hall of mirrors between our brain theaters, regarding group hierarchies. With primates in the wild, clever individuals find sneaky ways to mate or feed by deceiving their higher ups, signaling a false threat or not reacting to newly found food. In humans, such functional deception evolved into intentional deception and even *self-deception* to prevent the unconscious "leak" of a disadvantageous signal through motor and emotional contagion (Trivers). This uncertainty of self-deception relates, in my view, to the powerful hyper-theatricality of humans. We play many roles in our social networks, using various characterizations of self and other from our inner theaters, continually readjusted by the hall of mirrors in others' reactions to us. Research has found signs of deceit in children as young as two and half, involving right frontal cortex activity (with the inner *mime improviser*), although in a performative, "as if" mode, rather than intentional deception, which comes later. Thus, our playfulness with as-if scenarios on stages and screens throughout our lives, involving imaginary objects, becomes distinct from the functional Machiavellian deception of other apes.

Primate troops also show aspects of "social proteanism," relating to trickery in our human groups. The alpha male might punish whimsically, creating uncertainty and obedience among subordinates. Likewise, females in various primate species use protean promiscuity or concealed ovulation, to keep males guessing as to whether an offspring is theirs. This reduces infanticide by males and increases their desire for long-term pair bonding. In our evolution from animal to human, females developed both promiscuity and hidden ovulation, unlike most primates. Thus, loyalty to mates and to larger group hierarchies increased in human evolution, as our social networks evolved in size and flexibility far beyond those of other apes. The competition for alpha power by selfish egos and the lack of work by free loaders, disruptive of group cohesion, was counterbalanced by external conflicts, with *altruistic* groups surviving beyond others that had less unity. Today, we use "social categories" to define our remnant instincts of survival, reproduction, and territory, often in binary stereotypes of "you're with us or against us." Yet this also involves a dynamic continuum, fueled by testosterone aggression (more so in males) and oxytocin bonding pleasures (in romantic pairs and with group camaraderie), through protean, transformative mirroring.

People can be *primed* unconsciously with social categories like "elder" and "professor," so that they walk more slowly or perform better on tests. Social psychologists have found not just a "looking-glass" effect of others used as mirrors for the Self, but also a "spyglass" effect of *merging* the self with the other through "behavior matching." This extends kinship identification to larger, sometimes contending groups—through *mimetic* desires and rivalries. Thus, humans *ape* even more than other apes. Children over-imitate adults, copying unnecessary steps, whereas chimps emulate behavior, finding shortcuts to a goal. Starting at nine months, children share *joint attention* with adults through *gaze following*, using the whites of our eyes, the distinct sclera that is lacking in other animals. Such a mimetic mechanism in us, extending from our mammalian ancestors, also involves oxytocin—as a "tend and befriend" hormone in females and a "tend and defend" neurotransmitter in males, which increases their ability to read subtle social cues, but also their ethnocentrism. Mimetic desire in the hall of mirrors between our brains, while fostering care bonds within a family or extensions of kinship to ethnicity, nationality, and religion, can also lead to rivalry and *reciprocal vengeance* between individuals and groups.

Oxytocin is involved with the alternatives of "liking" or disgust, mediated by the insula and VMPFC (as superego or inner stage manager), showing the relationship of moral and physical feelings, with personal and group attraction or repulsion. We *like* those we see as *like* us and reject those who are not, often projecting negative feelings and stereotypes onto those others. However, "wanting" involves dopamine reward circuits, which can become addictively attached to certain people, objects, or rituals—with compulsive, repeated actions even when *disliked*, as with mimetic rivalry and reciprocal vengeance.

Humans inherit powerful grouping mechanisms of herd contagion, kinship bonds, and pack/troop (seeking and hierarchy) coordination. Our inner theaters absorb outer reflections of contending identity models, categorical projections, social rules, and nurturing networks. But we also experience profound *alienation*—when misrecognized or rejected in the hall of mirrors between our brains. Chimpanzee group members groom each other for much of the day, negotiating favors in the social hierarchy that also involve food and sex, as nurturing bonds. Yet sometimes, they go on border patrols to capture a female or to chase, beat to death, and even eat the flesh and drink the blood of a lone chimp who might have been a former kin (as they also do when hunting monkeys).[3] Humans have done far worse, however, when inspired by the "Social Darwinism" of racism, rationalizing the survival of the *cruelest* as the fittest, as a super-natural Übermensch beyond morality. This is a potential in all of us, individually and collectively, reflected by the movie screen's undead vampires and feral werewolves, as lone shape shifters or in clans and packs.

Chapter 4 explored the development of vampires and werewolves across cinema history, with references also to earlier folktales. Nosferatu, Count Dracula, the Werewolf of London, and the Wolf Man, as foreign or exotically infected adults of the 1920s–40s, eventually turned into local teenage vampires and werewolves, female as well as male, in the 1950s. (An Eastern European Dracula also appeared in a series of British films, produced by Hammer Pictures from 1958 to 1974.) Such lone shape shifters returned to the screen in the 1980s as students in high school and college or travelers infected in Europe, again showing the desires and fears of alienated, tragicomic, hormonally charged youths. Black vampires appeared onscreen in the 1970s and 1990s (also with a black vampire slayer in a series that continued into the next decade). Homoerotic vampire couples and families, from the Old and New Worlds, with theatrical and musical groups, appeared in the 1980s–2000s. A high-kicking *female slayer* of vampires, with her teacher, crew, and tragicomic twists, came to movie and TV screens in the 1990s. Faux historical figures (Jesus Christ and Abraham Lincoln) became vampire hunters, too, in the new millennium. The lonely Romanian vampire, Vlad the Impaler, was also revived in 1992 and 2014. A lone werewolf appeared again, in today's work world or an older setting, in 1994 and 2010. A lone girl's menstruation also made her lycanthropic in 2000 and 2004. Yet clans of vampires and packs of werewolves became increasingly popular since the 1980s, with the *Underworld* and *Twilight* series having both, as enemies or allies (drawing on their preexisting videogame and young adult novel fans).

Thus, the animal-to-human staging of consciousness within us and the hall of mirrors between our brains (and groups) became reflected in various transformations onscreen—with vampires as undead, bloodsucking, yet charmingly ideal egos and with werewolves as moon-struck, hairy, fanged, and savagely tormented ids. Across the decades they also expressed various modern and postmodern fears, involving superego clashes in the mass movie audience, through conflicts of culture, sex, race, and age identifications: as foreign gentlemen and exotic ladies, homoerotic families, black criminals, rebellious teens, or terrorist tribes. They showed the primal drives of dominance, attraction, and trickery—behind our insatiable desires, social demands, and yearning for a sense of destiny, alone and in groups. Especially when these monsters are portrayed sympathetically, along with their "chosen" slayers, through ironic twists onscreen, a catharsis of particular *rasas* might occur in viewers' inner theaters. Beyond fear, contagious panic, vengeful rage, and lustful seeking, a cognitive reappraisal can be evoked, with intimacy and yet distance, refining such feelings into *new perspectives*. But that also depends on how we watch.

Chapter 5 explored how the hall of mirrors between our brains, exemplified by such films with undead vampires disappearing in mirrors and werewolves changing in the moonlight, involves fun-house-like distortions of the body, displacements of identity, and illusions of immortality exposed by current research on real-life subjects (with rubber hand, enfacement, and head-mounted video experiments). This also reflects our hyper-theatrical "ape ego," developing through the stages of human childhood with shifting growth spurts and distinctive functions in the right and left hemispheres. Chapter 5 thus set up the investigation of lab-created hybrids and simian civilizations in the next chapter on movies. It also developed further details of the brain's inner theater with various metaphors that relate to the outer theaters of stage, screen, and everyday life.

First, the "gene-based" logic of infanticide, territoriality, and promiscuity was explored in various mammal species, regarding our biocultural evolution of cuteness nurturing, aggressive/defensive, and lustful passions. Next, the mirror awareness of one's own face and body by most simians was related to its gradual development in the human child—toward social emotions, Self memories, and theories of the other's mind (involving "mind games" that have also been observed in wild chimps). Growth spurts occur in right and left cortices: the right in the first 18 months, then the left until age 3, the right again until 5, the left until 7, the right until 11, and the left again after that, though to a lesser degree with each swing (Cozolino, *Human* 70–71). This gives evidence for specific functional shifts, as they develop in the human child and become aspects of our inner theater throughout life. Especially the greater growth spurts in the early years reflect the development of Imaginary, Self-Other awareness functions in the right cortex and then Symbolic, verbal, social identity, and self-deception functions in the left—building to the "terrible twos" of the human toddler's ape ego.

Even 1-year-old humans have an "intentionality detector" about others' desires and goals, like other primates. At 15 months, babies expect fairness in food distribution and demonstrate altruistic sharing, perhaps with reciprocal expectations, again like other primates. But in humans this shows a further development of *left*-cortical, prosocial networks, along with self-assertiveness toward age 2. Self as actor/character becomes redefined by the 2-year-old, who tests the theatrical frame of one's double in the mirror, challenging others' desires and demands with a repeated "No" and temper tantrums. Yet, children at that age start to console others in pain, as seen also with some monkeys and apes. Children at age 2 enjoy "pretend play," showing a sense of others' beliefs, which relates to the Machiavellian deceptions performed by primates in the wild.

At age 2½, with a shift toward *right*-cortical growth, children become active deceivers (also showing activity in the right frontal cortex). At age 3, they create as-if scenarios, asking others to join in, and form in- and out-groups, with more empathy toward those like them (even with just the same color of T-shirt). At that age, the child's impulse control can be increased with Symbolic abstraction (using numbers for candies in a bowl), as shown also in tests with chimps. Yet at age 4, children have a more holistic, right-cortical awareness, a "lantern" of attention rather than a spotlight, more like that demonstrated by chimps who can recollect a pattern of numbers on a checkerboard better than adult humans, using eidetic (photographic) memory. At that age, children also form "best friends" with real and imaginary playmates, while interpreting personality traits in tests with puppet animals.

As early as 18 months, but increasingly toward age 5, children overimitate adults in tests of following a series of ritual actions to get a prize—*unlike* chimps who emulate the goal instead and discover a quicker route to the treat. Here we see a crucial difference in the super-natural ape ego of the developing child: a mimetic inclination, along with symbolic powers of language and abstraction, which greatly increases the human learning of cultural behaviors and thus the *flexibility* of our species in adapting to and transforming new environments. Yet, the protean mimicry of children, upgrading to new levels of creativity, faces a challenge in the teen years.

Hormonal passions evoke primal emotions in the young animal-human, like in the werewolf onscreen, just when the social rules and roles of family and school may come into conflict with peer group pressures. Media ideals of muscular or curvaceous sexiness challenge the teen's changing self and body images. Oddly like the testosterone-driven *female hyena*, with a strangely enlarged clitoris that looks like a penis, young human males often exhibit physical aggression to handle confusing changes in gendered identity, while females specialize in psychosocial conflicts. (The hyena-swine-man character, as leader of the tragic rebellion against Dr. Moreau in the 1977 film, expresses both the physical and psychosocial drive for meaningful identity in human adolescents.) Unlike most mammals, the sexual receptiveness of human females, through estrogen and progesterone, also involves testosterone, as if with a degree of male urgency. Thus, the adolescent girl is akin to the hyena in her mix of hormones fueling different passions. Yet testosterone, with its gendered receptors in the human brain, still distinguishes males from females. Men have a more continual sex drive while women are cyclical and mysterious in theirs, especially with hidden ovulation.

Vasopressin, in humans and other animals, also shifts the brain/body toward more *masculine* pushiness and competitiveness, while oxytocin evokes more *feminine* tending and befriending—even with today's changing cultural interpretations of such identities. Indeed, oxytocin fuels nurturing and protective behaviors in *both* genders, along with confidence, trust, parental play, gloating, and the ability to interpret facial expressions. Thus, testosterone and vasopressin help viewers, both male and female, to identify with masculine warriors onscreen, even in ape-planet movies. But oxytocin helps connect viewers to more feminine characters and psychosocial scenes, which also occur in such films, especially with the female chimp doctor in the original series and the male parenting of an infant chimp in the 2011 version. (The female-dominant, alternatively sexual culture of bonobos has not been shown in the series, even with a nominal character of that species in the 2011 and 2014 versions.)

Brain mapping studies with male college students who were shown *erotic* film clips revealed arousal activity in limbic and subcortical areas, especially on the right side (Beauregard et al.). But when these males were asked to become detached observers of such erotic scenes, their right lateral prefrontal cortex became active, an area that I describe as the "inner character" of the brain's theater, because other studies also show that it is involved with mirror self-awareness (on both sides). Hence, awareness of one's appearance to others *and* of one's primal sex drive activates the lateral PFC, especially on the right side. In another study involving female viewers and *sad* film clips, the right ventromedial PFC (the "inner stage manager" in my model) also became involved, along with the lateral PFC, in the detached observation of one's own mimetic feelings. A similar study with *children* showed five times more activation in the LPFC on both sides, suggesting more difficulty for them in detaching from mimetic feelings while watching tragedy onscreen. Teenagers also lack the fully insulated wiring of emotional detachment and self-control networks in the PFC, which develops later in life. This shows the importance for growing brains of mindfulness training in general and of particular *rasa*-refining films, as well as the danger of cathartic backfire with melodramatic action movies (or videogames), shaping the viewer's sexual and sadness circuitry.

Like but far beyond other mammals and apes, humans have an advanced "mentalizing system" for theatrical perspective taking, in becoming aware of *others' viewpoints* (Matthew Lieberman). I call it the "inner director" because the director's role in staging a play is to give the actors, designers, and technicians a sense of cohesive aptness—as the creative Other watching their performance choices, prior to the audience becoming involved. This brain network includes hubs in the dorsomedial prefrontal cortex (DMPFC), temporal parietal junction (TPJ), precuneus (PC), and posterior cingulate cortex (PCC). It is also called the "default network," active when the mind is *not* focused on a particular task, which shows its importance for our social survival, imagining how others might view us and everything else, like and unlike our own view. Additionally, the brain's septal area, deep in the limbic system, with many oxytocin receptors and direct ties to the DMPFC, shows a relation between personal pleasure and *empathy* in helping others, through understanding the other's viewpoint.

However, the medial prefrontal cortex or MPFC (between the DMPFC and VMPFC) is a distinctive hub for another network, involved with *self-knowing* based on the prior influence of others. I call this network the "inner actor" because an actor onstage, like a person in everyday life, has a sense of self from past experiences that is distinct from the character being performed at a certain moment. (This is related to Baars's metaphor for the staging of consciousness, but that involves *any* percept/concept becoming a *player* in the spotlight of awareness, which might include the inner actor.) In a study showing erotic film clips, as mentioned earlier, the MPFC and LPFC were active on the *right* side when subjects imagined a detached view, distancing themselves from the arousal. But in another study, with subjects talking themselves into a distanced viewpoint of aversive pictures, the *left* LPFC was active. This shows how the right hemisphere more generally, as "mime improviser/scene designer" in my model, connects with the inner character and actor (LPFC and MPFC) in imagining a detached perspective for oneself, while the left hemisphere as "critic/scripter" ties with the inner character (LPFC) in verbalizing such detachment. But there are also many other functions of the right and left hemispheres, taken from Ian McGilchrist's meta-analysis, which relate to these terms in my model, as listed in Table 5.2 in Chapter 5.

Another study on the cognitive reappraisal of *aversive* (disgusting) photos showed the DMPFC as active (Beauregard et al.), thus relating to the mentalizing or default system as "inner director" in my model, with subjects reflecting on their feelings through the Other within. A study involving young girls showed the MPFC, LPFC, and VMPFC as active when they distanced themselves from *sad* film clips, observing their own reactions. This involvement of the inner actor, character, and stage manager also relates to the tragicomic twists explored in lab-hybrid and ape-planet movies in Chapter 6, which may distance the viewer toward cognitive reappraisal. Other studies show that the *right* VMPFC (or OFC) integrates memory, attachment, emotion, bodily regulation, and social cognition (Siegel, *Developing* 40, 303)—as inner stage manager, I would add, especially in its Imaginary (right cortical) dimension, involving the inner mime

Table 7.1

Summary of Inner Theater/Cinema Elements (simplifying Table 5.1)

Performance Element	Cognitive/Affective Function	Neurological Hub
Applying both theater and filmmaking terms to Self/Other brain functions and areas, through Lieberman's and Zacks's neuroscience research (as well as Baars's cognitive model):		
actor	inner self, yet influenced by others as mask/prism	MPFC network
director	mentalizing system, imagining others' views of oneself	DMPFC network
stage/production manager and cinematographer	moral and goal behavior monitor	VMPFC/OFC network
character	outer self, recognized in mirror	LPFC network
stage itself and film editor	working (short term) memory, rational attention, time chunking, and inhibiting of self-interest	DLPFC network
light/sound operator and filmmaking grip	controller of impulses and attitudes	rVLPFC network
audience	language, movement, emotional color, intuitional memories	temporal lobes and insula

Table 7.1 (Continued)

Performance Element	Cognitive/Affective Function	Neurological Hub
More generally, with McGilchrist's meta-analysis & Panksepp's research:		
critic/scripter	abstract rules and verbal focus	left cortex (senses/actions on right side)
mime improviser and scene designer	holistic, contextual, visual, spatial, & tonal openness to new percepts	right cortex (left side of body, stronger ties to limbic system and brainstem areas)
stagehands	social emotions and pleasure/pain of primary emotions (seeking, rage, fear, panic, care, and play), plus survival/reproduction drives	limbic system (including temporal lobes) and brainstem areas

improviser/scene designer. Also, the MPFC, with its ties to the VMPFC, creates a meta-cognitive representation of one's feelings, as inner actor, regarding social and moral behavior, including others' views. (See Table 7.1.)

Matthew Lieberman's studies show that teenagers, more than adults, use the DMPFC mentalizing system (inner director), along with the MPFC network (inner actor), when reflecting on the Self, showing the importance of peer and authority pressures on the developing brain. Lieberman even argues (and I would agree) that despite the Western stress on an independent, True Self, as inner self-knowing actor, such an identity *masks* a "social construction" developing especially in our childhood and teenage years through others' influences. As I added in Chapter 5, the masks of the inner actor are *prisms*, absorbing conscious and unconscious scripts and images from others, for self-knowing, self-reflections, and mentalizing projections upon others—in the hall of mirrors between us. There are many potential selves (and ghostly influences of others) within each of us, as beast-people at various levels of our brain, from natural to virtual, tied to our past, current, and projected social networks.

Lieberman finds a shift from left to right VLPFC activity in the "reappraisal" of emotions, involving *calmness and mental clarity*, plus a diminished amygdala (fear/rage) response. He contrasts this with the "suppression" of emotions, in distracting the brain to avoid feeling and looking distressed, which involves a later VLPFC effort in restraining one's facial expressions. One might relate this distinction to the Freudian notion of "working through" difficult emotional memories, rather than just "repressing" them. My theory of Artaudian/Brechtian *rasa*-catharsis also benefits from these distinctions.

Initially, the viewer's *right*-cortical (Imaginary) mime improviser/scene designer becomes *immersed* in a scene, with Artaudian openness to its emotions while identifying with the main character (in alliance and allegiance), extending what is perceived onstage or within the film frame to larger contexts through personal associations. Then, an ironic, tragicomic twist in the scene (through plot and/or presentation), or a choice by the viewer in how to see and feel it, triggers a shift toward the *left*-cortical (Symbolic) critic/scripter, reappraising and redrafting the scene with its personal extensions inside the mind, through a *distanced* Brechtian perspective (and perhaps a laugh). Thus, the scene's *rasa* flavors, resonating with the viewer's (Real) *bhava* drives, offer calmness and clarity through the cognitive reappraisal of emotions, in a shift back from *left to right* VLPFC activity, from immersive openness to distance and then to global awareness—not just a VLPFC suppression of emotions. (Further details of *rasa*-catharsis are given in Table 7.2.) This is also distinct from the more common *melodramatic* immersion in emotional identifications (and good versus evil stereotypes), experienced as fear, sorrow, and rage in theater or cinema, but often "suppressed" in everyday life. The conventional notion of catharsis, with vicarious participation and then crying or laughter to "purge" such emotions, relates to the simpler sense of VLPFC activity—as the fictional experience and then *suppression* of remnant feelings, especially with violent entertainment as a distraction from deeper concerns, which may *backfire* if the stereotypes are evoked again in everyday life.

Chapter 5 explored the different functions of brain hemispheres that we share with most animals: intense emotional responses in the *right*, with a wider awareness for danger or mating, often inhibited (or transformed) by the *left's* narrower focus on goals such as food gathering or hunting—and abstract, analytical, linguistic operations in humans (McGilchrist). In toads, horses, and humans, *new stimuli* activate the right hemisphere (with ties to the left side of the body) until they fit into (or alter) the left's categories. Most emotions are more active in the right cortex, with its stronger ties to the limbic system and brainstem areas. But *anger* is stronger in the left cortex, with dopamine pleasure circuits and a binary operator (in the left parietal lobe)—which relates to the power of *melodramas* onstage and onscreen, evoking viewer identification with the vengeful rage of the hero against the villain. The left hemisphere, from animals to humans, is also involved with *objectifying* prey and using tools—as in the objectifying of a

Table 7.2

Possible Steps of *Rasa*-Catharsis

1. The playful space around a stage or screen shifts the viewer's background emotion (mood) from malaise toward well-being and from stress to calm, yet with fictional malaise and stress.
2. *Sympathetic identification is* engaged, plus *pleasure/pain* with the main character(s), through initial plot conflicts, facial expressions and gestures, vocal tones, scenic images, sound effects, and music. This involves *mirror/intuition neurons and emotional contagion* between the viewer's inner actor/character/director/stage-manager/operator/audience/stagehands and scenes onstage or onscreen—also involving the inner improviser/designer and critic/scripter.
3. Primal emotions are evoked, fictional yet real, such as *fear* at the villain's threats, *sorrow* and *rage* at the victim's suffering, and *courage* in the hero taking revenge or stopping the villain, against overwhelming odds. Social and moral emotions may also be evoked. (See Table 1.1.) But this involves paradoxical contradictions as the viewer sees *tragicomic flaws* in the hero's efforts or *sensible motivations* in the villain's acts.
4. Such ironic edges, along with the viewer's choices in how to watch, *alter* the viewer's perspective and feelings (through plot twists and recognition scenes, according to Aristotle). This occurs from the hero's initial situation, challenges, conflicts with villains, ties to victims, and commitment toward a course of action, through increasing conflicts and potential changes in the hero, toward the climax and resolution of the overall conflict with the villain/antagonist.
5. The viewer's inner theatre changes via moments of awareness at sympathetic impulses with those onstage/onscreen, through *immersion* in the fictional action and yet a *distancing* effect at ironic twists (as in the theories of Antonin Artaud and Bertolt Brecht).
6. Shifts occur from right-cortical emotion to left-cortical analysis, grasping new sensations and perspectives, with a shift again toward right-cortical openness, letting go of prior concepts, especially in-group/out-group projections.
7. There is a refinement of the viewer's *bhavas* through such shifts of hot/cold emotional perspectives, evoking inner mimesis and yet critical questions, with resonance & distance in the savoring of *rasas* (as aesthetic flavors and cognitive reappraisals)—perhaps also with Brechtian motives for social as well as personal change.

monster that must be slayed with certain tools (like a stake in the vampire's case), unless the film also evokes sympathy for the beast-person, through tragicomic twists and a new right-cortical awareness.

Studies have found that the *right* hemisphere is used more for facial expression and recognition, involving the left side of the face and left visual field, in certain species of birds, sheep, and monkeys, as well as humans. This also relates to facial expressions and other subtle gestures on the movie screen, evoking the viewer's mirror neuron mimicry and emotional contagion. (Pain mimicry specifically involves the right anterior cingulate cortex.) Babies are more often held on the left side by chimps, gorillas, and humans, increasing the right-cortical, emotional *attunement*. Such attunement might also occur between the viewer's inner theater and the screen's emotional seductions. Regarding the limbic system, this would involve more of the left amygdala for conscious emotional representations and the right for unconscious processing.

The *left* hemisphere's "conscious verbal self" tends to rationalize or inhibit the right's unconscious intuitions (Timothy Wilson). It also tends to be overly optimistic, while the right is better at perceiving deceit, with more attention to the other's eyes (while the left attends to the mouth). So the eventual triumph of the hero over the villain, despite great odds against him, in action movie *melodramas*, confirms left-cortical, stereotypical, optimistic pleasures—and their "stickiness" against the right's perceptions (as McGilchrist puts it). But a complex film with *tragicomic* twists might connect more with the right's suspicions and alternative perspectives. Right frontal delay and left frontal distancing may also be evoked, in relation to circuits toward the back of the viewer's brain. Right and left, front and rear areas of the cortex become significant, too, regarding depression and mania in humans, related to the freezing and fleeing instincts in animals.

Even 10-month-old babies show greater *left* prefrontal activity when watching a movie actress smiling and more *right* when watching her crying. This happens with adults also: more left with amusing or uplifting videos, evoking detachment, and more right with frightening or disgusting videos, triggering a sense of danger or immorality. Yet Richard Davidson's research on this found great *variability* between viewers, which he eventually defined as different "Emotional Styles," involving degrees of Resilience, Outlook, Social Intuition, Self-Awareness, Attention, and Sensitivity to Context. Davidson has demonstrated that Resilience and Outlook are malleable through mindfulness training, which increases *left* prefrontal activity. I would relate this to the potential for *rasa*-catharsis to alter a viewer's emotional style in theater and cinema as well. This involves a crucial separation yet interplay between brain hemispheres, in the spreading "inner tree" of consciousness, especially left-cortical, prosocial controls versus right-cortical, alternative freedoms—which often play out in political theaters and on movie screens, as a clash of moral tribes.

Chapter 5 concluded with recent research on the "rubber hand illusion" and "body swapping." Such experiments suggest how the movie viewer's body image and sense of Self might *merge* with the fictional figures onscreen. Once the illusion is created of a rubber hand, or a tabletop, as an extension of one's body (involving right frontal and parietal brain areas) through synchronous sight and touch, then bending the rubber finger backward or hitting the table with a hammer evokes an automatic nervous system response, as if damaging to one's Self. Although the subject knows the rubber hand is just rubber, or the tabletop is not a part of one's body, the reshaping of identity disrupts that left-cortical, categorical knowledge. This shows how readily a movie viewer's sense of Self might be affected, consciously or not, through *identifications* with screen characters, their bodies, and their animal-human transformations, as well as physical and emotional wounds. Even more akin to movies, a head-mounted video display can make the subject feel that a manikin or another person, sliced by a knife, is one's own body, through a simple exchange in visual viewpoints—like a cut between angles onscreen.

Experiments also show that implicit antiracial bias can be *altered* in subjects who experienced the illusion with a rubber hand of a different skin color. This suggests that cross-racial identifications by movie viewers may change their in-group/out-group prejudices, depending on how stereotypes are challenged by ironic twists in plot and character expectations. Yet, such expectations are sticky (with left-cortical presumptions) as shown in an experiment that caused one group of viewers, given a stimulant, to laugh more at a short comedy, but found that they reported it as not being any more enjoyable than others who got a placebo or tranquilizer—with their ratings influenced by prior experiences with that type of film and its star.

Identification with another person's face onscreen, or "enfacement," has also been measured in the laboratory. It occurs more readily with beautiful faces. Thus, the sticky prejudices of beauty and racial ideals, or their binary opposites, might *evoke or block* face identifications by movie viewers. Yet the mostly unconscious (right cortical) malleability of one's body image and sense of Self offers an opportunity for viewers to engage with characters onscreen, *across* in-group/out-group and beauty/ugliness stereotypes. This can happen especially with beast-people films that evoke a mix of fear and desire, along with other paradoxical emotions, at *rasa*-cathartic moments, building toward transformative challenges and changes of the viewer's inner theater. But the details of that depend on numerous personal associations, in the particular memory, fantasy, and projection networks of each brain.

Chapter 6 explored possibilities of such *rasa*-cathartic twists in beast-people films that reflect fears of a godlike scientist, manipulating animal-human subjects in a lab, or of "apes" doing that someday to humans, in a vengeful clash of our evolving cultures. It considered recent experiments that show the predilection of young children to attribute agency, design, and purpose to all living things, intuitively believing in a "creator." The films based on H. G. Wells's *The Island of Dr. Moreau* reflect this "natural" inclination toward super-natural belief, but also a fear of science, with a "genius" conducting ruthless experiments as creator-parent. The 1932 version shows Depression Era fears of class revolution, colonial hubris, and biotechnology—even prior to World War II and Auschwitz—with sympathy evoked for the subhuman creations of the mad doctor, even as they rebel violently against his "Law." Despite the expected framework of a melodramatic adventure movie, the viewer's identifications are shifted, through tragicomic twists. Through the hero's noble fight against a ship captain, his romance with the island's Panther-Woman, and then his disgust with her as a beast-person and with Moreau as a cruel experimenter, the viewer may experience a fear of, yet sympathetic rage with the creatures that rebel, first against the Sayer of the Law and then against their godlike creator. Various *rasas* are evoked and potentially refined, especially through reflections onscreen of the viewer's inner actor/character, director, and stage manager, with beautiful and odious creatures, a fearful scientist-puppeteer, and his morally troubled assistant and visitor.

In the post-Vietnam, 1977 version, ironic sympathies and fears are elucidated, even about the island's visitor, as heroic surrogate for the movie viewer, and yet as bestial. The visitor's love interest is not a Panther Woman, but she does have a wild cat as a pet. The visitor also meets a Bear-Man in Moreau's lab, whose "natural instincts" are reduced and thus his body transformed by the doctor's injections, from animal to human, in an anti-werewolf direction. But when the visiting hero rebels against Moreau, mating with his daughter and threatening to leave the island with her, the scientist uses an opposite serum against him, making him bestial with full-body hair like a werewolf. After the beast-people rebel and kill Moreau, the animal-human hero tries to raise his corpse like a scarecrow with faux immortality. But that only delays their destruction of his compound and of themselves.

In the post-Rwandan-genocide, 1996 version, the beast-people are initially sympathetic but become even more monstrous, evoking a range of *rasas* for the viewer. Yet the island's visitor also kills at the start of the film, in a struggle to survive against other humans, with dwindling supplies on their lifeboat, before he is rescued. This is ironic, too, because the visitor was on a UN peace mission when his plane crashed in the ocean and he became a brutal survivor. Moreau's assistant (Montgomery) acquires perverse traits as well, as drug user, drug pusher, and Moreau-god mimic. Moreau himself appears as clownish, yet with scenes showing his calm fatherly compassion for the frightened visitor, for the troubled teen

Panther-Woman (who develops sharp teeth and claws like a werewolf), for his other house-servant children, and for the rebellious fieldworker beast-people, even the reckless Hyena-Swine. Thus, various *rasas* are stirred throughout the film, especially with Moreau's death at his creatures' hands, tearing him to pieces offscreen after they learn (from Hyena-Swine) to tear out the pain chip that their "father" implanted in their bodies as an obedience device, and with further cruelties in their revolution. Such scenes elicit sorrow, disgust, awe, courage, fear, and anger, through different perspectives, involving viewers' VMPFC *stage manager* and limbic *stage-hands* (especially the ventral striatum, insula, and dorsal ACC) for fairness and pleasure/pain networks, both social and physical. This challenges viewers' body-swapping and hierarchical ape-ego identifications from earlier in the film, toward a *reappraisal* of their own chimeric impulses.

In a parallel with the mad scientist Moreau and his Panther-Woman daughter, the 1942 film, *Cat People*, posits a backstory of Serbian village folk who use satanic magic to turn themselves into wild animals. The film's femme fatale, Irena, evokes tragic *sympathy*, as well as audience *fear*, as a victim of her predatory ancestry, in the multicultural lab of a modern New York City, when she changes into a panther through rivalry with another woman for her man. The 1982 remake presents brother and sister were-panthers, shifting the setting to the black magic or "voodoo" aura of New Orleans. This suggests also the threat of the Black Panther movement in the 1970s, although both were-panthers are white. The older sibling, Paul, wants to have sex with his sister, Irena, because he is cursed with becoming a panther whenever he feels lust. He must kill then, as the beast, in order to return to human form. Rejected by her, the Panther-Paul is killed by a female zookeeper. But Irena is turned into her panther form by a male zookeeper who makes love to her, while she is tied up, and then cages her for her own good. Thus, the zookeepers become slayers and Law-Sayers, containing the wild beast-people, through romantic rivalries and passions. This reflects the inner (left cortical) *critic/scripter* of movie viewers inhibiting, yet also renegotiating its powers with the (right cortical) *mime improviser/scene designer*. But the 1982 version shows different fears from the 1942 original: of the animal-Other within one's own culture and family, not from an exotic Eastern European country with its folk magic and world war threats.

Further tragicomic ironies are shown involving left-cortical, obsessive, objectifying functions with the mad scientist in *The Fly* series, especially in the 1958 original and 1986 remake (though not as much in their sequels). In the 1958 version, a meticulous inventor ignores family distractions to perfect his teleportation device, but accidentally swaps bodies with a fly, producing two hybrids with human and insect parts. The one with a human body but fly head and hand pleads for help from his wife to be fed, understood, and then mercifully killed as he fails to reconstitute his ape-human ego and body. Many of the eight *rasas* may be evoked along the way, as the viewer takes the wife's as well as the scientist's perspective: romance, awe, fear, disgust, heroic courage, rage, sorrow, and even humor (at the human-headed fly, an inverted double of the genius-inventor with a fly head, pleading for help at the end, while caught in a spider web, instead of asking heroically for death). These *rasas*, refined by ironic juxtaposition, raise a cognitive awareness of the viewer's *own* emotional impulses through new perspectives—depending on *how* each viewer watches the film.

The 1986 version focuses more on the inventor's horrifying hubris, propelled by a new mate whom he meets at a party, takes to his lab, and has sex with. This inspires his teleportation error, though he sees it as improving his body and brain toward super-natural skills, in a radical alteration of left-cortical rationality, obsessiveness, and optimism, right-cortical openness, and limbic/brainstem drives. His former girlfriend, a science journalist, agrees to videotape his gradual transformation into a Kafkaesque wingless bug, becoming more vicious, gene driven, and monstrous as a humanoid insect—until she gets too disgusted and fearful to be with him. He then tries to mate with her again, so as to produce a further chimeric offspring. He also dismembers a male human rival, showing ruthless rage. Yet the scientist-inventor retains his courageous wit while struggling with a changing body and hybrid powers, which eventually incapacitate him, evoking sympathetic sorrow and dark humor, as well as fear, disgust, and awe, in movie viewers. The film also reflects increasing fears in the 1980s of sexual wildness, AIDS, and technological monsters—although the inventor tries to view his disease as having a "purpose" that will benefit mankind.

The godlike purpose of human science and technology, to discover nature's mechanisms and improve on them, is questioned further in the 2009 film, *Splice*. A couple of scientists, who are romantically involved, defy their corporate bosses to persist in their biotech discoveries. They create a new hybrid species and nurture the plant-animal-human infant until it changes its form and gender, from big-brained but strange-legged child with a barbed tail to dangerous femme fatale and then to primal demonic threat for both parents. This also involves the female scientist's haunting memories of being abused as a child by her mother—and the male scientist's betrayal of her while drawn to his hybrid daughter as she grows toward sexual maturity, before changing into a male. The "mom"-scientist even operates on her offspring (like Moreau), amputating the barbed tip of her tail after it killed a pet cat. But is this operation pragmatic (with a genetically modified monster), or vengeful (against the sexy daughter who also mated with the scientist dad), or repeating a childhood trauma?

Such multiple dimensions in the scientist-parent and were-animal motifs shift audience identifications in radical ways. We may sympathize with the *courageous* bio-engineers fighting the big bad company to discover medical cures but *fear* their tragic hubris in taking the godlike experimenting too far. We might also fear the wild monster they create, especially when it seems like an invasive alien, yet feel an oxytocin-fueled *love* for it, as a child in a dress. At times, we may view it with *disgust* but then also with *awe* as it transforms, like and unlike a human. We might experience *rage* and yet *sorrow*, along with the two scientists, as she battles the maternal demon inside her, as he struggles with incestuous desires, and as they fight each other and then their super-natural teen. These shifts evoke and reflect our inner-theater elements, our own ape egos, and our multi-mirrored morphing relationships of animal-human signaling and transgression—which can create personal and group rivalries, but also *rasa*-cathartic compassion.

The *Planet of the Apes* series explored the ape-ego rivalries of our neural and political theaters more directly: initially with five sci-fi movies in the 1960s to 1970s and then with three in the 2000s to 2010s. In the Civil Rights Era, the first film (*The Planet of the Apes*, 1968) turned the tables on the Western domination of others by having an American astronaut, Taylor, crash land on a planet where he becomes a colonized subject. He also finds primitive humans there that are mute slaves and experimental subjects controlled by talking gorilla warriors, chimp scientists, and orangutan ministers, who have early modern weapons and religious beliefs. His shipmates from the spacecraft are even more objectified: a black man killed and stuffed in a museum and another white man made obedient with a large surgery scar on his skull. Taylor escapes, though, with the help of a female chimp scientist who realizes he can speak intelligently, but hides that from most of the other apes. The astronaut, barely covered by animal skins (or shown naked from behind) fights off the racist apes, gathers a native girl with him, kisses the chimp scientist on the lips in parting, and discovers to his horror that he is actually on a future earth, destroyed by nuclear war. This surprise ending, with a mostly buried Statue of Liberty, recycles the viewer through the various allegorical warnings of the film, engaging specific inner-theater networks, personal associations, and collective *rasas*, as if saying: evolve your culture in a better way or this could be your descendants' fate.

The four sequels in the 1960s to 1970s played upon nuclear annihilation and communist revolution anxieties as well as civil (animal and human) rights issues. They also evoked tragicomic twists through further ape sympathies and fears. And they continued a trans-historical question given at the end of the first film. Are we humans an aberrant species, "a devil's pawn," that will inevitably destroy the world, because of our brilliance as a wild yet technological "beast" which, alone among the apes, "kills for sport or lust or greed"? In other ape species, males may indeed kill out of genetic lust or greed, especially when taking another male's mate and destroying her infants, or when taking power as alpha. Females in many mammalian species also kill their infants to conserve resources. But we humans have extended this genetic rationale of self and kin, of survival and reproduction, into vast cultural aims to sacrifice oneself for the extended kinship of clan or country or religion and to take territories across the globe, through genocide and slavery in many historical periods, even our own. Yet we also have highly reflective, inner theater and outer social networks to alter ourselves, regarding the first film's warning: "Beware the beast, man. . . ." Movies today may help us or harm us in *playing with* that bio-cultural evolution of animal-human morality.

The second movie (*Beneath the Planet of the Apes*, 1970) showed an advanced society of radiation-scarred, mutant, telepathic humans who live underground, in the ruins of New York City, and worship a totemic nuclear missile from the film's own period. Sexual and social revolutions are also suggested with the primitively dressed Nova and with young chimps holding signs in protest of the gorilla armies marching into the Forbidden Zone where the underground humans live (also suggesting prehistoric cave artists). But the movie turns cynical about such 1960s ideals, as gorilla warriors dominate the ape agenda and the mutant telepaths, with the astronauts' help, destroy the earth in a nuclear explosion—shown then to the movie viewer at a godlike, outer-space distance. It seemed as if the movie series might have to end there, too, with a raging, fearsome, and awe-inspiring *rasa* warning.

Yet the filmmakers found a way to continue, returning the series to earth. In the third film (*Escape from the Planet of the Apes*, 1971), Zira and Cornelius, along with the chimp-genius Milo, somehow rebuild the American spaceship that crashed on their planet and travel to the earth of our time. Ironically, the chimps are at first kept in a zoo by the humans, like the astronaut in the future had been by them, and Milo is strangled to death by a gorilla in the cage next to theirs. But soon, Zira and Cornelius become the toast of New York City, as intelligent, speaking chimps from the future. Forces become allied against them, however, showing our current intolerance for wisdom in other species and races. Authorities fear a consciousness-raising revolution, with these advanced apes, that will send the human world in the direction of annihilation, which the chimp scientists have foreseen (along with movie fans). Zira and Cornelius are sacrificed but their baby survives, hidden in a circus cage as an ordinary, subhuman ape. This encourages further identifications of viewers' inner theaters with the chimp characters and their human allies, against the government villains, involving *rasas* of sorrow (at Milo's death), joy (at the chimps' initial celebrity), courage (at their struggle to survive with their offspring), rage (at their killers), and tragicomic sorrow/joy again (at their demise yet the baby's first spoken word at the end of the film, "Mama").

The fourth film (*Conquest of the Planet of the Apes*, 1972) showed the rise of that baby as "Caesar," the Messiah-like leader of an ape revolution against the humans in the near future. His only human allies are Latino and Black, again suggesting Civil Rights alliances. However, as his name suggest, Caesar becomes tempted toward imperial power, especially when he and the other apes want revenge on the humans, through their successful, collective violence. This offers an ironic twist for viewers cheering the apes' melodramatic fight against fascist oppressors—and connects, through tragicomic *rasas*, with viewers who fear the dangers of righteous vigilantes or communist revolutionaries, on either side of the political spectrum (or terrorist insurgents today). Yet Caesar becomes cathartically aware of the apes' rage toward vengeance with the help of a female chimp who gazes at him with an imploring "No," the first word by any of the other apes in the film. Caesar tells the apes to put down their weapons and promises that they will dominate the humans humanely, with "compassion and understanding"—evoking the ninth *rasa* of peace.

The fifth film (*Battle for the Planet of the Apes*, 1973) again involves a melodramatic battle between good apes and evil humans, yet also another between the apes. This time the humans are "mutants" who survived a nuclear war by living underground, but then attack the apes to acquire their fertile territory. After Caesar leads the apes in repelling the mutant humans, his rivalry with a certain gorilla general resurfaces. The militant gorillas almost assassinate Caesar, but he maintains his alpha "savior" role with the help of a wise orangutan, who reveals that the rival general killed Caesar's son, breaking the sacred law of "Ape does not kill ape." Caesar kills the general, getting heroic revenge, which movie viewers might cheer. But he also stops the apes from getting further revenge against his innocent human friends, who had been imprisoned by the general and almost killed. This ironic twist might make viewers pause, too, and consider in their own lives the emotions of vengeful violence (sorrow, rage, fear, and courage), even if it seems righteous—with more regard for the victims' perspectives, whether innocent or not.

Chapter 6 also explored the three remakes of the *Planet of the Apes* series in our new millennium, regarding sci-fi extensions of our ape egos, territorial rivalries, vengeful violence, and yet kinship kindness. A new *Planet of the Apes* was released in 2001, after Y2K fears and just two months before the 9/11 terrorist attacks in the United States. It again involves an astronaut crash landing on a planet in the far future (but this time it is "Ashlar" in 5021, where everyone somehow speaks his language). He tries to escape bondage, along with some primitive human friends, in an ape-run society with elements of our own, including a decadent upper class. Again, there is a sexy, animal-skin clad, native human female for the astronaut and a female chimp love interest who helps the humans escape. However, in this version the native humans can speak and the apes move in more natural or super-natural ways, running on all fours and with powerful leaps. The ape warriors again ride horses, but wear heavier armor and fight without guns—in a premodern style. There is a comical orangutan slave-trader and a bellicose chimp general with a hunchback like that of Richard III. He functions in the film's melodrama as a somewhat sympathetic villain and trickster, fighting brutally, yet cleverly to continue his alpha role over the gorilla warriors, wealthy apes, and escaped humans. He is also shown in a poignant scene with his dying father and then later trapped in a crashed space station, as ape temple, when he falls from power.

More ironically, another chimp (played by an actual chimp) appears as a *deus ex machina*, like Caesar as Messiah to the apes in the earlier series. But here, the apes soon lose their awe at the god's return and the humans must fight for freedom with technological, instead of divine weapons. In a further, tragicomic twist, the human astronaut returns to what appears to be earth, crashing near the Lincoln Memorial, but in a different time-space parallel, on a planet where the chimps have created an Abe-like historical god (putting a further ironic spin on the film's trickster-villain). Such twists complicate the typical action-movie *rasas* of sorrow, rage, romance, fear, and courage, spicing the viewer's awareness with humor, disgust, and ironic awe. Thus, the film offers not only postmodern winks at the spectators' prior knowledge of the series, but also a cathartic reappraisal of the emotions circulating between our inner theaters and the silver screen.

The playfully perverse, somewhat satirical "reboot" of the *Planet* series in 2001 turned more respectful with a brain-enhancement, animal-rights version of the rise of Caesar, in 2011 and 2014, involving viral epidemic and culture clash fears from recent decades. Like the *Island* films, the 2011 movie, *Rise of the Planet of the Apes*, starts with scientists experimenting on animals for biomedical reasons. As in *Splice*, a scientist nurtures a baby lab creature, rebelling against a corporate demand that the animal's life and such experiments be ended. But the baby chimp's brain has been enhanced with a viral drug the scientist was developing, which it absorbed from its dead mother. The scientist nurtures it with his father's help, although his father is suffering from Alzheimer's dementia. The scientist tries the drug on his father, without the company's permission, when he sees its effect on the developing brain of their chimp child. Thus, oxytocin-fueled bonding and nurturing circuits are shown in two male mothers with a male chimp offspring, saved from death at the lab, evoking *right*-cortical empathy from viewers regarding animal treatment and unconventional family structures, along with hopes for *left*-cortical scientific cures.

The use of computer graphics and motion capture technologies, to create more realistic and expressive ape characters, in this film and its sequel, alters viewers' empathic, ape ego, body-swapping ties to the screen. We see little Caesar turn gestures into sign language, while interacting with the facial expressions of the human characters around him. The virtual yet realistic images of a baby, then adolescent, and then adult chimp become extensions of the viewer's Imaginary interactions with the movie, through mirror-neuron mimesis and emotional contagion. With their brains signaling to mimic Caesar's actions and facial emotions, especially through responses by the parental actors onscreen, viewers accept that the animated chimp could be real—in scenes that were filmed without it or with an actor in a motion-capture suit—adding a greater reality to the screen via their inner theaters.

The adoptive, cross-species parenting performed by the scientist and his father in this film has been found in real-life animals as well. But it is usually done by females, even by predators who nurture a baby rather than eating it, after killing its mother (Hrdy, *Mothers* 209–11). Within our species, unlike nonhuman apes, we evolved a "cooperative breeding" style, with the birth mother often getting help from others, especially kin, but also from "wet mothers" as servants (30, 206). Thus, *oxytocin* circuits (evoked in viewers with the chimp development scenes in this film) are crucial to our high capacity for collaboration with others, through shared intentions, goals, and action plans that reach far beyond our fellow apes', as does our sharing of childcare (9–11, 138). Likewise, *dopamine* reward centers are active when mothers look at photos of their smiling babies (213)—and probably within male and female viewers when the infant Caesar smiles onscreen.

This pleasure at a smiling infant also involves activity in the OFC/VMPFC *stage manager*, even when adults are shown pictures of unfamiliar infants (Hrdy, *Mothers* 220). It suggests, too, that human morality, in the huge and highly complex social groups of our ape species, may be based in our "attachment" experiences as infants, as well as our ability to care for others' children. Various studies have shown that babies with multiple caretakers grow up feeling secure, with "more enhanced capacities to view the world from multiple perspectives" (132). Movies, as I have been arguing throughout this book, might also help us to feel secure enough to engage with fictional conflicts and increase, through *rasa*-cathartic twists, our perspective-taking capacities.

The first half of this ape-planet film focuses on the brain-enhanced Caesar, developing from child to adolescent, in the care of the scientist and his new girlfriend, as well as his father. But a conflict arises when Caesar (with white sclerae in his eyes, signaling more expressively like us) sees that he is similar to a dog on a chain, while on a walk with his human caretakers. He asks existential questions and learns from the scientist how his mother was killed in the lab. Another conflict occurs when Caesar tries to protect the scientist's father (whose dementia is increasing), in a confrontation with an angry neighbor, and the chimp's protective *rage* emerges as bestial in public. The scientist-parent is forced to put Caesar in a "shelter" where he eventually takes an alpha role, administers the stolen viral drug to his fellow apes, and speaks with them through sign language (subtitled in the film). He leads them in escaping from the shelter and its abusive handlers, in freeing other apes in a zoo and the lab, where they become like terrorists to the human authorities, taking a violent vengeance on building structures. But Caesar also leads the apes in crossing the Golden Gate Bridge toward the redwoods, despite a fierce battle with humans trying to stop them. Thus, viewers' *rasas* are altered across various perspectives, involving emotional *resonance* and yet critical *distancing*, with and against Caesar: (1) in his territorial conflict with the neighbor and alpha competition/cooperation with other apes, (2) through his trickery in staying at the shelter instead of going home with the scientist and in stealing the viral drug to enhance his fellow apes' brains, (3) in his outwitting and punishing of human handlers at the shelter, and (4) with his leading the apes as terrorist invaders of San Francisco, yet also as criminal fugitives or displaced refugees seeking a true shelter in the redwoods.

The eighth film, *Dawn of the Planet of the Apes*, was released in the summer of 2014, a few months before Ebola outbreaks in West Africa and elsewhere on the globe, carried by airline travelers. Coincidentally, a similar but more extensive plague was shown at the end of the 2011 film with the fictional, brain-enhancing, viral drug becoming deadly to humans. The sequel in 2014 begins with Caesar's tribe (10 years later) hunting deer with spears and living in a beautiful, orderly, forest village—although with dangers, too, as a bear attacks Caesar's son and they are both saved by the bonobo Koba. This romantic ideal of the "noble savage" contrasts with how the humans are shown, also in a small tribe, but as traumatized survivors of the worldwide viral plague, sheltered in the ruins of San Francisco, and running out of the technological power they need to survive. Territorial threats occur, with incursions from each side, as the humans enter the ape area to fix a hydroelectric dam—recalling to viewers, perhaps, the battles over territory and oil reserves in recent Middle East wars. The apes, with painted faces, also offer a "show of force" by entering the ruined city on horseback—reminiscent of Indians evoking fear and awe in white settlers and their viewers during Wild West movies from earlier decades.

But despite the melodramatic stereotypes and escalating battle scenes, alpha rivalries are shown within each tribe, complicating the viewer's alignments and allegiances. Malcolm argues with Dreyfus about developing trust with the apes, to solve the human techno-power problem, and this works at first. Yet Dreyfus's defensive measures prove right, after distrust arises on each side, emerging in specific ways. Koba (scarred by human lab tests and not acting like lovemaking bonobos in the wild) saves the lives of Caesar and his son, but then fights with that alpha-friend about loving humans more than apes. Koba avenges the shooting of another chimp by a human, but later kills that same chimp (Ash) when he feels betrayed—and betrays his friend, Caesar, as their rivalry grows. Koba may initially charm the viewer with his laughing antics at the armory, just as he tricks two humans while stealing their weapon and shooting them, thus shocking the viewer at his cruel cleverness. When he also makes Caesar fall, tricking the other apes into blaming the humans, movie viewers see his further villainy. And yet, there is a tragic twist in Caesar's revenge against the villain. He must rationalize his own violation of a sacred rule (shown at the start of this film as well as earlier in the series): "Ape not kill ape."

In another ironic twist, Dreyfus turns into a suicide-bomber against the Koba-led apes, destroying the metal tower that they both killed many others to control. This might provoke movie viewers to think again about their *rasas* in the us-against-them battle scenes (or in current cultural parallels). And that is precisely what I have tried to identify with the many film examples in this book, using inner theater and *rasa*-catharsis models. Even in popular, money-making, "escapist" entertainment, there may be tragicomic, Artaudian-Brechtian, emotional resonating yet critical distancing twists that engage various *rasas* and inner-theater elements in ironic ways. If we value such movies (and TV shows and videogames) with *tragicomic edges*, encouraging their challenges to us, with how we watch and discuss them, a spreading ripple effect in our culture is produced. This may evoke a more mindful awareness across many brain theaters about the primal alienation, violent impulses, and collective signaling in our evolutionary successes and current dangers as dominant, attractive, and tricky beast-people.

CATHARTIC BACKFIRE AND MORALITY PLAY

Wildness in humans involves *objectifying* one another beyond the limiting patterns of animal instincts, even while fueled by remnant instinctual drives. Whether idealized through erotic desire or demonized with horror and rage, or ridiculed with laughter, humans may appear as bestial objects that attract and repel us—as vampires, werewolves, lab hybrids, and civilized apes onscreen, entertaining us at a safe, enticing distance. Camaraderie is also evoked in the audience, in the fight against such beast-people, with a romantic duo or "slayer" team (or scientist-creators as lawgivers) versus the demon or pack, even when involving sympathetic monsters, like the Cullens, against others that seem worse. But slayers versus beast-people in melodramatic battles may confirm stereotypes of heroes and villains, activating emotional circuits of our inner theaters: from brainstem, limbic *stagehands* (with fear, rage, and in-group/out-group passions) to left-cortical *critic/scripters* (projecting binary, good against evil schemas).

Movies have been doing this, in various genres, for over a century. For example, in 1915, D.W. Griffith's silent film, *Birth of a Nation*, presented blacks as beastly villains and white-robed Ku Klux Klan members on horseback as heroic slayers, saving other whites. This may promote identifications with such slayers outside the movie theater, who fight against the fearful terrorist monsters, *exercising* rather than exorcizing violent emotions in viewers' brains. (*Birth of a Nation* was a mass-audience hit when it premiered and was used for decades thereafter as a recruiting tool by the KKK.) The conventional attempt at a *cathartic venting* of emotions, in theater, cinema, TV, or videogames, can *backfire*, as studies show, increasing aggressive tendencies, especially in children (Craig Anderson et al.; Zacks 113–35). Through melodramatic and participatory formulas, outer theaters may capitalize on the *addictive* habits of viewers and gamers, extending primal drives of lust and survival—with oxytocin, dopamine, and opioid networks transforming the inner theaters of those watching, even to the point of *acting out*, beyond the screen.

Such "cathartic backfire" involves a realignment of inner-theater elements—from the easy, entertaining, yet addictive violence of melodramatic good-versus-evil onscreen to its ripple effect of aggressions in real life, in minor mimetic acts or major copycat crimes. Viewer identifications may shift (in Murray Smith's terms) from "alignment" with righteous slayers to "allegiance" with their vigilante violence—through right-cortical *mime-improviser* attunement. This could increase the brain circuits for acting impulsively, or acting out vengeful fantasies, given similar stereotypes of victims and villains, beyond the movie's dreamlike rehearsal spaces.[4]

Yet tragicomic twists onscreen may provide an antidote to cathartic backfire for many viewers (or videogame players). Such ironic, antiheroic twists might provoke a further shift of viewer's inner-theater circuits, balancing the self-conscious *character* of left-LPFC verbal and right-LPFC imagistic reappraisals of limbic emotional *stagehands*, triggered by the action onscreen. This may spur a more mindful, reflective thinking, with a "dismantling" of the brain's top-down, left-cortical stereotypes, as "automatic classifications," because of a shift to right-cortical "imagery-based processing" through the present flow of working memory, in the DLPFC *editor*, instead of relying on long-term memory patterns (Siegel, *Mindful* 248–51). Such a metacognitive "de-automatizing" awareness (Kang et al.), along with personal associations in watching a film (or playing a videogame), might also involve the MPFC *actor* reassessing one's alignment/allegiance—in relation to the VMPFC *stage manager's* monitoring of moral behavior and the DMPFC *director's* sense of others' views, as reflected by the screen.

According to psychologists Jonathan Haidt and Craig Joseph, there may be an inherited, animal-to-human basis for fundamental aspects of morality. Their research shows that both humans and nonhuman apes make intuitive moral judgments about minimizing suffering, supporting fairness, respecting hierarchical authorities, being loyal to group bonds, and maintaining cleanliness. These "universal moral modules" in today's human brain probably evolved through natural and cultural selection, from our ape and hominin, hunter-gatherer ancestors (Gazzaniga, *Who's* 172–73). They involve right-cortical perceptions of others' motives, through emotion simulations and expectations, which are then rationalized by left-cortical judgments (174–76). In reaction to moral intuition and reasoning (in the right and left cortices), a specific brain area, the right dorsolateral prefrontal cortex, inhibits self-interest for the sake of *fairness*—especially as children learn particular social norms (176–78; Steven Anderson et al.). Thus, the inner "film editor," as I have called it, *redefines* one's sense of self, regarding personal appetites and desires, yet also fairness to others (involving the inner stagehands, stage manager, actor, character, and intuitive audience networks).

Recently, Haidt has added "liberty" to his list of moral foundations, derivable from ape observations, and put them into abstract terms with binary opposites. The list also involves characteristic emotions, as shown in Table 7.3, mostly using Haidt's terms, but with a few added in relation to Table 1.1, in Chapter 1. I would also note that the animal instinct to fight for freedom changes greatly in various human contexts, depending on how one feels trapped or how the ideal of freedom seems threatened. Humans can feel free while in a cage if believing that their sacrifice is for a higher good. Humans can also experience too much freedom, without a meaningful purpose in life, as a trap. Haidt has argued, with data from online surveys involving tens of thousands of volunteers at YourMorals.org, that libertarians are more concerned about the first two binaries in this list, liberals with the first three, and social conservatives with all six (*Righteous* 297–306).

Haidt values the "breadth" of conservatives' strong concern for all six moral foundations, although he calls himself a "Durkheimian utilitarian" not aligned with the Republican Party (*Righteous* 372n36). He suggests, with section titles in his book's final chapter, that liberals are "yin," while libertarians and conservatives are "yang." He does not explore these gendered or Daoist associations further, except to recommend that we shift from Manichaean, good-versus-evil polarities "to a more respectful and constructive yin-yang disagreement" (312). But Haidt's characterization of *liberals* as yin relates to what I have been describing in this book as

Table 7.3

Haidt and Joseph's Moral Foundations Theory (with added elements in brackets) adapted from Haidt and Joseph's "Intuitive Ethics," plus Haidt's *Righteous Mind* (125)

Ape [or Animal] Observation	Human Emotion	Moral Binary (Good/Evil)
[fighting for freedom]	[rage]	liberty/oppression
supporting fairness	anger, gratitude, guilt	fairness/cheating
minimizing suffering	compassion	care/harm
being loyal to group bonds	group pride, rage at traitors	loyalty/betrayal
respecting hierarchical authorities	respect, fear [and awe]	authority/subversion
maintaining cleanliness	disgust [and awe]	sanctity/degradation

the more nurturing, holistic, oxytocin-fueled functions of the right-cortical mime improviser/scene designer, with stronger ties to limbic emotional stagehands than the left. This might be viewed as the more *feminine* or yin side of our animal-human (super)nature, with the long-evolved role of women as primary caretakers and with women on average being better able than men to interpret facial emotions using their right hemispheres. Libertarians could then be seen as more *masculine* or yang, especially regarding the moral module of "liberty." They are sensitive about threats to it (using right-cortical wariness) like liberals, but more eager to *fight* against any social controls or service institutions that might reduce it—using the left-cortical critic/scripter and its limbic rage circuits. Conservatives can also be viewed as more yang-oriented and abstract rule based, not just as balanced across the six "moral foundations." They may be holistically sensitive (through the right cortex) about threats to all six, yet conservatives manifest strong left-cortical critic/scripter and limbic stagehand networks in order to, as Haidt puts it, "fight back ferociously when they believe that change will damage the institutions and traditions that provide our moral exoskeletons (such as the family)" (305).

Of course, we all use both sides of the brain all the time (unless there is damage to certain areas) and liberals fight for their beliefs, too. But the conservative versus liberal split in today's politics, although it gets different names elsewhere in the world, is often characterized as "right" versus "left."[5] This again suggests a connection between *conservatives* and the rule based, narrowly focused functions of our left cortex, which controls the *right* side of the body—or between *liberals* and the detail oriented, openly aware, right cortex, tied to the *left* side. Another way to regard our still evolving bio-cultural moral spectrum involves parental roles. George Lakoff describes liberals as using a "Nurturing Parent" (feminine, yin, maternal) model of government while conservatives employ a "Strict Father" metaphor for their moral politics, believing in people's inherent evil, which must be contained by surveillance, threats, or rewards, and retributive justice. One might add that such a figure of good trying to contain evil often appears in melodramatic horror and action movies, especially in vampire films with a "slayer" (who might be female, like Buffy, but is still taught by an older male) or in lab-hybrid movies with a scientist as controlling parent. Sometimes, the monster is a Strict Father to itself, trying to subdue its own bestiality or protect others from it, as in various werewolf films, or to stop the spread of evil by others of its kind, as in films with vampire or werewolf clan battles. There is also a violent rivalry for alpha male, Strict Father rule in ape-planet movies, often with Nurturing Parent (or scientist) conflicts as well.

Psychologist Joshua Greene sees a flaw in Haidt's research about the broad spectrum of conservative moral foundations, arguing that his online surveys, finding clustered responses by certain political types, were already "designed with five clusters in mind" (*Righteous* 386n). Greene reveals a danger in Haidt's admiration for the conservative breadth of morals,[6] which also suggests, to my mind, a danger in Lakoff's envy, as a liberal, of conservatives' successful appeal to the country-as-family metaphor (19). Greene argues that we should not resort to such "tribal" intuitions with automatic moral emotions. Instead, we should cultivate effortful perspective taking, or what Daniel Kahneman calls slow, "System 2," deliberative thinking, as a way to improve our understanding of others' views and to alter our fast, "System 1," intuitive, tribal-based morals. This is akin to the potential of theater and film, even when evoking tribal, System 1, good versus evil, hero against villain identifications, to challenge viewers' perspectives with ironic, tragicomic twists, through emotional immersion and yet critical reappraisal, as System 2, perhaps with a dose of playful laughter.

Greene points to neural areas involved in this shift, which correspond to my inner-theater model. His brain-mapping research shows more activity in the amygdala and MPFC, plus parts of the VMPFC, with a *personal* moral dilemma, involving System 1, automatic feelings—as compared with an *impersonal* dilemma, which has more activity in the DLPFC, using "explicit decision rules" that can "override competing impulses" (*Righteous* 120–22). As dilemmas for test subjects' brains, Greene used a fictional "trolley car" story, offering an impersonal "utilitarian" choice of saving five people's lives but killing one person, by throwing a switch to change the tracks and divert a runaway trolley, or a more personal version of pushing a person off a footbridge over the tracks to divert the trolley. Most people in the lab could do the first easily (when asked to imagine this) but hesitated or declined doing the second, even though the numbers saved and killed were the same—showing the influence of System 1, automatic, intuitive, tribal-based morality.

Further evidence for this "dual process" of moral judgment,[7] involving DLPFC utilitarian decisions competing with MPFC-VMPFC-amygdala emotions, comes from brain-damaged patients. Those with VMPFC damage were much more likely (five times more in Damasio's experiments) to agree, at least fictionally, to kill someone with a personal push off a footbridge in order to save five others below, even to kill a family member in order to save a larger number of strangers (Greene 125).[8] But in normal subjects, tests also show that inducing mirth or evoking skepticism about one's intuitions with tricky math problems "increases utilitarian judgment" (126–27).

Adding my inner-theater terms to Greene's, the amygdala *stagehands* signal "an initial alarm bell," the VMPFC *stage manager* "integrates" emotional signals in monitoring one's potential behavior (causing hesitation in the footbridge version of the trolley experiment) and the MPFC *actor* forms a decision, which might be overridden by the DLPFC *film editor.* This dual process, faster or slower PFC system also relates to Joseph LeDoux's finding, as considered in Chapter 1 here, of a quicker, "low road" of sensory cortex to thalamus to amygdala circuits for fear, bypassing the conscious "high road" of PFC circuits, in signaling the body to act with animal survival instincts. Thus, it takes practice for brain circuits to *shift* from the fast, "emotional automatic settings" of thalamus-amygdala survival/mating emotions *and* amygdala-VMPFC moral foundations (or immoral temptations) *to* the slower, System 2 roads of outer-inner, top-down, DLPFC "manual-mode reasoning" (quoting Greene's terms, 349).[9] But insightful, *tragicomic twists*, even of popular action-adventure, romantic comedy, or melodramatic horror scenes in films and videogames, offer such rehearsals for the viewer's (or gamer's) inner theater, as might the cognitive reappraisal practice of "mindfulness" therapy. Moral foundations and reasons are still *at play* in our bio-cultural evolution, as each of us alters our neural circuits of tribal allegiances, impulsive actions, and cognitive reflections—with potential ripple effects of conserved archetypes or altered stereotypes—through what and how we watch.

ATTACHMENT FIGURES MADE MINDFUL

Philosopher Patricia Churchland argues that "morality originates in the neurobiology of attachment and bonding," through the oxytocin/vasopressin (in-group trust) network that we inherit as mammals (71). Yet such personal attachment and group bonding networks, setting up various moral foundations with tribal emotions and utilitarian ideals, involve more influences than just a Nurturing Parent or Strict Father. As mentioned earlier, regarding the ape-planet movies, humans evolved a form of childcare distinct from other primates. "Alloparenting," with multiple caretakers beyond the mother and father, enables our young to develop in complex and challenging social worlds, with diverse perspectives (Hrdy, *Mothers* 138).

And yet, the primary caretaker is extremely important to human development, especially with our very early birth, premature in comparison with other apes. Through genetic predispositions and early childhood experiences, we develop a certain "attachment style," which continues into adulthood: secure, avoidant, anxious, or disorganized/disoriented

(Mikulincer and Shaver 25).[10] One-year-old infants, in a strange lab room without their mother, show a "secure" behavior in holding her briefly and yet continuing to explore the space at her return, "avoidant" with little distress but also not trusting her for comfort when she returns, "anxious" (or ambivalent/resistant) with extreme distress at separation and conflicted responses at her return, or "disorganized/disoriented" with high levels of both avoidance and anxiety. Mothers of *avoidant* infants, in home observations, "tend to be emotionally rigid, and as well as angry and rejecting" (26). With *anxious* infants, home interactions "are characterized by lack of harmony and lack of caregivers' consistent responses." *Disorganized* infants oscillate between strategies or they do "bizarre" things in the lab room with the mother's reappearance, such as lying face down on the floor or sitting passively under a table. The "disorganized, unpredictable, and discomfiting" behavior of parents is likely due, "research shows," to their also suffering from unresolved losses or "attachment-related traumas." Vampires, werewolves, and negatively nurtured lab hybrids, show extreme attachment disorders onscreen, with bizarre behaviors conveyed across generations.

Genetics and family environment[11] set up attachment styles and group bonds, *orienting our moral and political foundations*, through the biochemistry of our brain's inner theater—with more of a Strict Father model when others are perceived as untrustworthy or evil. For example, a study showed that oxytocin levels were significantly lower in women who suffered "abuse or neglect" in their childhood (Churchland 80). Their mammalian, in-group, trust network may be more fragile, and seek firmer footing, than people with secure attachments in childhood. Yet *disorganized* attachment emerges not just with abuse, but also in children whose parents show "frightened, dissociated, or disoriented behavior" (Siegel, *Developing* 108).

Attachment ideals in politics and entertainment, through our mass media, become significant for all of us, but especially for people who did not have secure attachments in childhood. People with *avoidant* attachment may be stronger individuals in certain types of work and play, using more of the left-cortical critic/scripter to enforce and define rules in new or threatening situations. People with *anxious* attachment, on the other hand, might help a group with their alertness to danger through their right-cortical mime improviser/scene designer. Such avoidant or anxious attachment styles also relates to Brechtian distance or Artaudian immersion in viewers' participation with the scene onstage/onscreen. All of us, as animals, need a source of security, especially with our evolved extension of youth and playfulness as humans. For better or worse, this derives from parental influences that we absorb, rebel against, and renegotiate throughout our lives. Such influences relate also to the Nurturing Parent and Strict Father modes of government, involving ideals of liberty, fairness, and care for liberals *versus* additional ideals of loyalty, authority, and sanctity among conservatives.

Child psychiatrist Daniel Siegel describes how each of us goes through "natural oscillations" of an internal or external focus, of solitude or connection with others, and thus the importance to a developing child's brain of the caretaker's attunement with such patterns (*Developing* 101). This involves a "mutual co-regulation of resonating states" between child and parent brains, with the disengagement and reengagement of "alignment" (88)—akin to what we rehearse with the movie screen or live theater stage. Research shows that children with *avoidant* attachment, who are disengaged from others, have parents who are "emotionally unavailable, relatively insensitive to their children's state of mind, imperceptive of their children's need for help, and not effective at meeting those needs once perceived" (92–93). Children with *anxious* attachment, who are not easily soothed when the parent returns to the lab room, had parents who were "inconsistently available" and sometimes intrusive with "emotional invasions into the infant's state of mind" (100). This might include apparently positive acts, such as grabbing a happy child and showering it with kisses, though that disrupts its focus in play. Such children and later adults (in my view) may seek a better Other for emotional resonance, with engaged and disengaged insights, through various entertainment screens, as well as theatrical play, games, and sports.

Siegel explains how there is "an internal image of the parent" within the child's mind by about 9 months of age (*Developing* 102). This "virtual other" becomes a "filtering process" for perceiving and representing the parent and others throughout one's life. Here, one might see a correlation to the prismatic mask/lens of the inner theater, creating a "hall of mirrors" between brains: I see you through others in my past while projecting/perceiving how you see me through your others, as we interact. The filtering of perceptions by the *virtual parent-Other* inside one's brain thus involves the temporal lobe's audience of intuitive memory associations, the DMPFC's mentalizing director, the VMPFC's moral stage manager, and the MPFC's inner performing actor, which is also a social construction, having absorbed scripts and images from others.

Sometimes, we might even hear the ghostly *voice* of an influential person inside our heads, or see the Other's image, absorbed into memory traces that form the Self, which also has an inner voice. For many people, this involves a religious dimension, with the virtual Other of attachment figures extending to saints or gods or God. According to psychologist Lev Vygotsky, the inner voice of Self (vis-à-vis the Other) occurs especially through early childhood experiences, as we shift, at about age three, from image thinking to verbal thinking by internalizing our learned behaviors in verbalized interactions with others. I would also relate this to the narrative "voice-over" in some movies, a ghostly persona usually identified with the main character (as with Louis or Bella in their vampire films), mediating scenes between the screen and the viewer's inner theater, with its associated images and voices.

In daily life, the image or voice of a *virtual Other* in our heads involves not just the birth mother and primary caretaker, but also various *alloparents*, from strict and kind father figures to various relatives, teachers, coaches, and peers (perhaps also older siblings). They, along with our mothers, help us to realign our Imaginary bonds with the primal (m)Other toward further Symbolic networks and narrative identities. The play space around "transitional objects" (in D. W. Winnicott's terms) is especially significant in this development of external ties and inner-theater elements, from childhood to adulthood. Movies are transitional attachment objects of *identity play*, as we integrate and change our shared reality/fantasy perceptions. Thus, the teacher's alloparenting role gains particular value, at various developmental levels, when films are used in the *classroom* to evoke emotional engagement and yet cognitive reappraisal, as *rasa*-catharsis. This might involve the home and peer-group environments, with scripts and characters absorbed by the child's inner theater and then brought to school.

Research shows that resilience in adults is affected by "the level of security established with the primary caregiver during the first two years of life" (van der Kolk 161). Parents and alloparents, including teachers, may pass on to children the insecure attachment they experienced and still carry within as a virtual Other—or their resilience (and "Post-Traumatic Growth") if they become more mindful themselves. A study showed infants raised by severely depressed mothers developed a shift toward *right* frontal-cortex activity similar to their moms'. And yet, an eight-week Mindfulness-Based Stress Reduction (MBSR) program caused a shift toward *left* frontal-cortex activity in both long-term meditators and new practitioners, which was correlated in the latter with better immune system functioning (Siegel, *Mindful* 220–21). Mindfulness also helps to develop VLPFC-limbic circuitry that may be deficient with manic-depressive bipolar disorder (281–82). As Siegel puts it, "parents who come to 'make sense' of their lives can actually alter their attachment status and raise children who thrive" (204).

Siegel defines a "resonance circuit" in the insula and temporal lobes (superior temporal sulcus), also involving the prefrontal cortex, especially the MPFC and right VLPFC, plus the mirror neuron system—with mindfulness skills enhancing both compassion and affect regulation in these areas (*Mindful* 354–55). He theorizes that *intrapersonal* resonance within one's brain will increase *interpersonal* resonance between brains, especially between parent and child. Such resonance between neural areas may involve the inner audience, actor, and light/sound operator (or film grip) of one brain becoming more mindful and thus altering the attunement with another person's brain. *Resonance* between a viewer and a movie might likewise involve intra- and inter-personal changes, with limbic emotions and PFC reappraisals, through some degree of *attachment* to certain screen figures and yet tragicomic twists of *awareness*. Especially with horror movies, this could evoke monstrous identifications and powerful emotions, as with traditional Tibetan images in Vajrayana deity meditation,[12] but evolving then into compassion for and from the virtual Other, through personal memory networks of the temporal-lobe and insula audience.

Various therapies, such as MBSR, LKM (Loving-Kindness Meditation), CFT (Compassion-Focused Training), and CBCT (Cognitively Based Compassion Training), have translated Eastern meditation traditions into Western medical contexts, mostly without religious elements. Many studies demonstrate their therapeutic benefits.[13] But they can sometimes have dangerous side effects, including the appearance of personal nightmarish images during meditation, as with other forms of "sensory deprivation and perceptual isolation" (Lindahl et al.).[14] Yet these effects may be akin to hallucinations that prehistoric humans experienced, through sensory deprivation and perhaps other media, in the caves of France and Spain, where they left images of large herbivores, predators, and animal-human figures—as emotional pictures, moving in firelight. These might have become a very early form of *cinematic meditation*, sharing such pictures with others, between inner theaters, as a way of ritually containing and yet exploring their super-natural dreams and nightmares as evolving humans.

MOVIES AS ANIMAL-HUMAN THERAPIES?

Today's horror films about vampires, werewolves, and other beast-people attract viewers for various reasons. Some may just want the adrenalin rush of fear at a safe distance or to share that with a partner as sexually arousing. But some may seek relief from the pain of personal traumas and subsequent anxieties. Psychiatrist Bessel van der Kolk found that Vietnam veterans withstood pain 30% longer while watching a violent film clip from *Platoon* (1986, dir. Oliver Stone) than when watching a peaceful film clip (32–33). The stress of strong emotions and memory associations in veterans' inner theaters, while watching the violent scenes from that film about the Vietnam War, produced endorphins in their brains that blocked the pain of holding one's hand in a bucket of ice water, equivalent to 8 milligrams of morphine. "This suggested that, for many traumatized people, reexposure to stress might provide a similar relief from anxiety." Thus, horror movie stress, at a safe distance, might also relate to *exposure therapy* for people with phobias for certain things, such as elevators or spiders, gradually increasing their ability to experience what is feared, and to exposure and response prevention (ERP) with obsessive-compulsive disorder.

In later experiments with *trauma victims*, van der Kolk found that their *right* cortex reacts, when reminded in daily life of what they suffered, as if it were happening again in the present, without the *left* cortex making them aware of the threat being in the past (45). When triggered in this way, they become "furious, terrified, enraged, ashamed, or frozen" and may blame somebody else for it. I would add that beast-people movies, especially with tragicomic twists, might help all of us with our traumas from the past, with specific fears or general anxieties, in being reexposed to such stress—not just with an adrenalin or endorphin boost, but with a greater awareness, in the shift of neural networks between right and left hemispheres.

In *severely traumatized* people, adrenalin-fueled stress is triggered easily and does not return readily to normal levels. This causes memory and attention problems, irritability, sleep disorders, and long-term health problems (van der Kolk 46). For these viewers, as for those with an anxious attachment style, *movie stress* may backfire—if it simply triggers right-cortical, mime-improviser, bottom-up, emotional associations. It may even become *addictive* as an adrenalin (exciting) and endorphin (pleasurable, morphine-like) high. Van der Kolk describes the danger of addictions and self-harm practices when trauma victims seek to relieve bodily anxiety in the wrong way (32). But I would argue that beast-people films with ironic twists might also shift viewers toward left-cortical, critic-scripter, top-down, cognitive reappraisals, through cathartic *rasas*. This may help us to practice a more mindful awareness about our everyday stresses and anxieties, as *not* being so life threatening to the animal within.

According to van der Kolk, some trauma victims "go into denial" and ignore messages from their emotional brain, but their organs still get the stress hormones and their bodies "keep the score" (46). With such viewers, like those with avoidant (and secure) attachment to their virtual Other, the monster onscreen may offer insights as an object of fear *and* subjective identification, evoking emotional messages and yet altering cognitive viewpoints. This might involve right-cortical improviser/designer threat signals, otherwise ignored, *and* a left-cortical critic/scripter narrative awareness, with compassionate perspectives connecting the hemispheres in new ways.

As a sympathetic creature and yet object of fear, the beast-person onscreen could represent to viewers both a traumatized *child* and the abusive (or non-attuned) *parent,* who may have been traumatized earlier in life—a tragic cycle across generations reflected, too, in the blood infection curse of vampires and werewolves. Movie vampires often have wildly *disorganized* attachment styles, with avoidant and anxious aspects, like undead babies in adult bodies (or even wilder beings in immortal childish bodies). They appear as charmingly aloof, exotic strangers, yet also sharp toothed and blood sucking. At times, they are avoidant of touch and sunlight; at others, they anxiously cling to and bite their victims, like a child desperate for more than the parent's life can offer. However, to the seduced victim onscreen or to the movie viewer, in awe of the vampire's super-natural energies and immortal powers, taking and giving life, that animal-human might appear like a suddenly enraged or joyfully intrusive *parent* to a tiny child. The vampire might evoke deep memory circuits in the viewer like those in the disorganized child whose caregiver is "simultaneously necessary for survival and a source of fear" (van der Kolk 117).[15]

Likewise, werewolves onscreen transform into voracious beasts, beyond self-control, akin to a monstrous parent, yet also a virtual Other inside the child, teen, or adult, with animal-human emotions sometimes erupting and stresses changing the body. Werewolves may represent, especially to sympathetic viewers, the "denial and dissociation" that can emerge with parental "disengagement and misattunement during the first two years of life," as the later teen or adult does not "feel real inside" and resorts to extreme behavior "to feel something" (van der Kolk 121). Lab-hybrid films reveal such parenting problems more directly, with a genius inventor or a scientist couple failing to handle a new monster (even if it is himself), or a clan of such beast-people, seeking the right moral balance of nurturing mother and strict father, as godlike source of life and yet Lawgiver.

Many of the vampire and werewolf films considered here also show tragically troubled, or comically twisted, romantic and family relations—with humans choosing whether to slay the beast-person or mate with and become one, sometimes creating monstrous offspring. Even ape-planet movies have recently focused on the difficulty of raising an animal-human child in the right way, with alloparenting problems tied to larger social conflicts and cross-species clan wars. Such films may evoke, for some viewers, the radical uncertainties of an abusive or intrusive parent, which they experienced early in life and retained as a virtual Other in unconscious memory networks. But they also reflect the animal-human emotions and bodily symptoms of stress in all of us, as we renegotiate gender, age, class, and race projections in our social networks, through inner images and voices from our family lives.

Van der Kolk refers to Paul MacLean's notion of the rational brain as a "more or less competent rider" on the "unruly horse" of the emotional brain, related to that scientist's model of rational neo-mammalian, emotional paleo-mammalian, and primal reptilian levels of neural circuitry (64). Haidt puts the rational rider on an emotional elephant (*Righteous* 48–50). To my mind, this rider on a horse/elephant (and reptile/dragon) recalls Plato's ancient allegory of the human "soul" or mind as a flying chariot with reason as the charioteer, holding the reigns of two winged horses: a noble, spirited, white horse flying upward toward heaven and its dark, pleasure-seeking opposite plunging downward toward earth (*Phaedrus*). In Book 9 of *The Republic*, Plato also describes the human soul as "a multitudinous beast, having a ring of heads of all manner of animals, tame and wild." He says that the "supporter of injustice" feeds the beasts and starves the man, but the "maintainer of justice" strengthens the man by using his "lion heart" to bring together in unity the "many-headed hydra."

But van der Kolk gives another model, with new metaphors, for the animal survival mechanism that often goes awry in human brains—through misattunement in infancy that might set up later anxieties, avoidances, and differing views of justice, with moral foundations of nurturing or strict government (in Haidt's and Lakoff's terms). With van der Kolk's *castle kitchen model*, the thalamus is a "cook," the amygdala system a "fire alarm" (both in the limbic system), and the MPFC a "watchtower" that can calmly observe, predict, and make conscious, mindful choices (60–63). The thalamus-cook takes in and blends perceptions, along with expectations, sending *slower* signals to the watchtower-MPFC or *faster* to the amygdala as fire alarm. Anxious-attachment people have a hypersensitive alarm system; avoidant people have a watchtower that ignores the smoke; secure people have a better balance between them.

This cooking metaphor suggests, to my mind, the ancient Indian notion of *rasas* in art as a mix of emotional flavors between brains. But it focuses on the *alarm system* of the brain's theatrical kitchen. As shown with LeDoux's description of fear activating a quicker, "low road," the amygdala fire alarm, with feedback from hippocampus memory circuits, signals a *survival threat* through stress hormones and the autonomic nervous system (ANS) "to orchestrate a whole-body response" (van der Kolk 60). This also involves "polyvagal" ties between mirror neurons, the body's various organs, and our herd-like sensitivity to danger or safety signaled by other humans or animals around us, with subtle facial cues and other physical gestures (78; Porges).

In relation to the model I gave earlier, trauma victims with a hypersensitive alarm system have a temporal-lobe (amygdala/hippocampus) *audience*, along with subcortical, ANS/polyvagal *stagehands*, yelling "fire" in a crowded inner theater. They are thus "misinterpreting whether a particular situation is dangerous or safe" (van der Kolk 61–62). Sometimes, especially in people with an avoidant attachment style, the executive "watchtower" ignores the smoke signals from the kitchen, though they still alarm the body, causing stress and health problems. At other times, especially with PTSD anxieties, the conscious MPFC *actor* is overwhelmed by signals from the sensory *stagehands* and memory-projection *audience*. According to van der Kolk, people with intense fear, sadness, and anger have increased subcortical (alarm system) activity and *reduced* frontal (watchtower) *mindfulness* (62–63). Thus, "mindfulness practice, which strengthens the MPFC, is a cornerstone of recovery from trauma" (96).

With theater or filmmaking, the actor onstage or in the camera frame, like the inner MPFC watchtower, is partly *watching* his or her performance, aware of saying the correct lines and moving mostly as planned (left cortically), though also with room for improvisation (right cortically). Different schools of acting, such as the various American offshoots of Stanislavski's "Method," emphasize more of the left-cortical planned technique for finding and sharing something Real from the "given circumstances" of the script (and with "outside-in" bodily training) or stress right-cortical "emotional recall" (as less watchful, "inside-out" spontaneity). But many acting theories, especially those derived from Stanislavski's system, value "relaxation" techniques to help the actor be mindful onstage, even when playing an emotionally stressed-out character. Of course, the *outer* director and stage/production manager also help the actor's *inner actor and character* (along with other inner-theater elements) to prepare and be aware, while performing, of how the *outer* character might appear onstage or, at later points, onscreen.

In van der Kolk's model, if people are hypersensitive due to trauma and attachment disorders, there are "two ways of changing the threat detection system," top-down and bottom-up (63). Top-down alterations involve "strengthening the capacity of the watchtower to monitor your body's sensations," through certain meditation techniques and talk therapies, with self-awareness and new perspective taking (63). The bottom-up way helps to explore the score that the body has kept, even in avoidant personalities, or to calm its alarms, in anxious persons, with mindful "breathing, movement, and touch." Van der Kolk discusses various bottom-up techniques, such as EMDR with rapid eye movement (REM), neurofeedback, and psychomotor group therapy (related to Internal Family Systems therapy). In the last of these, participants act out roles from a patient's inner theater. The patient then becomes a protagonist/director, to "go back into the movie of your life and rewrite the crucial scenes," especially through "non-verbal, right-hemisphere realms" of communication (256–61, 281, 297–99, 305, 352). Both ways are included in "Trauma Drama" and other theater workshops for at-risk youth, which van der Kolk also describes (334–47).

For movie viewers, *rasa*-cathartic mindfulness may be mostly top-down, with the body seated in stillness. Yet the viewer's mimetic and intuitive mirror neurons (as visual and acoustic systems) also signal vicariously with the action onscreen, especially actors' facial expressions and gestures, plus the emotive soundtrack, camera movements, and editing cuts, toward empathic immersion much of the time, and yet distanced awareness at certain points. Perhaps this relates to the REM technique in EMDR therapy, which the patient performs while reimagining and narrating traumatic memories. According to van der Kolk, this may involve cross-cortical stimulation and be akin to REM sleep, during which dreams activate "distant associations" that integrate recent experiences with long-term patterns, forging "new relationships between apparently unrelated memories" (260–61). Likewise, viewers' eyes move rapidly across the screen and between angles. While watching horror and action scenes, this may evoke a dreamlike, yet narrative *integration* of "fragmented sensory and emotional traces" from traumas in their own lives (176), associated with the fictional figures onscreen. As with IFS and psychomotor therapies, this might involve the inner theater of virtual Others as well, through top-down and bottom-up, as well as cross-cortical networks, tied to the beast-people onscreen, as objects of fear and yet also forces inside us.

Each of us plays a role in our species' animal-to-human *journey of consciousness*. We might stress different "moral foundations," advocating a Nurturing Parent or Strict Father form of government. But these feelings involve brain structures that we all have in common, along with various degrees of security, anxiety, avoidance, and disorganization in our childhood to adult attachments. If we feel secure about our past and present attachments, yet sympathetically anxious about the dramas and traumas of others (as replayed by our news media), we may feel more Nurturing Parent *compassion*, beyond our family and tribal bonds, extending kinship to other classes and races, even across the globe. We might advocate the moral foundation of *care*, with extensions of *liberty*, *fairness*, and *loyalty* to those foreign to us, against our oxytocin-fueled, evolutionary need to belong and cooperate within our own clan, in competition with others. This also relates to our human drive for a personal mission and meaning in life beyond just survival, reproduction, and territory. The ideal of spreading "freedom" to others can be misplaced, however, especially when mixed with territorial anxiety or survival vengeance—as with U.S. difficulties in reproducing its sense of democracy in other cultures, from Vietnam to Afghanistan and Iraq.

If we are anxious about our attachments, or avoidant about such feelings, or oscillate between these positions, we would be more likely to see certain people as threats, based on our in- and out-group stereotypes. The amygdala-hippocampus fire alarm (as inner audience), drawing on ANS-polyvagal body networks (as stagehands), might send frequent smoke signals to the MPFC watchtower (as inner actor). Hence, the moral foundations of liberty, fairness, care, and loyalty would have a different sense, restricted to kin and allies, with projections of their opposites onto others. More weight might be given to *authority* and *sanctity*, with the inner watchtower/actor under constant threat and with right-cortical fears of *subversion* and *degradation* from within, sometimes projected outward, activating left-cortical *rage* from the kitchen-castle.

Beast-people movies evoke, but also challenge these primal emotions and moral ideals, with characters changing from human to animal (vampires and werewolves), or showing bestial elements in families, inventions, and societies (lab hybrids and simian civilizations). They often align viewers melodramatically, with or against the slayer-team, scientist-creator, monster, and ape-warrior, through moral binaries and related feelings (such as those shown in Table 7.3). But in my view, the best beast-people movies mix the *rasas* of awe/disgust, fear, anger, and courage with romantic love, sorrowful sympathy, and humor, for tragicomic twists of cathartic immersion and mindful detachment.

Movies, like other forms of storytelling, help us to integrate the traumatic memory fragments within each of our brains, which cause suffering for alienated individuals and between conflicting groups (Siegel, *Developing* 320–36). However, we should be careful about what we watch onscreen—and how we view it, value it, discuss it, and teach it to our children and students. If movie images and narratives consolidate stereotypes of certain people as beasts, drawing on primal emotions of the viewer's virtual Other and focusing them as an evil versus good binary, as a monster that must be slayed, then the potential cathartic healing may *backfire*. Anxieties could increase, especially in child and teen viewers, as undead animal-human characters, even if slayed onscreen, mate with deeper memory fragments and haunt the viewer's inner theater after the film. With adult viewers, too, personal and collective *myths of vengeance* may form, or be reconfirmed, with melodramatic victims, heroes, and villains onscreen. Groups often need such myths about the *evil other* to project their internal mimetic rivalries upon an outer enemy. Such cooperation within a group, through fantasies of righteous vengeance and vigilante justice, may help the traumatized to feel less alienated. But it also fuels competition between groups, causing much bloodshed throughout human history. Individuals sometimes feel compelled to sacrifice themselves and others, with *deceptive morals*, when "terrorists" appear. Ironically, they become like the beast-person they fear, in seeing others as that threat.[16]

Animal-human monsters and slayers onscreen, when reflecting such ironies, encourage us toward a tragicomic awareness about traumas of the past, affecting our inner theaters and how we perform in everyday life, as parents, teachers, coworkers, and morality players. They also reflect how the staging of human consciousness continues to evolve, inside our brains and in the "hall of mirrors" between us—with each of us in a full-body prismatic mask/lens, involving various animal scripts and personal ghosts. Most of the signaling between us, by the inner stagehands, is subconscious while a small raft of Self/Other awareness tries to focus all the outer refractions and inner waves onto a momentary screen, with left-cortical frameworks and right-cortical flashes to related memories and fantasies. Our ape egos aspire to be godlike, yet our remnant instincts and traumatic traces make us go wild at times, wilder than beasts, individually and collectively. But each of us can choose how we cling to suffering, with survival and mating anxieties, causing others to suffer *or not*—through how we participate, each day, in the animal drives of dominance, attraction, and trickery, reproduced and reshaped by cultural trends. Beast-people movies are thus a way to join popular culture's stream of consciousness, with its collective daydreaming of unconscious desires. Yet we might also alter it, by selecting what to value in, question about, and take from the seductive, computer-enhanced, body-swapping illusions onscreen.

So with each film you watch, especially those about beast-people, ask yourself and others around you, or on the Web, some questions like these.

- What does the film reflect in our animal to human evolution?
- Where is that process headed with our current super-natural technologies, including our screen media, as suggested by the film's fantasy or sci-fi monsters?
- Which melodramatic stereotypes entertained you?
- What tragicomic twists or ironic moments do you recall?
- Which *rasas* did such moments evoke, through resonance or distancing?
- How did subsequent developments, or the film's ending, make you realign your ideas and feelings about earlier plot points and various characters?
- How did the film engage your inner-theater elements? For example, how is the beast-person from the screen a part of you *now*, related to your inner audience of memory traces?
- How does it also interact with your inner actor, character, director, stage/production manager (or cinematographer), film editor, light/sound operator (or film grip), critic/scripter, mime improviser/scene designer, and deeper animal-based stagehands?

Notes

INTRODUCTION

1. French theorist Jacques Derrida deconstructs this binary in relation to various forms of "Otherness," through philosophy and linguistics, in ways that resonate with this book's use of neuroscience, theater, and film.

2. The notion of a fundamental human alienation from nature, through language and culture, as a "lack of being" or "want to be," comes from psychoanalytic philosopher Jacques Lacan.

3. I capitalize "Self" here and throughout this book when I mean the illusory sense of a whole, integrated identity that we often try to maintain, despite the many selves in our brain's theater. For details, see my book, *Ghosts of Theatre and Cinema in the Brain*, which explores the Self as a ghost, reflected in certain plays and films. I capitalize "Other" as meaning the social network that defines our sense of Self, which is sometimes idealized in divine figures, especially parents, pop-stars, and gods. For more on that, see my book, *Inner Theatres of Good and Evil*. The term "Other" also relates to Lacanian psychoanalytic theory and cultural studies.

4. See acknowledgements for this book.

5. On the philosophical idea of an inner "Cartesian theatre," with the *cogito* as spectator or playwright, critiqued by Daniel Dennett, but refined by Slavoj Zizek, see Pizzato, *Ghosts* (15–19).

6. Imaginary, Real, and Symbolic are terms from Lacanian cultural studies, which will sometimes be capitalized in this book, as a reminder of that reference. For Jacques Lacan, the Imaginary order (of personal memories, dreams, fantasies, perceptions, and projections) and the Symbolic order (of social rules and language) make up "reality" as we know it. But the Real is another order, repressed by, yet erupting through our mind's Imaginary and Symbolic sense making. I related these three orders to neuroscience in my previous books and will do so again in this one, but that connection is rare in Lacanian psychoanalysis.

7. The terms in this sentence come from cognitive psychologist Bernard Baars's "global workspace" model. But see also neuroscientist Iain McGilchrist: "The attentional 'spotlight' . . . is a function of the left hemisphere" (224).

8. I use the term "unconscious" here regarding the "adaptive unconscious" studied by cognitive and social psychology, but I will also refer to the "dynamic unconscious" in psychoanalytic theory, which I view as very much related to it. For a similar view and definition of these terms, see Timothy Wilson 9–15.

9. Eagleman argues that our legal system should replace the notion of blameworthiness with "modifiability," due to how criminal behavior may be produced by brain abnormalities, stemming from genetics and bad environments (192).

10. Plantinga draws on Lamarque, who defines *katharsis*, regarding ancient tragedy, as a working through of emotions, involving "self-knowledge"—due to a double perspective offered the viewer, "internal" imaginative engagement and "external" moral interpretation (68–69).

11. For psychological and sociological research on media violence increasing fear (according to "cultivation theorists") and mimetic aggression ("effects theorists"), see Cantor; Centerwall; Condry; Gerbner et al.; Hough and Erwin; Huesmann; Huesmann and Miller; Kalamas and Gruber; Krcmar; Lyon; Moise and Huesmann; Murray; Philo; and Shrum. See also Pizzato, *Inner* 310n9–16.

12. Throughout this book, I use the term "melodrama" in this sense, deriving from the stage genre that emerged in the 18th century and became dominant onstage in the 19th and then onscreen in the 20th and 21st—not in the narrower sense of emotional films about women, as the term is sometimes used in film studies. See Singer. See also Plantinga, who describes melodrama as "one of Hollywood's dominant modes, extending across many genres" (188).

13. Marco Iacoboni, who researches mirror neurons, mentions that the correlation between people watching media violence and having aggressive behavior is closer than that of inhaling passive smoke or asbestos and getting lung cancer (208–9). See Bushman and Huesmann for more details.

14. Media researcher Dolf Zillman argues, somewhat like I do, for the potential cathartic effects of violence onscreen, as a "cure" for anxiety in the viewer, through heroes who develop a "tragic flaw" and villains who show an ironic, "positive side" ("Anatomy" 160; "Mechanisms" 48–49).

15. This is not the same as Noel Carroll's theory of "clarificationism," which argues that films merely clarify what a viewer has already experienced in life. For a critique of that theory, see Grodal 86. See also Nussbaum 388–91, on alternative translations of Aristotle's *katharsis*, from *The Poetics*, not as "purging," but as purifying or clarifying emotion, "since the word obviously has a cognitive force" (389).

16. See Calderazzo 100, on defensive laughter in stage melodrama, with a cognitive shift from danger to relaxation, as the audience protects itself from fear, especially in the work of musical theater artist Stephen Sondheim.

17. See my first book, *Edges of Loss*, with chapters on Antonin Artaud and Bertolt Brecht, in relation to the psychoanalytic theories of Jacques Lacan and Julia Kristeva, exploring belief and perversion at the edge of the theater stage. See also

Hakemulder 150–51, on "catharsis" as the retrieval of repressed memories, a relief from their negative emotions, and a reduction of destructive impulses—through "identification with fictional characters" as an exploration of possible selves in *reading* literature. Experimental studies are cited that show positive effects "on moral development and empathic ability," including "enhancement of critical thinking and reduction of anxiety."

18. "Brechtianism" is an exception. It has greatly influenced semiotic and Marxist film studies. It has also been critiqued in the cognitive rebellion against such theories since the 1990s. See Murray Smith, "The Logic." He charts a lineage of Brechtian film studies from Roland Barthes and Louis Althusser to various *Screen* (journal) theorists. Smith critiques this camp as lacking "notions of consciousness and agency" and he offers the terms "*alignment*" and "*allegiance*" as better ways to understand the viewer's identification, empathy, or distancing (139–41).

19. See Raz et al., who call for an interface between "affective neurocinematics" and film studies, which would also involve Lacanian "continental theories" (300–305). Cf. Grodal. Although critical of psychoanalytic theory in film and cultural studies, he uses Freudian terms and argues that "psychological explanations based on evolutionary and biological considerations are highly compatible with historical and moderate culturalist, moderate constructivist, and antireductionist approaches to film studies" (18). See also Greg Smith's critique of Freud and psychoanalytic film theories for focusing on drives, subject positions, and desires, but neglecting emotions (174–94).

20. Psychologist Steven Pinker, while arguing that violence has *declined* in recent human history, still states that "most of us . . . are wired for violence" (483).

21. 97% of right-handers and 70% of left-handers show "left-hemisphere speech and language localization" while others show right-hemisphere or bilateral language representation (Toga and Thompson 38).

22. See Cozolino 24–25, for a brief overview of the problems with MacLean's model.

23. Zacks points to midbrain structures, interacting with the posterior parietal cortex, especially its medial (interior) areas, as key to the shifts in visual attention and eye gaze when we watch the movie screen (203).

24. Our evolutionary heritage of forebrain, midbrain, and hindbrain may even relate us to insects, extending back to a common ancestor of insects and chordates (Hirth et al.).

25. MacLean's notion of the visceral brain or "limbic system," for emotional circuits between the upper brainstem and neocortex, especially the prefrontal cortex, is still used by neuroscientists today—but sometimes questioned or redefined. I will continue to use the term with this caveat. Cf. LeDoux, *Emotional* 96–103.

26. See Lynch's theory of popular beliefs as "thought contagions," related to Dawkin's and Blackmore's notion of "memes" as gene-like components of ideologies that compete for survival in human minds and cultures.

27. For an application of Damasio's neuroscience to theories of acting onstage, see the books by Blair and Kemp. See also Lutterbie, who touches on neuroscience

while applying cognitive psychology and "Dynamic Systems Theory" to acting. Rokotnitz, too, stresses a cognitive and embodied approach to theater with a few references to neuroscience, such as the orbitofrontal cortex in relation to the Freudian id (5), mirror neurons (5–7, 72), and the "second brain" in the gut (81–82). Likewise, Cook only uses neuroscience briefly (despite her title) while applying "cognitive blending theory" to Shakespeare.

CHAPTER 1

1. See Díaz 739, for a phenomenological approach to consciousness as a "cinematic or narrative stream of explicit mental events" regarding the inner monolog.

2. See the Discovery Channel video, directed by Carol Fleisher, *Why Dogs Smile and Chimpanzees Cry*.

3. See also Solms and Panksepp, who align the "core self" with Freud's idea of *id*, as the "font of all consciousness," and the "declarative self" (or autobiographical self) with Freud's *ego* (167).

4. First discovered in monkeys in the 1980s, mirror neurons fire when a primate, including humans, watches as well as performs a certain action, such as picking up a peanut. So when watching another primate perform, the spectator's mirror neurons send signals to mimic the action, but they are usually inhibited by other circuits. Mirror neurons, along with canonical neurons (firing with typical objects for certain actions) and intuition neurons (firing when emotions are shared), form "simulations" within one's brain of what might be going on inside another person's brain, in each social situation. See Iacoboni. See also Jacob and Jeannerod 230–34 and Ramachandran, *Tell-Tale* 117–52.

5. For a counterargument with research by dualist, nonmaterialist neuroscientists, see Beauregard and O'Leary.

6. See also Dehaene et al. on specific neural circuits involved in the brain's Global Workspace.

7. Though he published this statement in 1997, Baars recently confirmed to me, by email, that it is still valid.

8. It is not only certain brain areas but also the rate of neuronal groups firing in synchrony that stages levels of consciousness. Gamma waves (oscillations of around 40 Hertz or higher) relate to highly interconnected thoughts. Beta waves (13–30 Hz, i.e., cycles per second) appear with normal waking consciousness, at the higher range with more active, alert, and focused thinking. Alpha waves (8–13 Hz) or mu waves (with mirror neurons at that frequency) relate to calm, flowing thoughts and feelings. Theta waves (4–8 Hz) appear with deep meditation and dreaming sleep. Delta waves (below 4 Hz) relate to deepest meditation and dreamless sleep. Neuro-feedback videogames are now being used, along with meditation techniques, to help people alter brain waves (Gabriel). See also Egner and Gruzelier; Monastra et al.; Nauert; and www.brainworksneurotherapy.com.

9. Recently, Kosslyn has argued for a new model of the brain with top and bottom circuits as more or less active in certain people, forming four cognitive

modes: mover (high activity in both), stimulator (high top, low bottom), perceiver (low top, high bottom), and adaptor (low in both). "Top" is defined as the top part of the frontal lobe, including the dorsolateral prefrontal and medial frontal cortex, plus the parietal lobe and anterior cingulate cortex. "Bottom" is the lower part of the frontal lobe, including the ventromedial prefrontal cortex, plus the temporal and occipital lobe. People who think in the mover mode implement plans and register consequences. Stimulators create and execute plans but often fail to register consequences. Perceivers interpret experience, put it in context, and understand implications, but do not initiate complex plans. Adaptors go with the flow. See Kosslyn and Miller 32, 87–90.

10. Goleman is also known for the notion of "emotional intelligence," which he ties to the right somatosensory cortex, right insula, right amygdala, anterior cingulate, and ventromedial prefrontal cortex (*The Brain* 13).

11. See Nadal et al. 111–12, on the greater expansion of the "how" pathway than the "what" pathway in human evolution, regarding artistic appreciation today.

12. See the *Journal of Neuro-Psychoanalysis*. See also Sacks, "Sigmund," on the Freudian foundation of current neurology, especially regarding regression and drives. Cf. Doidge 233–34, 379–80. See Kandel, *In Search* 385–88, on neurological evidence for the Freudian unconscious. See also Corrigall and Wilkinson; Cozolino, *Psychotherapy*; Kandel, *Psychiatry*; Green; Kaplan-Solms and Solms; Solms and Turnbull; Vaughan; Wexler 122–29—plus Panksepp, "Emotions," and the commentaries that follow that article.

13. Recently, Solms has revised his use of Freudian terms with the neuroscience finding that human Self-consciousness is based in the arousal, emotional, and pleasure/pain mechanisms of the upper brainstem, especially the periaqueductal grey (PAG), as an "animal within," or an "id consciousness," from which "ego consciousness" of the bodily Self derives ("Interview"). Consciousness, says Solms, "is not inherently perceptual; it is inherently affective. And in its primary manifestations, it has less to do with cognition than with instinct" ("Conscious" 12). Solms gives the example of hydraencephalic children, born without a functional cortex, whose subcortical networks still make them "emotionally functional human beings" (Solms and Panksepp 161). Basic consciousness thus involves survival and reproduction instincts in three categories: (1) monitoring interior bodily states, such as hunger and thermoregulation, (2) sensory affects to regulate those drives, such as taste and warmth, and (3) emotional networks with action tools (157).

14. Bilateral damage to the ventromedial prefrontal lobes (a.k.a. orbital frontal cortex) causes various symptoms that Solms and Turnbull relate to Freud's notion of id-driven, primary process thinking: paradoxical logic, timelessness, projection of internal imagery onto external reality perceptions, and the delusory transference of feelings between objects (103–4).

15. Panksepp, "Emotions" 31, considers the VMPFC and other brain areas in relation to Freudian therapeutic issues: anterior cingulate areas regarding depression and obsessive compulsive disorders, or lateral and medial temporal lobes and amygdala zones, plus frontal areas, regarding anxiety and anger.

16. Cf. Solms, "Consciousness" 13: "cortical representations are unconscious in themselves; however, when consciousness is *extended onto* them (by 'attention'), they are transformed into something both conscious *and* stable, something that can be *thought* in working memory." Solms calls these representations: "mental solids." When activated by emotions, they "enable the id to *picture* itself in the world and to think. But mental solids also threaten to obscure all else from view, even in primary-process cognition. One is reminded of Plato's cave" (14). See also Solms and Panksepp 165: "cortex transforms the fleeting, fugitive, wave-like states of consciousness into mental solids. It generates objects," involving both short- and long-term memories. In a similar but simpler way, Graziano points to the superior temporal sulcus and temporal parietal junction as "nodes" for consciousness, with his "attention schema theory" (166).

17. See also Keenan 49–52, for doubts about dolphins and elephants. But cf. Reiss, who found that they do pass the "mirror test." In 2010, rhesus monkeys were also found to look at themselves in a mirror when familiar with it from infancy (Keim). Pigs, pigeons (when elaborately trained), and European magpies show signs of this form of self-awareness, too (Broom et al.; Epstein et al.; Prior et al.). The language-trained gorilla, Koko, has also passed the mirror test (and uses mirrors extensively), so have a few other gorillas in zoos, but some have not (Wise 222–24). Many individuals from the other great ape species (bonobos, chimpanzees, and orangutans) do show mirror self-recognition.

18. See Murray Smith's theory of the spectator's "engagement" with a character onscreen, through alignment and allegiance, which are not always morally conjoined, especially with "perverse allegiance." See also Plantinga 106–8, who comments on these terms regarding the star system.

19. See Schechner, *Performance Theory* 266–73, on correspondences in Ekman's study of universal facial expressions for certain emotions with the tradition of Indian dance theater, based in the *Natya-Shastra*.

20. According to researchers, the fast fear route is activated within 100 milliseconds and the later, "high" pathway, which "provides object-identification information to the visual cortex," takes 300 milliseconds. See de Gelder.

21. See Sapolsky for further effects and risks to the body with prolonged stress, in relation to our animal heritage. He also points out that the rat and human hippocampus degenerates with age, increasing glucocorticoid levels and decreasing a person's ability to recover from stress—causing a "cascade" of "glucocorticoid neurotoxicity" (240–46).

22. See also Cozolino, *Human* 60, on the orbital medial prefrontal cortex (OMPFC) as inhibiting amygdala emotions, through conscious awareness. But when we are very frightened and have high levels of amygdala activation, "the OMPFC becomes inhibited, and we have a difficult time being rational, logical, and in control of our thoughts."

23. As brain researcher David Linden puts it, humans "reserve significant brain growth and synapse formation until after birth. This . . . allows sensory experience to guide the fine-scale wiring of the brain" (144).

24. See Miall 235–39, on the brain's "*defamiliarization–reconceptualization* cycle," as offering cathartic insights to the reader of literature, involving the left hemisphere's interpretive process and the right's connotative, predictive, and integrative functions. Also, reading an "action word" may trigger the brain's mirror neuron system, signaling the inner theater to simulate that action—as with the viewing of an action performed by someone else (242).

25. In *Theatres of Human Sacrifice*, I consider Martin Scorsese's *Taxi Driver* (1976) and Oliver Stone's *Natural Born Killers* (1994) as films that inspired copycat violence.

26. Cf. also Gerow.

27. Sen differs from Virtanen in arguing that catharsis is *not* homeopathic, according to the ancient Greek medical tradition. Virtanen assumes it is: "Aristotelian *katharsis* emphasizes removal, discharge—which is brought about homeopathically—while the Indian concept of *rasa* is based on relishing, though the consequents are not ignored. *Katharsis* returns the experiencer back to a normal state, but adds unity and harmony; *rasa* is considered to change the whole awareness into an abnormal state by lifting it to an extraordinary, transcendental (*alaukika*, *lokottara*), and joyful level" (62). And yet, Virtanen states: "The principle of *katharsis* seems to comprise a part of the principle of the emotional process belonging to *rasa*, because the aspect of [homeopathic] removal is also present in the realization of *rasa*." My view is more aligned with Sen's here, but I appreciate the distinctiveness of *rasa* as involving transcendent joy. Cf. also Chaudhury.

28. See Virtanen 60, on *rasa* as freeing the spectator from worldly concerns: "the aim of art is both to instruct . . . [and] to give extraordinary aesthetic delight, directed towards *moksa* (liberation). . . ."

29. There are also 33 transitory feelings, "despondency, weakness, apprehension, envy, intoxication, weariness, indolence, depression, anxiety, distraction, recollection, contentment, shame, inconstancy, joy, agitation, stupor arrogance, despair, impatience, sleep, epilepsy, dreaming, awakening, indignation, dissimulation, cruelty, assurance, sickness, insanity, death, fright, and deliberation," plus eight *sattvika* states, "paralysis, perspiration, horripilation, change of voice, trembling, change of color, weeping, and fainting" (Richmond 81). See Scheff, who as a sociologist in the 1970s extended Aristotle's notion of catharsis to further emotions, but without considering the ancient Indian *Natya-Shastra* tradition. On *rasa* theory (and Artaud's Theatre of Cruelty) in relation to Indonesia theater, see Gillit 196–97.

30. See Kale 109–12. The eight *bhava-rasa* pairs are (with some of the related "inconstant" feelings listed by Kale): pleasure-erotic (guilt, weariness, anxiety), mirth-comic (impudence, lethargy, jealousy), wrath-violent (quick temper, excitement, arrogance, vengeance), vigor-heroic (poise, arrogance), disgust-repulsive (paroxysm, turbulence), wonder-wondrous (excitement, perplexity, joy, madness, poise), sorrow-pathetic (distress, exhaustion, dejection, misery), and fear-fearful (guilt, misery, excitement, quick temper).

31. Cf. Virtanen 73–74. See also Gnoli, especially xv–xvi.

32. See Kale 173, on differences between the ancient Indian spectator as "devotee" and the Greek theater's citizen—with the tranquility, serenity, and rapture that *rasa* brings, like yoga, as unlike the "equilibrium" of Aristotelian catharsis, *if* understood as restoring "emotional balance." Cf. Schechner, "Rasaesthetics" and "Thoughts," on *rasa* as distinct from the Aristotelian tradition. However, Sen argues that catharsis and "*Rasa-srota* (as a result of purification from *Bhava*)" are parallel concepts (90).

33. See also Bharata-Muni 6.15–17. Cf. Byrski 136–49, on the relation of dramatic structure in the *Natya-Shastra* to Hindu sacrifice and the idea of life (or multiple lifetimes) as a cycle, with regard to *rasa*.

34. See Pizzato, *Ghosts*.

35. This might also relate to the structural archetypes that many myths have in common, across various cultures, as argued (somewhat differently) by Carl Jung, Joseph Campbell, Mircea Eliade, Claude Lévi-Strauss, and their followers. Cf. Adam Frank 208–15.

36. See also Sapolsky 218–22, on "learned helplessness" in experiments with rats as a way to understand human depression, caused by inconsistent, uncontrollable stressors.

37. See Shiller on the "feedback loop" of speculative economic bubbles (60–68).

38. In his recent work, Panksepp defines these seven mammalian emotional systems together as "SEEKING (expectancy), FEAR (anxiety), RAGE (anger), LUST (sexual excitement), CARE (nurturance), PANIC/GRIEF (sadness), and PLAY (social joy)" (Panksepp and Biven 2).

39. The subgenual area of the anterior cingulate is also a hub, along with the temporal pole, for the brain's courage network (Nili et al.).

40. Solms (writing apart from Turnbull) has subsequently changed his view of the neuro-anatomy of the conscious ego. See note 13 for this chapter.

41. See Winkielman et al. who point to "evidence that the somatosensory cortices and the insula might jointly contribute to emotional experience by generating a model of the current body state" (347).

42. For example, rodent sexual behavior is restricted "almost entirely" to the period when the estrogen level peaks in the female (Sapolsky 123). But such behavior is much less restricted with primates, especially humans, chimps, and bonobos, who use sex for social, not just reproductive purposes.

43. The seeking network extends from the ventral tegmental area of the brainstem, where dopamine is produced, to the medial (inner) surfaces of the limbic system and neocortex (Solms and Turnbull 39). It includes the hypothalamus, nucleus accumbens, anterior cingulate, and amygdala.

44. The dopamine reward system, and the "valuation" of choices, also involves other areas of the basal ganglia, with a feedback loop to the midbrain (deeper in the brainstem), and the frontal cortex (Glimcher 305–13).

45. Cf. Ariely, *Upside* 175–76, on the "hedonic treadmill" of consumer culture, to buy new things and "keep up with the Joneses," without realizing how the initial happiness soon wears off and we need the next new thing.

46. See Linden 147. Most female mammals advertise their pre-ovulation with swollen areas, certain odors, or postures and sounds. Sex is usually restricted to such times (except with dolphins and bonobos). Humans, along with orangutans and vervet monkeys, are rare among primates in having a hidden ovulation cycle. Human females are also rare in experiencing menopause and living beyond their declining fertility, unlike other animals.

47. The brain's humor network has recently been defined as involving the perception of incongruity in temporo-occipito-parietal areas, plus reward and salience processing in the inner, cortical-limbic, dopaminergic system and amygdala (Vrticka et al.).

48. The rage system may also involve the cerebellum at the back rear of the brain (Panksepp 204–5).

49. See van der Kolk 31–33, on trauma survivors as thrill seekers and on a study with Vietnam veterans showing that watching a violent war movie may act like an addictive drug, blocking pain receptors in the brain, especially for traumatized people.

50. See Ramachandran and Blakeslee 191.

51. Left neocortical structures are predisposed to regulate language in about 90% of people today, but right-hemisphere and subcortical areas are involved with specific language functions as well (such as the subcortical basal ganglia, which sequence automatic word-usage, building on earlier evolved, movement processes). Also, the right brain may assume a greater role in language after left-brain damage. Such lateralization of left and right hemisphere communication functions, with distinct verbal areas in humans, relates to other species, too. Complex aspects of birdsong and frog vocalization are regulated by a dominant brain side (usually left). See Philip Lieberman 144–45, 158–59.

52. Research shows that English-speaking mothers name objects more, using more nouns and fewer verbs, when talking to their 18-month-old babies, compared with Korean-speaking mothers who talk more about actions (Gopnik et al. 89). Likewise, the children speak likewise and English-speaking children also categorize objects earlier, while Koreans solve action problems earlier, like getting something out of reach by using a toy rake (90).

53. McGilchrist makes a similar argument as a neuroscientist, exploring European history and art in great detail, with comparisons to Asian cultures.

54. It was also found that women's brains were activated more than men's, in a test using women's "stress sweat," while men's had an equal effect on both genders (Radulescu and Mujica-Parodi).

CHAPTER 2

1. Most of the prehistoric cave art has been found in Europe. But recently, a handprint was found on the island of Sulawesi in Indonesia, dated to 40,000 years ago, with a "pig-deer" figure near it, dated at 35,000 (Aubert et al.).

2. See also Pizzato, *Inner*, for details on another dozen caves that I visited in 2009.

3. For a critique of such a "magico-religious" theory, see zoologist R. Dale Guthrie 8–9, 36, 413, 427. In support of the theory, with a focus on Chauvet, see Whitley. See also Clottes, *Cave*, for its photos and timeline.

4. On the relation of cave surfaces to prehistoric paintings, see Ogawa.

5. For photos of the Chauvet cave, see Clottes, *Chauvet*, and Curtis.

6. Cf. Lewis-Williams, "Of People," 146, 150.

7. But see McNamara 1: 65, who views such Venus figurines as being created by women, "to promote female political alliances," unlike the shamanic cave paintings.

8. Cf. McNamara 1: 19–26, on the possibility of Neanderthal bear cults, using cave rites, as the origin of human religion.

9. Red and black paint were also combined, in an unusual way, for purple shading at Tito Bustillo. And there is a figure that may be a whale. For those traveling to the cave, across the river Sella from Ribadesella, it has a nice interpretation center and tickets may be booked in advance by phone. It is also within a 60–90-minute drive of the two other Asturian caves described in this chapter, El Pindal and El Buxu, but they are more difficult to reach, through winding, narrow, mountain roads, sometimes shared by grazing cattle, and with long walks to the caves from small parking areas. Reservations should be made, especially at those more remote caves, since guides are not always present.

10. See Beltrán.

11. Cf. Lewis-Williams, *Mind* 36–37. He also details other chambers of the Altamira cave.

12. See Pizzato, *Inner* 39.

13. Cf. Lewis-Williams, *Mind* 204.

14. It is difficult, yet possible, to visit all five Cantabrian caves described in this chapter in just one day, with 60–90 minutes of travel time needed between them, except for the adjacent caves, El Castillo and Las Monedas. Those two have a large parking lot and interpretation center, but the other caves are located on narrow mountain roads near small villages (though El Pendo is also near the large city of Santander). Tour reservations may be booked online.

15. Cf. Montelle, *Palaeoperformance* 85–86, on his experience of "fragmentation" in the Gargas cave.

16. Cf. Bataille 27–36.

17. Cf. Renfrew, who finds evidence of "artistic patterning" in pierced shells, which were probably used as beads 70,000 years ago in southern Africa. Those may also show the emergence of Self-awareness in a new way.

18. See Mithen, *Prehistory* 140–42, 163–65, 208. Lewis-Williams, *Mind* 107–11, relates Mithen's theory to his own about cave art.

19. Unlike Merveilles, the Gargas cave (with a small museum) is difficult to reach, on winding mountain roads, west of St-Gaudens, in the area of Aventignan.

20. The Villars cave is about an hour's drive north of Perigueux, near the tiny village of Le Cluzeau.

21. Curtis sees it as a provocation (118). Comparing this with another bison and man scene in Lascaux, he calls it a "universal myth."

22. There are about 75 anthropomorphic figures in all of prehistoric cave art, a very small percentage of the total figures discovered. See Lewis-Williams, *Mind* 277. He makes comparisons with southern African (San) and Native American rock art, which involve many human figures. "[S]ome societies focus on animals as sources of power or as helpers, while others emphasize anthropomorphic depictions of shamans and spirits."

23. See Hatfield et al.; Papousek et al.; Sullins 166–74; Gallese and Goldman 493–502; and Blakeslee and Blakeslee.

24. See McGilchrist 189 on the "holistic or *Gestalt* nature of thought associated with the right hemisphere . . . [and] the analytic nature of thought in the left hemisphere," regarding gestures and verbal language.

25. See Menary.

26. Even today, "Hölle" is the word for both "cave" and "Hell" in German. Yet, Our Lady of Lourdes, a nineteenth-century vision by a girl in a grotto in the Pyrenees near Gargas, is still valued by many Catholics as a miracle-working image and site.

CHAPTER 3

1. See Ekman on the universal nature of human facial expressions and vocal tones, signaling certain emotions, despite different cultural display rules or personal attempts to mask them (with micro-expressions still leaking out).

2. However, even at age three, children do not integrate past and present selves, with a sticker on the face in a video taken just minutes before—and refer to the child on the tape by name, not as "me" (Gopnik 46).

3. Chimpanzees follow gazes, and use them for pointing, but do not exhibit joint attention (Fernandez-Duque and Baird 83).

4. Even at age three, children have difficulty realizing how their own perspective changes, just a few minutes after acquiring new knowledge (Gopnik et al. 46). They "make the same mistakes whether they are reporting their own mental state or predicting the mental states of other people" (47).

5. Barr and Keysar find, however, that the "egocentrism" of children prior to age four, lacking an appreciation of others' beliefs, and in adults with frontal lobe damage, is also "present in normal adults" (273).

6. Chimpanzees have also been found to refuse a food reward if another gets a better one for the same task. Chimps in unstable, short-term groups respond more to such inequity than others (Brosnan). Some researchers argue, however, that it is the desire for better food, rather than the other animal getting it, that leads to the refusal: "expectation" instead of "inequity aversion." See Bräuer et al., who also experimented with orangutans, gorillas, and bonobos. They found that chimps do more begging, rather than food throwing, which may show expectation, not

refusal. And yet, dominant chimps are less likely to accept low-quality food if a subordinate gets something better. See also Jensen et al., who refute Brosnan's study, finding that "in a food-acquisition context, chimpanzees are essentially indifferent to differential payoffs for a conspecific" (1021).

7. Wolves, coyotes, and dogs that do not play fairly are sometimes excluded from play (Brosnan 155). Ravens show "third-party enforcement" of the social norm that "a food possessor can maintain possession" of its food (168).

8. Bird references in de Waal's model indicate parallel developments on a very different branch of the evolutionary tree, though with similar forms of social cooperation and competition in survival and reproduction.

9. This distinction relates to LeDoux's research, mentioned in Chapter 1 here, on a quicker "low road" network of fear, directly through the amygdala, and slower "high road" network through the prefrontal cortex, with the possibility of "educating the low road," as Goleman, *Social* suggests (97–100). See also Panksepp and Biven 222–33.

10. See van der Kolk on various performative therapies that deal with the mental and physical effects of early traumas: internal family systems (IFS), Pesso Boyden system psychomotor (PBSP), eye movement desensitization and reprocessing (EMDR), yoga, sensorimotor psychotherapy, and theatrical exercises for at-risk and convicted youth. See also Ekman 215, on "an individual's *emotional profile.*"

11. Recent research shows that human neoteny includes the late development of our most advanced brain area, the prefrontal cortex, through genetic differences from other primates, increasing the influence of culture (Somel et al.).

12. See Raffaele on recent observations of bonobos in the wild, finding that the alpha male may dominate, as in chimpanzee groups, but with less violence. He determines where the troop eats and sleeps, and when it moves, or defends it against leopards and pythons, like a military "general." But the alpha female still makes final decisions, like a "queen."

13. See Timothy Wilson 121–22, on an experiment with homophobic men who had erections when shown homosexual porn videos, supporting the psychoanalytic notion of a "reaction formation," with repressed desire related to aggression against those who represent it.

14. See Greeson, for a meta-analysis of research from 2003 to 2008 on mindfulness training in relation to mental and physical health.

15. Baumeister also uses this rationale to explain why female genital mutilation is conducted by women against girls (358).

16. For a summary of the critiques of an earlier theory of "group selection," by V.C. Wynne-Edwards in the 1960s, see Trivers 79–85.

17. Cf. Gómez for a distinction between pretense about "intentional relations with non-existing objects or properties," which he says apes lack, and about "intentional relations with existing, although not necessarily currently perceived, objects," which apes have (586).

18. See also Tomasello, *Why* 27–28.

19. See Oughourlian for a development of Rene Girard's theory of mimetic desire and rivalry regarding hysteria, spirit possession, and hypnosis.

20. See also Averbeck.

21. For further details, see Berridge et al.

22. See also Wicker et al.

23. See Dawkins, *Selfish*, for his metaphor of the "selfish gene." See also Dawkins, *Extended*, Blackmore, and Aunger, on "meme" theory, which extends the selfish gene to the realm of ideas using human minds as vehicles. For the more specific, biological notion of genome "intentionality," see John Maynard Smith, especially 144.

24. See Zhu et al., on the different cultural shaping of medial prefrontal cortex activity in Western and Chinese subjects, regarding ideas of "self" and "mother" (more connected in that brain area with Chinese).

25. See Haidt, *Righteous*, for a social psychologist's view of secular conservatism as different from "orthodoxy" and valuing "moral capital," which he says is a "blind spot" in liberalism (288–94).

26. Consciousness itself may be a collective experience—through connections with others, existing or imaginary (Noë).

27. On waves of "information" in quantum physics, biology, and theology, see Davies and Gregersen.

CHAPTER 4

1. This also relates to Friedrich Nietzsche's notion of the Dionysian chorus as the womb of ancient tragedy (in *The Birth of Tragedy*) and to Julia Kristeva's idea of the psychological and linguistic *chora*, which I explain in Chapter 2 here, regarding prehistoric cave art.

2. See Pizzato, *Inner* 169–70, and "Soyinka's."

3. There were few references to vampires in literature prior to this, but also not many vampire novels after it, just 65 published until the 1960s (Day 14–15). But there were early stage versions in England and France in the 1820s based on Polidori's novella (Jenkins 71–72).

4. Lesbian vampire films inspired by Le Fanu's novella, *Carmilla*, include *Vampyr* (1932), *Blood and Roses* (1960), *The Vampire Lovers* (1970), *Lust for a Vampire* (1971), *Twins of Evil* (1971), *Vampyres* (1974), and *Lust for Dracula* (2005).

5. The word "entertainment" suggests this, from the Latin terms (*inter* and *tenere*) for "between" and "to hold." Unfortunately, some in the mass audience seem to be inspired toward copycat (or copy-vampire) violence by role-playing games, novels, and films about vampires (Jenkins 28, 32–33).

6. See Gelder, *New* 36–43, on the neighborliness of vampires in several films not considered in detail here: *Frostbitten*, *Let the Right One In*, and *Let Me In*.

7. "Capital is dead labor, that, vampire-like, only lives by sucking living labour, and lives the more, the more labour it sucks" (Vol. 1, Chapter 10, Section 1). For a critique of Marx's view, as unfair to vampires because it makes them, along with capitalism, purely evil and their slayer just heroic, see McClelland 152–53.

Friederich Engels also referred to the middle class as a "vampire" bleeding the workers (Jenkins 83).

8. Holte calls Schreck's Nosferatu a "supernatural human rat" (33).

9. This racist stereotype can still be found today, sometimes in relation to *Nosferatu*, on White Supremacist websites such as http://www.stormfront.org/forum/t169895/ (posted on Dec. 6, 2004).

10. The eugenics movement to sterilize the "feeble minded," so that their children would not be a burden to society and weaken the gene pool, began in the United States prior to the rise of Nazism in German and continued after the war—until 1974 in the state where I live, North Carolina. More than 60,000 people were sterilized in the United States.

11. Cf. Gelder, *New* 13–15. He uses a comment on this film by Slavoj Zizek to remark that vampires are both cinematic and pre-cinematic, modern and ancient, with "nothing much to do with reality at all, because they are Real," in a Lacanian psychoanalytic sense.

12. See Wise who categorizes honeybees, with their intelligent dances communicating the direction and distance of food, water, and potential new hive locations, as possessing "sufficient autonomy for liberty rights," along with dogs, elephants, and parrots, though on the low end of the spectrum (241).

13. Thousands of TV episodes have involved vampires, too, from *Dark Shadows* as a daily daytime series (with Barnabas Collins) in the late 1960s to *Buffy*, *True Blood*, *The Vampire Diaries*, and *The Originals* as weekly nighttime shows more recently.

14. Auerbach calls Langella's Dracula "erotically easy in his animal self," with the director allowing him to play himself as bat and wolf, instead of giving artificial transformations (145).

15. A similar story about Vlad the Impaler as Dracula the Vampire fighting Turkish Muslim invaders forms the core of the recent film, *Dracula Untold* (2014, dir. Gary Shores).

16. See Gelder, *New*, about such a cinematographic moment in this and other vampire films.

17. Holte refers to Coppola's remark about Dracula as a fallen angel, like Milton's Satan, showing that the director saw the Count as a "complex hero from the beginning of his planning for the film" (83).

18. Holte sees this as redemptive with Dracula going to "heaven" (85–86). I am not so sure.

19. See Abbott, who gives a longer list of vampire types and calls the 1970s "The Vampire Decade" (4, 75–87).

20. See Creed, *Monstrous* 62, on the vampire as a "menstrual monster."

21. See Pizzato, *Edges*, on Antonin Artaud's theory of a "theatre of cruelty" in relation to his mother.

22. Cf. Zunshine's cognitive theory of "mind-reading" in visual art and cinema (29).

23. See Gelder, *New* 80–83, 107–15. He also considers international films with Asian or Jewish vampires (or about vampire filmmaking), such as *Irma Vep* and

The Breed (51–57, 124–27). For more on Asian vampires, see the six essays by various scholars at the end of Browning and Picart's volume.

24. The feared diseases porphyria and pellagra may be other sources (Jenkins 46–48).

25. Also, in *House of Dracula*, the Count's blood is found to contain an unknown parasite and Larry Talbot learns that his werewolf symptoms may be due to pressure on the brain, suggesting a neural disease like rabies, though the film's doctor wants to treat him by reshaping his skull.

26. I thank my former student, Carley Foster, for her insights about this and other vampire and werewolf films.

27. Abbott acknowledges that in the last couple decades vampire films have involved computer-generated images, a development that "reinfuses the vampire with the spectral and . . . virtual" but she does not consider the spiritual and sacrificial dimensions of the postmodern vampire's human-animal hybridity (214).

28. For more on the witch and Wiccan elements of the TV series, see Winslade.

29. On Jungian archetypes in TV's *Buffy*, see Frankel. On various gender, race, and class issues in the series, see the volume edited by Levine and Parks. On the biology, chemistry, and physics lessons in the show, see Ouellette.

30. Newer technologies are often used in vampire-slaying films, such as *Van Helsing* (2004) and the *Blade* series (1998–2004), involving the viewer in videogame-like violence. This sometimes occurs in *Buffy* episodes, too, for example in season 2, episode 14, when Buffy uses a rocket launcher to destroy the villain.

31. See Jowett on the various non-heterosexual orientations that the TV series explores, with good and bad women and men, and certain characters complicating these simple melodramatic identifications.

32. I thank my colleague, Dr. Lisa K. Perdigao, for her insights about the *Buffy* film/TV series and the *Twilight* films.

33. See Cunningham on the various styles brought to the first three films by their different directors and DuBouis on the diminishing of Bella's narrative voice after the start of the first film, whereas she narrates most of the novels. For more on that comparison, see Kapurch.

34. See Murphy on the "frenzy" of teen female fans, and even moms, for Edward, through the novels, set up by Buffy's sensitive vampire boyfriend, Angel, on TV (58, 66). See also Bode, on critical views of the female teen and adult fans of the first film.

35. Cf. Erzen.

36. See Natalie Wilson 135–50, on the Mormon reinterpretation of the Fall of Man as not tragic, but "fortunate," leading to the human development, at least of white believers, toward becoming gods—in relation to Meyer as a devout Mormon and the *Twilight* novels.

37. Simple staking is not enough to kill vampires in the *Twilight* film series. See Jenkins 171–76, on medieval practices of decapitation and cremation of corpses to prevent witches or vampires from returning from the grave.

38. *Cronos* (1993) likewise combines hypodermic injection technology and the super-natural with a scarab-shaped clockwork device from 450 years ago that contains a strange insect whose sting transforms a human into a vampire.

39. A kinder view of a single child vampire in 1980s Sweden, in a budding friendship with her 12-year-old neighbor, yet needing blood from others, is shown in *Let the Right One In* (2008) and its American remake, *Let Me In* (2010).

CHAPTER 5

1. See the essays by David F. Bjorklund and Katherine Kipp, by Richard Byrne, and by Michael C. Corballis in Sternberg and Kaufman 27–54, 79–96, 117–44. See also Cheney and Seyfarth.

2. This is a general overview given in Cozolino. For specific growth spurts within each hemisphere, see Thatcher et al. and Chiron et al. Cf. Hervé et al. for more recent research.

3. Empathy involves limbic emotion circuits plus the right inferior parietal and right lateral prefrontal lobes, the latter "inhibiting the automatic tendency to espouse one's own point of view" (McGilchrist 57).

4. Regarding adult "prosocial goals," social psychologist Ervin Staub defines two "value orientations": (1) a *prosocial orientation* with "concern about the welfare" of others and (2) a focus on *duty and obligation*, societal rules, and abstract moral principles (111). I would associate the former more with the right cortex and the latter with the left.

5. In humans, "the left hemisphere favours analytic, sequential 'processing,' where the right favours parallel 'processing' of different streams of 'information' simultaneously" (McGilchrist 228).

6. "Eidetic imagery has been defined as the memory capability to retain an accurate, detailed image of a complex scene or pattern. It is known to be present in a relatively high percentage of normal children, and then the ability declines with age. Our present study shows that young chimpanzees can quickly grasp many numerals at a glance, with no decline in performance as the hold duration is varied. Moreover, the young ones showed better performance than adults in the memory task. Our study shows that young chimpanzees have an extraordinary working memory capability for numerical recollection better than that of human adults. The results fit well with what we know about the eidetic imagery in humans" (Inoue and Matsuzawa). Monkeys (rhesus macaques) likewise succeed at holistic addition and subtraction tests with a small number of spots (Dell'Amore). See also McGilchrist 83: "whenever an image is either only fleetingly presented, or presented in a degraded form, so that only partial information is available, a right-hemisphere superiority emerges . . . [but with] certainty and familiarity, a left-hemisphere superiority emerges. . . ."

7. See Nielsen and Tomaselli 734: "when an adult models, children automatically assume that the adult intends for them to learn something new, and hence they interpret the specific actions of the adult as being purposeful. A mind evolved

to construe actions in this way cannot be easily overridden, even when the actions demonstrated are clearly superfluous."

8. Cf. Whiten et al. 2427: "chimpanzees may be restricted in their capacity to modify and upgrade the first approaches to a task they learn, in contexts where children show a readiness to upgrade, facilitating a process of cumulative cultural change."

9. As Durham points out (106–7, 184–86), the female ideal of thinness is influence by Western cultural values that have shifted from mature plumpness, in the 1700s and 1800s (with corset and yet bustle) to thinner body sizes after the 1950s. Western beauty ideals of youth, slimness, white skin, and large eyes have also been exported through the media to many cultures worldwide.

10. Staub distinguishes blind patriotism, with "embedded self-concepts," from constructive patriotism with a connected self that can still be critical of the group (355). He also explores how helping is *less* likely with *more* bystanders due to "pluralistic ignorance" (people seeing others not reacting become inclined to see the event as a non-emergency) and "diffused . . . responsibility" (74).

11. I have added the term "theatrical perspective-taking" to Lieberman's argument. But his examples from video experiments suggest this, too, and he points to a study showing that people who "read fiction tend to have stronger mentalizing abilities, suggesting that engaging with fictional minds may strengthen this system" (154).

12. Matthew Lieberman tested three types of treatment and found that affect labeling (with phobics verbalizing their fear while seeing a real tarantula) worked best, especially with more negative labels. This worked better than exposure therapy (just seeing the spider) or reappraisal treatment (verbalizing an ideal, logical reaction, such as "the spider isn't actually dangerous," while seeing the tarantula).

13. The study was by Ochsner et al., published in 2002, but Schwartz et al. also refer to it. See also Ochsner and Gross.

14. See Cozolino, *Human* 71–72, on a related area, the "right-biased" orbital medial prefrontal cortex, as the first region in the frontal lobe to develop in children, during the first year of life, and as "the executive center of the right-hemispheric networks of attachment, social relationships, affect regulation, and higher-level input into bodily homeostasis."

15. Damage to the OFC leads to "distractibility, impulsivity, emotional outbursts, shallowness, argumentativeness, verbal and physical aggressiveness, hypersexuality," and other problems with immoral and risky behaviors (Beauregard et al. 171). Various studies show that the OFC is also involved in "decision making related to reward and punishment," plus "empathic and moral judgments" (172).

16. Here, Lieberman may be echoing, albeit ironically, Freud's notion about dreams as the "royal road" to the unconscious. To my mind, though, Lieberman's research demonstrates even more of Lacan's axiom: "one's desire is the desire of the Other."

17. From now on, I will use the term *inner actor* in this more specific sense than Baars's "player" onstage, which refers to any percept or concept that becomes conscious.

18. "Moral judgment involves . . . the right ventromedial and orbitofrontal cortex, as well as the amygdala in both hemispheres. Damage to the right prefrontal cortex may lead to frank psychopathic behaviour" (McGilchrist 86). This relates, in my view, to the theatrical "stage manager," who takes over control of the show after the director finishes the rehearsal process and "calls" the show by signaling the light and sound control operators, stagehands, and actors just prior to and during each performance, like an orchestra conductor, but verbally. It also relates to the "production manager" who coordinates the technical aspects of the show during the rehearsal (or film shooting) process.

19. See Roth et al. on MPFC activation with pride, more than with shame/guilt, in relation to various limbic, emotion-processing areas.

20. See Wagner et al. on the right OFC (VMPFC) as central to the brain's guilt network, also involving the DMPFC.

21. See Coricelli et al. on regret as involving the MOFC, ACC, and hippocampus.

22. With the term "scene designer" here I include not only the set, but also costume, lighting, and sound designers in theater, or the production designer and art director in movie making.

23. For details on the inhibition and excitation between hemispheres, see Hellige. He summarizes research and theories from the 1970s to 1980s on how "the corpus callosum tends to produce mirror-image patterns of activation and inhibition in the two hemispheres" (174). It mostly serves as an inhibitory barrier, "preventing maladaptive cross-talk between the processes for which each hemisphere is dominant" (175). McGilchrist adds that the "suppressive effect" of the left on the right is greater than vice versa (218).

24. While considering the research on split-brain patients, after their commissurotomy, McGilchrist states: "the two hemispheres have wills that may not always be in harmony" (219).

25. As McGilchrist puts it, the left hemisphere's sense of "belief" involves degrees of certainty, as in "I believe that is so." But the right's includes "care: it describes a *relationship*, where there is a calling and an answering, the root concept of 'responsibility,'" as in "I believe in you" (170).

26. Human infants who cry while the mother is absent have "higher baseline levels of right prefrontal activation" than infants who show less concern about abandonment (Davidson and Begley 38).

27. Greater Social Intuition involves higher activity in the fusiform gyrus and lower in the amygdala (Davidson and Begley 73). Attention uses the prefrontal and parietal cortices (238). Sensitivity to Context involves the hippocampus, which is in the temporal lobes on each side of the brain, like the fusiform gyrus and amygdala, but is used also to form long-term memories (76–77). In rhesus monkeys, the anterior part of the hippocampus, close to the amygdala, regulates behavioral inhibition in response to different contexts. People with post-traumatic

stress disorder (PTSD) have a loss of hippocampal volume and problems with social context, conflating normal situations now with dangerous traumatic memories (77). However, hyperactivity in the hippocampus can cause "excessive focus on context, which can inhibit emotional spontaneity."

28. See also Lane and McRae 105–6, for a model of implicit (brainstem, thalamic, limbic) to explicit (paralimbic, prefrontal) emotions, involving background feelings, focal attention to feelings, and reflection about feelings.

29. Cf. Broadhurst who relates this to "the mechanism of love" and to digital artwork (66).

30. The study used "memory confusion protocol" with errors in recall revealing whether subjects are categorizing target individuals into groups. "Subjects are asked to form impressions of individuals whom they will see [in photos] engaged in a conversation. They then see a sequence of sentences, each of which is paired with a photo of the individual who said it. Afterward, there is a surprise recall task: the sentences appear in random order, and subjects must attribute each to the correct individual. Misattributions reveal encoding: subjects more readily confuse individuals whom they have encoded as members of the same category than those whom they have categorized as members of different categories" (Kurzban et al. 15388). With verbal allegiance cues, reinforced by arbitrary visual ones (such as shirt color), subjects encoded a new dimension of coalition membership "far more strongly than race" (15389). The subjects were all college undergraduates and primarily Euro-, Hispanic-, and Asian-American. The people they were shown wore colored jerseys as members of rival basketball teams, involved in a fight earlier in the season, and were Euro- and African-American.

31. See Timothy Wilson 110–12, for evidence from social psychology that "on balance, a stranger knows as much as you do about the causes of your feelings, judgments, and actions."

CHAPTER 6

1. Cf. Packer, who defines a subgenre of sci-fi films, "neuroscience fiction films," including the Dr. Moreau and ape-planet movies considered in this chapter.

2. Barrett also summarizes research showing that 3 to 5-year-old Israeli children had a bias toward believing in God's immortality and 5 to 12-year-old Spanish children felt that a puppet-show mouse would still have desires and thoughts after death (116–19). With his own experiments, Barrett found evidence in adults, too, for an intuitive (unconscious) tendency to add human-like features to an abstract God, during a story-comprehension task, both in the United States and in India (160–65). Barrett agrees with anthropologist Stewart Guthrie that our ancestors evolved a "hyper-sensitive agency detection device" in their brains for its survival benefits, which became extended from animals in nature to the supernatural (208–9). Barrett also cites research showing that modern humans with a high proportion of "strivings" (day-to-day goals) involving God or religion

"tend to have better physical and mental health" with less conflict between goals (233–34).

3. See Jackson on *Splice* as following the Frankenstein trope, yet with a corporation as the mad scientist. See also McQueen for a Deleuzian "becoming-animal" approach to the film, as breaking the "Frankenstein barrier" of denied futurity. Hantke mentions that the names of the mad scientists in *Splice,* Clive and Elsa, may be a joking reference to the actors, Colin Clive and Elsa Lanchester, in the first two Frankenstein films (109). He also says the film is more like Mary Shelley's original novel in presenting a Bildungsroman of the monster's point of view.

4. In Wells's original novel, the feminine puma created by Moreau functions differently, killing him and bringing climactic chaos to the island, leading also to the death of Montgomery—representing feminist politics in the 1890s, as well as a "feminized landscape and nature that eventually returns to wreak disaster on the masculine scientist" (Cole 148).

5. See also Zack for a comparison of racist elements in the three feature films made from Wells's novel.

6. See Grant 156, for a simpler view of the film's ending: "The beastliness within us is clearly repressed in the film. . . . "

7. Lammes sees further elements of Moreau's Britishness, yet also finds that the American Parker "decontaminate[s]" Moreau's "bad science" (73, 81).

8. There was also a French test of a nuclear bomb near Brando's island home during the shooting of the film—and many conflicts between the initial writer/director, Richard Stanley, his replacement, John Frankenheimer, and the actor Val Kilmer. Many of the scenes were reshot, rewritten, and improvised. The finished film was a failure at the box office, losing money for the studio.

9. A sequel was made in 1944, directed by Gunter V. Fritsch and Robert Wise, with Irena as a ghost or imaginary friend to Oliver and Alice's daughter.

10. In the original short story, by George Langelaan, on which the film was based, Helene commits suicide at the end.

11. See Beard 153, on the distinctively un-Hollywood, Cronenbergian elements in this Hollywood remake.

12. See Creed, "More," on the womb envy of Seth Brundle as a Frankenstein-like scientist.

13. See Rochat and Striano, "Social-Cognitive" 13, about human infants: "It appears that newborns up to approximately 6 weeks behave in many ways as externalized fetuses."

14. See also Rochat and Striano, "Social-Cognitive," for details on infants' primary intersubjectivity, as dyadic and "contemplative," from 2 to 7 months of age, and then secondary intersubjectivity, as triadic and "intentional," from 8 to 12 months.

15. From 3 to 18 months of age infants only follow head turns, but after that they start to follow eye movements also (Moore 254).

16. Cf. Hogle 163, who describes the final Dren as "a vampiric male werewolf-figure with bat wings."

17. See Eric Greene 25–28, on the influence of Serling's initial screenplay on the subsequent film's adaptation of the novel, introducing the theme of nuclear destruction even though Boulle objected. Also, the apes were not as strictly segregated according to class in the original novel (32).

18. See Kirshner who praises the film's "sustained assault on religion" during the Cold War (43).

19. Others have written about racial issues in the original film (Susan McHugh) or the initial five-film series (Eric Greene in great detail, Sandy Rankin more critically, regarding capitalism as well), but not in relation to the brain's inner theater, as explored here.

20. Eric Greene sees the chimps in this film as "analogous to Jewish people in the United States," marginalized and discriminated against, although considered white. He also mentions a brief reference to "quotas" by Dr. Galen, a chimp colleague of Zira (31–32).

21. Screenwriter Michael Wilson suffered a similar suppression of speech and human rights as a victim of blacklisting during the McCarthy Era, a decade before the film was made (Eric Greene 38).

22. Eric Greene sees a political association here with a familiar image in the 1960s: the water hose as a "tool to enforce racial supremacy and quell resistance" (36).

23. In June 2013, the U.S. National Institute of Health announced plans to "substantially reduce" the use of chimpanzees in biomedical research (Moses). NIH director Francis Collins stated that chimps' "likeness to humans" has made them "uniquely valuable" in such research but also "demands greater justification for their use" (Moses). At that time, the United States was "the only remaining nation in the developed world" that used great apes as research subjects. But monkeys were still commonly used "in the US, Europe and many other parts of the world."

24. Eric Greene sees Caesar as a Christ figure: born in a manger-like place, with parents hunted by a ruler, while "figured as a mystical presence with uplifting powers for his people," and with his later ape followers naming him as a "savior sent by their god" (92). But Greene also sees him as having "more extensive" parallels to Moses in the Exodus tale.

25. Eric Greene observes that the apes are darker skinned in *Conquest* than in previous films in the series, with Caesar looking like a mixed race African American (99). He describes its political appeal to black audiences in the 1970s, as with "blaxploitation" films (84–85). And he comments on the irony of white supremacist groups in the 1980s referring to the NAACP as a "planet of the apes," showing their fear of a racial apocalypse (177).

26. See Eric Greene 115–17, 136–39, on the changes made by the Corringtons to Dehn's fatalistic story, with a Cain and Abel morality tale and a more hopeful ending that "empowers the victim to *choose* not to become the victimizer" (139).

27. Various producers, directors, and stars became involved in failed attempts to revive the series at Fox Studios in the late twentieth-century, including Oliver Stone, Chris Columbus, James Cameron, and Arnold Schwarzenegger (Argent).

See Eric Greene 152–86, on the TV shows, comics, toys, and other "comebacks" of the film series in the 1970s–90s.

28. Lindner compares the original novel, 1968 film, and 2001 remake: "Whereas both Boule's text and Shaffner's film feature animalized humans, Burton's version in turn re-animalizes the ape in terms of his natural behaviour as opposed to his humanized culture" (122).

29. Cf. Vint, on the "anxiety about the beast within" and the nihilism of patriarchal, militaristic humans in the original series and the 2001 remake, compared also to Stanley Kubrick's film, *2001*.

30. See Hersch 176–84, on such NASA animal astronauts in relation to ape and human sexual relations in the *Planet of the Apes* films.

31. Eric Greene calls Aldo a "totally unsympathetic . . . one-dimensional villain," more so than Zaius, Hasslein, Breck, or Kolp: "the filmmakers missed the opportunity [with Aldo] to explore more rigorously the complex and painful racial politics that inspired *Battle*" (120). He cites one of the screenwriters admitting that the gorillas in that film were "reverse racists," reflecting a white fear of black militancy (119). Leo in the 2001 remake is also a somewhat simplistic, melodramatic hero. "The unwavering depiction of Leo as hero means we are never encouraged to see the faults of ape culture as a commentary on human frailties" (Vint 244).

32. The statue is dressed like Lincoln, in a similar chair, but with Thade's chimp face, hands, and foot, plus an altered inscription behind it: "IN THIS TEMPLE AS IN THE HEARTS OF THE APES FOR WHOM HE SAVED THE PLANET THE MEMORY OF GENERAL THADE IS ENSHRINED FOREVER." See Lindner 124–30, for more on this ending, in the "tradition of" the original novel and film, but here "sacrificing the logical coherence of the plot" (130).

33. See, for example, Besel and Besel who find more "perspective by incongruity" in the original film as the reason for it being better than the 2001 version in exploring issues of class, race, and animal rights.

34. As Porter points out, the simians in this film "were created digitally without resorting to any live ape performances, a breakthrough that animal advocates have long hoped for" (498).

35. See Maisano 65, for more on Shakespeare and apes, with a brief tie to this film.

36. See Hughes, who praises this film for "discussing the ethics of cognitive enhancement drugs" (43).

37. David Denby calls this moment "interspecies pathos" and describes it as "ineffably, absurdly touching, the sweetest expression of family love seen in recent films" (315).

38. Comparisons have been made by others (Baum; Porter; Sorenson) to the ASL-taught, male chimpanzee, Nim Chimpsky (1973–2000), because a documentary about him, *Project Nim* (dir. James Marsh), was released in the same year as this film (2011). But Nim was shuttled between caretakers and homes, not raised in one family, like Lucy and Caesar. He also injured several of his teachers and caretakers, sometimes severely. And yet, Nim's experiences at the Institute for Primate

Studies at the University of Oklahoma, which also owned Lucy and where Temerlin was a psychology professor, may have been more like (and perhaps a basis for) Caesar's in the primate shelter. After that institute was closed in 1982, Nim was also used for biomedical testing like Caesar's mother and Koba, his bonobo colleague later in the film and its sequel.

39. An earlier chimp, Washoe, learned 250 words of ASL, also coining new words like Lucy (Hu). Bonobos, such as Kanzi, have learned to use hundreds more words with lexigrams (Hu) and understand thousands of vocalized words (Raffaele). Their conversations with humans show linguistic aspects of "turn-taking, negotiation, pauses and repetition, . . . [which] go far beyond information sharing" ("Great Ape"). Kanzi also vocalizes a few English words in conversation with his scientist-handlers (Greenspan and Shanjer; "Kanzi"). The gorilla Koko reportedly uses over a thousand ASL signs and understands two thousand words of spoken English (Fischer 26–28; Wise 216). But there are recent accusations, by concerned researchers and employees, about the maltreatment of apes at the Gorilla Foundation where he lives and at the Great Ape Trust (now renamed) where Kanzi lives (Hu).

40. Temerlin also points out that Lucy became more aggressive in the forest, on their ranch, and twice bit friends there due to her own excitement (78–81). Many aspects of Lucy's behavior, as a human-raised chimp, are not shown with Caesar in the film: her refusal to toilet train, her "extreme dependency" on mother figures (especially after losing her first, like Caesar), her drinking from the urine stream of humans, her disregard for the *Planet of the Apes* movie and for chimps on TV, her kissing people on the lips (including her "father" and newly met visitors), her concern at the death of her pet cat, her fascination with human dancing and with jungle sounds, her ability to tickle herself, her enjoyment of playing "keep away" and play fighting with sticks, her use of many human tools, her fondness for alcohol, her use of a water hose or vacuum cleaner for masturbation, her ability to lie, her making of art, her often signing "what?" but not "why?", her attempts at stopping sexual intercourse between her "parents," her mimicry of human vomiting, and her help as a "co-therapist" in Temerlin's psychotherapy group sessions (27–31, 58–61, 79, 83, 91, 93, 96–99, 101, 104–9, 123, 130, 167–72, 199).

41. Clover sees this as a pivotal moment in the film: "Caesar spots the leash in Will's hand and, seeming to reach out to him, he instead closes the door, locking himself in. No pretending that even the most enlightened bondage is more tolerable than an iron cage" (9).

42. Solms makes a similar connection between videogame playing and mirror-neuron, body-swapping, and rubber-hand experiments, regarding the "learnt nature of the external body" ("Conscious" 15).

CHAPTER 7

1. "As neural currents develop, variant individual experiences leave imprints such that no two brains are identical, even those of identical twins" (Edelman, *Second* 21).

2. Recently, Ramachandran and Seckel added "balance" to the list, to make ten laws, and also suggested "harmony" as the "use of multiple laws within a single image so that they don't clash" (386–87). This sounds to me like the *rasa* of "peace"—as a reflective awareness of other *rasas*.

3. See Calvin 11, who describes such chimps as looking like a "human gang [in] warfare" leaving a dying chimp with "its throat or genitals chewed out, great strips of skin pulled loose."

4. Videogames, news, and social media may also be involved in cathartic backfire—as shown in the recent case (August 2015) of the fired Virginia news reporter who got revenge by wearing a body camera and raising a gun within its frame, like a "first-person" shooter's view in a videogame, while killing two former colleagues, a reporter and another cameraman, "live" on the local TV news. There were many other personal and media aspects to this case, with the killer being black, his victims being white, his putting the body-camera video and statements about the killing onto the Web's social media, and his remarks about a "race war" in a fax to ABC News, referring to the South Carolina church killing of blacks by a white racist two months before (plus his oppression as a black and gay man, along with his admiration for the Virginia Tech and Columbine shooters)—before ending his own life.

5. As the source of this nomenclature, Haidt points to the tradition of the French Assembly from 1789, where "delegates who favored preservation" sat on the right side of the chamber, "while those who favored change sat on the left" (*Righteous* 277).

6. Directly refuting Haidt, Greene says that liberals' moral tastes, "rather than being narrow, may instead by more refined" (339).

7. Haidt developed a similar model, without the brain mapping, called "social intuitionism," with the metaphor of moral emotions or "automatic processes" as an elephant and reason or "controlled processes" as the rider, or an intuitive dog wagging a "rational tail" (*Righteous* 48–50).

8. Greene also mentions that low-anxiety psychopaths and people with alexithymia (a disorder reducing the awareness of one's own emotions) "make more utilitarian judgments," which suggests some caution about the moral philosophy that he advocates (125). Cf. Haidt, *Righteous* 62, on the danger to others of psychopaths with an "ability to reason combined with a lack of moral emotions."

9. See also van der Kolk 206 (diagram and caption), on the "rational, analyzing" DLPFC network, which has "no direct connections with the emotional brain, where most imprints of trauma reside," although the MPFC, "the center of self-awareness, does."

10. The term "resistant" or "ambivalent" is sometimes used for the anxious category and "controlling" as a further childhood development of "disorganized" (Howe 45, 164–65). With adults, different but parallel terms are sometimes used: secure/autonomous, dismissing, preoccupied, and unresolved/disorganized or "fearful avoidant" (Howe 58–64; Siegel, *Developing* 74). But I will consistently use secure, avoidant, anxious, and disorganized.

11. See van der Kolk 154, for experiments with monkeys regarding a "serotonin gene," shared with humans, showing that early-life experience with the mother "has at least as much impact on biology as heredity does"—and on symptoms of impulsivity, aggression, and severe depression.

12. See Amihai and Kozhevnikov, for a study showing activation of the body's sympathetic nervous (arousal) system and "an immediate dramatic increase in performance on cognitive tasks" with this type of meditation.

13. See Mascaro et al. for a recent summary of social neuroscience research on compassion meditation.

14. See Hay and Rocha. They each give journalistic accounts of recent research in the "Dark Night Project" (a.k.a. "Varieties of Contemplative Experience") at Brown University. See also Lindahl.

15. Cf. Siegel, *Mindful* 203, about disorganized attachment: "the parent is the source of terrified or terrifying behaviors that induce in the child a state of fear without resolution," involving brain circuits to flee the attachment figure and yet others to move toward her "for protection and soothing."

16. A tragic example of this appeared in the news media in June 2015, while I was finishing this book. Ten African Americans were shot at close range, 9 killed, in Emanuel AME Church in Charleston, South Carolina, by a young white supremacist, much influenced by Internet hate groups, as shown on his own website. (See also note 4 in this chapter, about a related killing in August 2015.)

Filmography

INTRODUCTION

The Animal (2001)

CHAPTER 4, VAMPIRE FILMS (SOME CONSIDERED AT LENGTH, OTHERS JUST BRIEFLY)

Nosferatu (1922)
Dracula (1931)
House of Frankenstein (1944)
House of Dracula (1945)
Abbott and Costello Meet Frankenstein (1948)
Blood of Dracula (1957)
Hammer *Dracula* series (1958–74)
Blacula (1972)
Scream Blacula Scream (1973)
Nosferatu: Phantom der Nacht (1979)
The Hunger (1983)
The Lost Boys (1987)
My Best Friend Is a Vampire (1987)
Near Dark (1987)
Def by Temptation (1990)
Bram Stoker's Dracula (1992)
Buffy the Vampire Slayer (1992)
Cronos (1993)
Interview with the Vampire (1994)
Vampire in Brooklyn (1995)

Immortality (a.k.a. *The Wisdom of Crocodiles*, 1998)
Blade series (1998–2004)
Jesus Christ Vampire Hunter (2001)
Queen of the Damned (2002)
Underworld series (2003, 2006, 2009, 2012)
Van Helsing (2004)
Let the Right One In (2008) and *Let Me In* (2010)
Twilight series (2008–12)
Thirst (2009)
Abraham Lincoln: Vampire Hunter (2012)
Dracula Untold (2014)

CHAPTER 4, WEREWOLF FILMS

Werewolf of London (1935)
The Wolf Man (1941)
She-Wolf of London (1946)
I Was a Teenage Werewolf (1957)
An American Werewolf in London (1981)
Full Moon High (1981)
The Howling (1981)
Teen Wolf (1985)
An American Werewolf in Paris (1987)
Teen Wolf Too (1987)
Wolf (1994)
Ginger Snaps series (2000, 2004)
Dog Soldiers (2002)
Underworld series (2003, 2006, 2009, 2012)
Blood and Chocolate (2007)
Twilight series (2008–12)
The Wolfman (2010)

CHAPTER 6, LAB-CREATED OR EXOTIC HYBRIDS

The Island of Lost Souls (1932)
Cat People (1942)
The Island of Dr. Moreau (1977)
The Island of Dr. Moreau (1996)
Cat People (1982)

The Fly (1958)
Return of the Fly (1959)
Curse of the Fly (1965)
The Fly (1986)
The Fly II (1989)
Splice (2009)
Men and Chicken (2015)

CHAPTER 6, APE-PLANET FILMS (AND OTHERS CONSIDERED BRIEFLY)

King Kong series (1933, 1976, and 2005)
Mighty Joe Young (1949 and 1998)
The Planet of the Apes (1968)
Beneath the Planet of the Apes (1970)
Escape from the Planet of the Apes (1971)
Conquest of the Planet of the Apes (1972)
Battle for the Planet of the Apes (1973)
The Relic (1997)
The Planet of the Apes (2001)
Percy Jackson and the Olympians (2010 and 2013)
Rise of the Planet of the Apes (2011)
Life of Pi (2012)
The Wolverine (2013)
Dawn of the Planet of the Apes (2014)

References

Abbott, Stacey. *Celluloid Vampires*. Austin: University of Texas Press, 2007.

Adamson, Lauren B., and Connie L. Russell. "Emotion Regulation and the Emergence of Joint Attention." *Early Social Cognition*. Ed. Philippe Rochat. Mahwah, NJ: Lawrence Erlbaum, 1999. 281–97.

Allen, Louise. "Monkey Business: *Planet of the Apes* and Romantic Excess." *Femspec* 3.2 (2002): 3–15.

Ambrose, Stanley H. "Late Pleistocene Human Population Bottlenecks, Volcanic Winter, and Differentiation of Modern Humans." *Journal of Human Evolution* 34 (1998): 623–51.

Ames, Daniel R. "Everyday Solutions to the Problem of Other Minds." *Other Minds*. Ed. Bertram F. Malle and Sara D. Hodges. New York: Guilford, 2005. 158–73.

Amihai, Ido, and Maria Kozhevnikov. "Arousal vs. Relaxation." *PLoS ONE* 9.7: e102990.

Anderson, Craig A., Leonard Berkowitz, Edward Donnerstein, L. Rowell Huesmann, James D. Johnson, Daniel Linz, Neil M. Malamuth, and Ellen Wartella. "The Influence of Media Violence on Youth." *Psychological Science in the Public Interest* 4 (Dec. 2003): 81–110.

Anderson, Steven W., Antoine Bechara, Hanna Damasio, Daniel Tranel, and Antonio R. Damasio. "Impairment of Social and Moral Behavior Related to Early Damage in Human Prefrontal Cortex." *Nature Neuroscience* 2.11 (1999): 1032–37.

Anspach, Mark. "Imitation and Violence." *Mimesis and Science*. Ed. Scott R. Garrels. East Lansing: Michigan State University Press, 2011. 129–54.

Apperly, Ian A., Dana Samson, Claudia Chiavarino, and Glyn W. Humphreys. "Frontal and Temporo-Parietal Lobe Contributions to Theory of Mind." *Journal of Cognitive Neuroscience* 16.10 (2004): 1773–84.

Argent, Daniel. "Evolution: The Development of *Planet of the Apes* (2001)." *Creative Screenwriting* 8.4 (2001): 44–46.

Ariely, Dan. *Predictably Irrational*. New York: HarperCollins, 2008.
Ariely, Dan. *The Upside of Irrationality*. New York: HarperCollins, 2010.
Aristotle. *Poetics*. Trans. S.H. Butcher. *Dramatic Theory and Criticism: Greeks to Grotowski*. Ed. Bernard F. Dukore. New York: Holt, 1974. 31–55.
Armel, K. Carrie, and V. S. Ramachandran. "Projecting Sensations to External Objects." *Proceedings of the Royal Society of London: Biological* 270 (2003): 1499–1506.
Armstrong, Karen. *Twelve Steps to a Compassionate Life*. New York: Knopf, 2011.
Aubert, M., A. Brumm, M. Ramli, T. Sutikna, E.W. Saptomo, B. Hakim, M.J. Morwood, G.D. van den Bergh, J. Kinsley, and A. Dosseto. "Pleistocene Cave Art from Sulawesi, Indonesia." *Nature* 514 (9 Oct. 2014): 223–27.
Auerbach, Nina. *Our Vampires, Ourselves*. Chicago: University of Chicago Press, 1995.
Aunger, Robert. *The Electric Meme*. New York: Free Press, 2002.
Averbeck, Bruno B. "Oxytocin and the Salience of Social Cues." *Proceedings of the National Academy of Sciences USA* 107.20 (2010): 9033–34.
Averill, Lindsey Issow. "Un-biting the Apple and Killing the Womb." *Theorizing* Twilight. Ed. Maggie Parke and Natalie Wilson. Jefferson, NC: McFarland, 2011. 224–37.
Baars, Bernard J. *In the Theater of Consciousness: The Workspace of the Mind*. Oxford: Oxford University Press, 1997.
Banissy, Michael J., and Jamie Ward. "Mirror-Touch Synesthesia Is Linked with Empathy." *Nature Neuroscience* 10.7 (2007): 815–16.
Bard, Kim A. "Neonatal Imitation in Chimpanzees (*Pan troglodytes*) Tested with Two Paradigms." *Animal Cognition* 10 (2007): 233–42.
Baron-Cohen, Simon. *The Science of Evil*. New York: Basic, 2011.
Barr, Dale J., and Boaz Keysar. "Mindreading in an Exotic Case." *Other Minds*. Ed. Bertram F. Malle and Sara D. Hodges. New York: Guilford, 2005. 271–83.
Barrett, Justin L. *Born Believers*. New York: Free, 2012.
Bartz, Jennifer A., Jamil Zaki, Niall Bolger, and Kevin N. Ochsner. "Social Effects of Oxytocin in Humans: Context and Person Matter." *Trends in Cognitive Sciences* 15.7 (2011): 301–9.
Bataille, Georges. *Lascaux*. Switzerland: Skira, 1955.
Baum, Bruce. "Apes, Humans, and Other Animals." *New Political Science* 34.1 (2012): 111–21.
Baumeister, Roy F. *The Cultural Animal*. Oxford: Oxford University Press, 2005.
Beard, William. "Thirty-Two Paragraphs About David Cronenberg." *North of Everything*. Ed. William Beard and Jerry White. Alberta: University of Alberta Press, 2002. 144–59.
Beauregard, Mario, Johanne Lévesque, and Vincent Paquette. "Neural Basis of Conscious and Voluntary Self-Regulation of Emotion." *Consciousness, Emotional Self-Regulation, and the Brain*. Ed. Mario Beauregard. Amsterdam: John Benjamins, 2004. 163–94.
Beauregard, Mario, and Denyse O'Leary. *The Spiritual Brain*. New York: Harper, 2008.

Beltrán, Antonio. *The Cave of Altamira*. New York: Harry N. Abrams, 1999.

Benning, Ashley. "'How Old Are You?'" *Theorizing* Twilight. Ed. Maggie Parke and Natalie Wilson. Jefferson, NC: McFarland, 2011. 87–101.

Berridge, Kent C., Terry E. Robinson, and J. Wayne Aldridge. "Dissecting Components of Reward." *Current Opinion in Pharmacology* 9.1 (2009): 65–73.

Besel, Richard, and Renee Smith Besel. "Polysemous Myth: Incongruity in *Planet of the Apes*." *Millennial Mythmaking*. Ed. John Perlich and David Whitt. Jefferson, NC: McFarland, 2010. 51–66.

Bharata-Muni. *The Natyasastra*. Trans. Manomohan Ghosh. Vol. 1. Varanasi: Chowkhamba Krishnadas Academy, 2007.

Bhat, G.K. *Theatric Aspects of Sanskrit Drama*. Poona: Bhandarkar Institute, 1983.

Blackmore, Susan. *The Meme Machine*. Oxford: Oxford University Press, 1999.

Blair, Rhonda. *The Actor, Image, and Action*. London: Routledge, 2008.

Blakemore, Sarah-Jayne, and Jean Decety. "From the Perception of Action to the Understanding of Intention." *Nature Reviews Neuroscience* 2 (2001): 561–67.

Blakeslee, Sandra, and Matthew Blakeslee. *The Body Has a Mind of Its Own*. New York: Random House, 2007.

Bode, Lisa. "Transitional Tastes." *Continuum: Journal of Media & Cultural Studies*. 24.5 (Oct. 2010): 707–19.

Boehm, Christopher. "Egalitarian Behavior and the Evolution of Political Intelligence." *Machiavellian Intelligence II*. Ed. Andrew Whiten and Richard W. Byrne. Cambridge: Cambridge University Press, 1997. 341–64.

Bogdan, Radu J. "Why Self Ascriptions Are Difficult and Develop Late." *Other Minds*. Ed. Bertram F. Malle and Sara D. Hodges. New York: Guilford, 2005. 190–206.

Bordwell, David, and Noël Carroll, eds. *Post-Theory: Reconstructing Film Studies*. Madison: University of Wisconsin Press, 1996.

Bosch, Oliver J., Simone L. Meddle, Daniela I. Beiderbeck, Alison J. Douglas, and Inga D. Neumann. "Brain Oxytocin Correlates with Maternal Aggression." *The Journal of Neuroscience* 25 (2005): 6807–15.

Bräuer, Juliane, Josep Call, and Michael Tomasello. "Are Apes Really Inequity Averse?" *Proceedings of the Royal Society B* 273 (2006): 3123–28.

Broadhurst, Susan. *Digital Practices*. New York: Palgrave, 2007.

Broom, D.M., H. Sena, and K.L. Moynihan. "Pigs Learn What a Mirror Image Represents and Use It to Obtain Information." *Animal Behaviour* 78.5 (2009): 1037.

Brosnan, Sarah F. "Nonhuman Species' Reactions to Inequity and Their Implications for Fairness." *Social Justice Research* 19.2 (2006): 153–85.

Brown, Steven, and Ellen Dissanayake. "The Arts Are More Than Aesthetics." *Neuroaesthetics*. Ed. Martin Skov and Oshin Vartanian. Amityville, NY: Baywood, 2009. 43–57.

Browning, John Edgar, and Caroline Joan (Kay) Picart, eds. *Draculas, Vampires, and Other Undead Forms*. Lanham, MD: Scarecrow, 2009.

Bunson, Matthew. *The Vampire Encyclopedia*. New York: Crown, 1993.
Burke, Brianna. "The Great American Love Affair." *Bringing Light to Twilight*. Ed. Giselle Liza Anatol. New York: Palgrave, 2011. 207–19.
Bushman, Brad J., and L. Rowell Huesmann. "Effects of Televised Violence on Aggression." *Handbook of Children and the Media*. Ed. Dorothy G. Singer and Jerome L. Singer. London: Sage, 2001. 223–54.
Buss, David M. *The Evolution of Desire*. New York: Basic, 1994.
Byrne, Richard W., and Andrew Whiten. "Machiavellian Intelligence." *Machiavellian Intelligence II*. Ed. Andrew Whiten and Richard W. Byrne. Cambridge: Cambridge University Press, 1997. 1–23.
Byrski, M. Christopher. *Concept of Ancient Indian Theatre*. New Delhi: Munshiram Manoharlal, 1973.
Cadinu, Maria Rosaria, and Myron Rothbart. "Self-Anchoring and Differentiation Processes in the Minimal Group Setting." *Journal of Personality and Social Psychology* 70.4 (1996): 661–77.
Calderazzo, Diana. "Theory of the Murderous Mind." *Theory of Mind and Literature*. Ed. Paula Leverage, Howard Mancing, Richard Schweickert, and Jennifer Marston William. West Lafayette: Purdue University Press, 2011. 93–104.
Calvin, William H. *A Brief History of the Mind*. Oxford: Oxford University Press, 2004.
Campbell, Anne. "Attachment, Aggression and Affiliation: The Role of Oxytocin in Female Social Behavior." *Biological Psychology* 77 (2008): 1–10.
Cantor, Joanne. "The Media and Children's Fears, Anxieties, and Perceptions of Danger." *Handbook of Children and the Media*. Ed. Dorothy G. Singer and Jerome L. Singer. London: Sage, 2001. 207–21.
Carlson, Stephanie M., Angela C. Davis, and Jamie G. Leach. "Less Is More: Executive Function and Symbolic Representation in Preschool Children." *Psychological Science* 16.8 (Aug. 2005): 609–16.
Carroll, Noel. *A Philosophy of Mass Art*. Oxford: Oxford University Press, 1998.
Carter, C. Sue, James Harris, and Stephen W. Porges. "Neural and Evolutionary Perspectives on Empathy." *The Social Neuroscience of Empathy*. Ed. Jean Decety and William Ickes. Cambridge: MIT, 2009. 169–82.
Carter, Rita. *Exploring Consciousness*. Berkeley: University of California Press, 2002.
Changeux, Jean-Pierre, Antonio R. Damasio, Wolf Singer, and Yves Christen, eds. *Neurobiology of Human Values*. Heidelberg: Springer-Verlag, 2005.
Chatterjee, Anjan. "Neuroaesthetics." *Aesthetic Science*. Ed. Arthur P. Shimamura and Stephen E. Palmer. Oxford: Oxford University Press, 2012. 299–317.
Chaudhury, Pravas Jivan. "Catharsis in the Light of Indian Aesthetics." *The Journal of Aesthetics and Art Criticism* 24.1 (1965): 151–63.
Cheney, Dorothy L., and Robert M. Seyfarth. *Baboon Metaphysics: The Evolution of a Social Mind*. Chicago: University of Chicago Press, 2007.
Chenier, Troy, and Piotr Winkielman. "The Origins of Aesthetic Pleasure." *Neuroaesthetics*. Ed. Martin Skov and Oshin Vartanian. Amityville, NY: Baywood, 2009. 275–90.

Chiron, C., I. Jambaque, R. Nabbout, R. Lounes, A. Syrota, and O. Dulac. "The Right Brain Hemisphere Is Dominant in Human Infants." *Brain* 120 (1997): 1057–65.

Churchland, Patricia. *Brain-Trust*. Princeton: Princeton University Press, 2011.

Clottes, Jean. *Cave Art*. London: Phaidon, 2008.

Clottes, Jean. *Chauvet Cave*. Salt Lake City: University of Utah Press, 2003.

Clottes, Jean, and David Lewis-Williams. *The Shamans of Prehistory: Trance and Magic in the Painted Caves*. Trans. Sophie Hawkes. New York: Abrams, 1998.

Clover, Joshua. "Fall and Rise." *Film Quarterly* 65.2 (Winter 2011): 7–9.

Cole, Thomas G. "'I Have Worked Hard at Her Head and Brain': Dr. Moreau and the New Woman." *The Sex Is Out of This World*. Ed. Sherry Ginn and Michael G. Cornelius. Jefferson, NC: McFarland, 2012. 134–51.

Colen, Bill D. "Making Old Hearts Younger." *Harvard Gazette*, 9 May 2013. Web.

Condry, John. *The Psychology of Television*. Hillsdale, NJ: Lawrence Erlbaum, 1989.

Connor, Steve. "Humans Evolved Monogamous Relationships to Stop Men Killing Rivals' Babies, Says Study." *The Independent*, 29 July 2013. Web.

Cook, Amy. *Shakespearean Neuroplay*. New York: Palgrave, 2010.

Cooley, Charles. *Human Nature and the Social Order*. New York: Scribner's, 1902.

Copjec, Joan. "Vampires, Breast-Feeding, and Anxiety." *October* 58 (1991): 24–43.

Coricelli, Giorgio, Hugo D. Critchley, Mateus Joffily, John P. O'Doherty, Angela Sirigu, and Raymond J. Dolan. "Regret and Its Avoidance." *Nature Neuroscience* 8.9 (Sept. 2005): 1255–62.

Corrigall, Jenny, and Heward Wilkinson, eds. *Revolutionary Connections*. London: Karnac, 2003.

Cozolino, Louis J. *The Neuroscience of Human Relationships*. New York: Norton, 2006.

Cozolino, Louis J. *The Neuroscience of Psychotherapy*. New York: Norton, 2002.

Cramer, Phebe. "Defense Mechanisms in Psychology Today." *American Psychologist* 55.6 (2000): 637–46.

Creed, Barbara. *The Monstrous Feminine*. London: Routledge, 1993.

Creed, Barbara. "More Human Than I Am Alone." *Screening the Male*. Ed. Steven Cohen and Ina Rae Hark. London: Routledge, 1993. 134–47.

Creed, Barbara. *Phallic Panic*. Melbourne: Melbourne University Press, 2005.

Cunningham, Mark D. "Traveling in the Same Boat." *Genre, Reception, and Adaptation in the "Twilight" Series*. Ed. Anne Morey. Farnham, Surrey, England: Ashgate, 2012. 99–214.

Curtis, Gregory. *The Cave Painters*. New York: Knopf, 2006.

Damasio, Antonio. *Descartes' Error: Emotion, Reason, and the Human Brain*. New York: Putnam, 1994.

Damasio, Antonio. *The Feeling of What Happens: Body and Emotion in the Making of Consciousness*. New York: Harcourt, 1999.

Damasio, Antonio. *Looking for Spinoza: Joy, Sorrow, and the Feeling Brain*. Orlando: Harcourt, 2003.

Damasio, Antonio. *Self Comes to Mind: Constructing the Conscious Brain*. New York: Pantheon, 2010.

Damasio, Antonio R., Daniel Tranel, and Hana Damasio. "Individuals with Sociopathic Behavior Caused by Frontal Damage Fail to Respond Autonomically to Social Stimuli." *Behavioural Brain Research* 41 (1990): 81–94.

Davidson, Richard J., and Sharon Begley. *The Emotional Life of Your Brain*. New York: Hudson Street, 2012.

Davies, Paul, and Niels Henrik Gregersen, eds. *Information and the Nature of Reality*. Cambridge: Cambridge University Press, 2010.

Davis, John. *Individuals and Identity in Economics*. Cambridge: Cambridge University Press, 2011.

Davis, Kenneth L., and Jaak Panksepp. "The Brain's Emotional Foundations of Human Personality and the Affective Neuroscience Personality Scales." *Neuroscience and Biobehavioral Reviews* 35 (2011): 1946–58.

Dawkins, Richard. *The Extended Phenotype*. Oxford: Oxford University Press, 1999.

Dawkins, Richard. *The Selfish Gene*. New York City: Oxford University Press, 1976.

Day, William Patrick. *Vampire Legends in Contemporary American Culture*. Lexington: University of Kentucky Press, 2002.

Deacon, Terrence W. *The Symbolic Species*. New York: Norton, 1998.

Decety, Jean. "Perspective Taking as the Royal Avenue to Empathy." *Other Minds*. Ed. Bertram F. Malle and Sara D. Hodges. New York: Guilford, 2005. 143–57.

Decety, Jean, and Thierry Chaminade. "When the Self Represents the Other: A New Cognitive Neuroscience View on Psychological Identification." *Consciousness and Cognition* 12 (2003): 577–96.

Decety, Jean, and Julie Grèzes. "The Power of Simulation." *Brain Research* 1079 (2006): 4–14.

Decety, Jean, and Claus Lamm. "Human Empathy through the Lens of Social Neuroscience." *The Scientific World Journal* 6 (2006): 1146–63.

Decety, Jean, and Jessica A. Sommerville. "Shared Representations between Self and Other." *Trends in Cognitive Sciences* 7.12 (2003): 527–33.

De Dreu, Carsten K. W., Lindred L. Greer, Michel J. J. Handgraaf, Shaul Shalvi, and Gerben A. Van Kleef. "Oxytocin Modulates Selection of Allies in Intergroup Conflict." *Proceedings of the Royal Society B*, 279 (2012): 1150–54.

De Dreu, Carsten K. W., Lindred L. Greer, Michel J. J. Handgraaf, Shaul Shalvi, Gerben A. Van Kleef, Matthijs Baas, Femke S. Ten Velden, Eric Van Dijk, and Sander W. W. Feith. "The Neuropeptide Oxytocin Regulates Parochial Altruism in Intergroup Conflict Among Humans." *Science* 328 (2010): 1408–11.

De Dreu, Carsten K. W., Lindred L. Greer, Gerben A. Van Kleef, Shaul Shalvi, and Michel J. J. Handgraaf. "Oxytocin Promotes Human Ethnocentrism." *Proceedings of the National Academy of Sciences USA* 108.41 (2011): 1262–66.

de Gelder, Beatrice. "Nonconscious Emotions." *Emotion and Consciousness*. Ed. Lisa Feldman Barrett, Paula M. Niedenthal, and Piotr Winkielman. New York: Guilford, 2005. 123–49.

de Gelder, Beatrice, Josh Snyder, Doug Greve, George Gerard, and Nouchine Hadjikhani. "Fear Fosters Flight." *Proceedings of the National Academy of Sciences USA* 101.47 (2004): 16701–6.

Dehaene, Stanislas, Michael Kerszberg, and Jean-Pierre Changeux. "A Neuronal Model of a Global Workspace in Effortful Cognitive Tasks." *Proceedings of the National Academy of Sciences USA* 95 (1998): 14529–34.

Dell'Amore, Christine. "Monkeys Can Subtract, Study Finds." *National Geographic News*, 28 October 2010. Web.

DeLoache, Judy. "Rapid Change in Symbolic Functioning in Young Children." *Science* 238 (1987): 1556–57.

Denby, David. *Do the Movies Have a Future?* New York: Simon, 2012.

Dennett, Daniel. *Consciousness Explained*. Boston, MA: Little, Brown and Co., 1991.

De Quervain, Dominique J.-F., Urs Fischbacher, Valerie Treyer, Melanie Schellhammer, Ulrich Schnyder, Alfred Buck, and Ernst Fehr. "The Neural Basis of Altruistic Punishment." *Science* 305.5688 (Aug. 2004): 1254–58.

Derrida, Jacques. *The Animal That Therefore I Am*. New York: Fordham University Press, 2008.

de Waal, Frans. *The Age of Empathy*. New York: Random, 2009.

de Waal, Frans. *Good Natured*. Cambridge: Harvard University Press, 1996.

de Waal, Frans. *Our Inner Ape*. New York: Penguin, 2005.

de Waal, Frans. *Primates and Philosophers: How Morality Evolved*. Princeton: Princeton University Press, 2006.

de Waal, Frans. "Putting the Altruism Back into Altruism." *Annual Review of Psychology* 59 (2008): 279–99.

de Waal, Frans. "The 'Russian Doll' Model of Empathy and Imitation." *On Being Moved: From Mirror Neurons to Empathy*. Ed. Stein Braten. Amsterdam: John Benjamins, 2007. 49–72.

Díaz, José-Luis. "A Narrative Method for Consciousness Research." *Frontiers in Human Neuroscience* 7 (2013): 739.

Díaz-Andreu, Margarita, Carlos García Benito, María Lazarich, Antonio Ramos, María José Cruz, and Gao Quian. "Archaeoacoustics and the Location of Rock Art in Spain." *American Indian Rock Art* 40 (2013): 661–70.

Dietz, Tammy. "Wake Up, Bella!" *Bringing Light to Twilight*. Ed. Giselle Liza Anatol. New York: Palgrave, 2011. 99–112.

Dimberg, Ulf, and Maria Petterson. "Facial Reactions to Angry and Happy Facial Expressions." *Psychophysiology* 37 (2000): 693–96.

Dimberg, Ulf, Monika Thunberg, and Kurt Elmehed. "Unconscious Facial Reactions to Emotional Facial Expressions." *Psychological Science* 11.1 (2000): 86–89.

Dinello, Dan. "Island of Lost Souls." *Quarterly Review of Film and Video* 27.5 (2010): 367–69.

Doidge, Norman. *The Brain That Changes Itself*. New York: Viking, 2007.

Domes, Gregor, Markus Heinrichs, Andre Michel, Christoph Berger, and Sabine C. Herpertz. "Oxytocin Improves 'Mind-Reading' in Humans." *Biological Psychiatry* 61.6 (2007): 731–33.

Donald, Merlin. *A Mind So Rare: The Evolution of Human Consciousness*. New York: Norton, 2001.

Donnelly, Ashley. "Denial and Salvation." *Theorizing* Twilight. Ed. Maggie Parke and Natalie Wilson. Jefferson, NC: McFarland, 2011. 178–93.

DuBois, Rachel. "Coming to a Violent End." *Genre, Reception, and Adaptation in the "Twilight" Series*. Ed. Anne Morey. Farnham, Surrey, England: Ashgate, 2012. 131–46.

Dundes, Alan. *Bloody Mary in the Mirror*. Jackson: University of Mississippi Press, 2002.

Durham, M. Gigi. *The Lolita Effect*. New York: Overlook, 2008.

Eagleman, David M. *Incognito*. New York: Pantheon, 2011.

Edelman, Gerald M. *Second Nature: Brain Science and Human Knowledge*. New Haven: Yale University Press, 2006.

Edelman, Gerald M., and Giulio Tononi. *A Universe of Consciousness: How Matter Becomes Mind*. New York: Basic, 2000.

Egner, T., and J.H. Gruzelier. "Ecological Validity of Neurofeedback: Modulation of Slow Wave EEG Enhances Musical Performance." *NeuroReport 14* (2003): 1221–24.

Ehrsson, H. Henrik, Nicholas P. Holmes, and Richard E. Passingham. "Touching a Rubber Hand." *Journal of Neuroscience* 25.45 (Nov. 2005): 10564–73.

Eisenberg, Nancy, and Natalie D. Eggum. "Empathic Responding." *The Social Neuroscience of Empathy*. Ed. Jean Decety and William Ickes. Cambridge: MIT Press, 2009. 71–83.

Ekman, Paul. *Emotions Revealed*. New York: Holt, 2003.

Epstein, Lanza, R.P. Skinner, and B.F. Skinner. "'Self-Awareness' in the Pigeon." *Science* 212 (1981): 695–96.

Erzen, Tanya. *Fanpire*. Boston: Beacon, 2012.

Euripides. "The *Bacchae*." *Three Plays of Euripides*. Trans. Paul Roche. New York: Norton, 1974.

Evans, Simon, Sukhwinder S. Shergill, and Bruno B. Averbeck. "Oxytocin Decreases Aversion to Angry Faces in an Associative Learning Task." *Neuropsychopharmacology* 35 (2010): 2502–9.

Farmer, Harry, Ana Tajadura-Jiménez, and Manos Tsakiris. "Beyond the Colour of My Skin." *Consciousness and Cognition* 21 (2012): 1242–56.

Farrer, C., and C.D. Frith, "Experiencing Oneself vs. Another Person as Being the Cause of an Action." NeuroImage 15 (2002): 596–603.

Fehr, Ernst, and Colin F. Camerer. "Social Neuroeconomics." *Trends in Cognitive Sciences* 11.10 (2007): 419–27.

Feinberg, Todd E. *Altered Egos: How the Brain Creates the Self*. Oxford: Oxford University Press, 2011.

Fernandez-Duque, Diego, and Jodie A. Baird. "Is There a 'Social Brain'?" *Other Minds*. Ed. Betram F. Malle and Sara D. Hodges. New York: Guilford, 2005. 75–90.

Ferrari, Pier F., Elisabetta Visalberghi, Annika Paukner, Leonardo Fogassi, Angela Ruggiero, and Stephen J. Suomi. "Neonatal Imitation in Rhesus Macaques." *PLoS Biology* 4.9 (2006): 1501–8.

Fischer, Steven R. *A History of Language*. London: Reaktion, 1999.

Fischer-Shofty, M., S.G. Shamay-Tsoorya, H. Hararib, and Y. Levkovitzb. "The Effect of Intranasal Administration of Oxytocin on Fear Recognition." *Neuropsychologia* 48 (2010): 179–84.

Fitch, W. Tecumseh, Antje von Graevenitz, and Eric Nicolas. "Bio-Aesthetics and the Aesthetic Trajectory: A Dynamic Cognitive and Cultural Perspective." *Neuroaesthetics*. Ed. Martin Skov and Oshin Vartanian. Amityville, NY: Baywood, 2009. 59–101.

Fleisher, Carol L., dir. *Why Dogs Smile and Chimpanzees Cry*. Writ. Paula Deats and Carol L. Fleisher. Devillier Donegan Enterprises, 1999.

Fost, Joshua. *If Not God, Then What?* Clearhead Studios, 2007.

Frank, Adam. *The Constant Fire*. Berkeley: University of California Press, 2009.

Frank, Lone. *Mindfield*. Oxford: One World, 2009.

Frankel, Valerie Estelle. *Buffy and the Heroine's Journey*. Jefferson, NC: McFarland, 2012.

Freeland, Cynthia A. "The Sublime in Cinema." *Passionate Views*. Ed. Carl Plantinga and Greg M. Smith. Baltimore: Johns Hopkins University Press, 1999. 65–83.

Freeman, Jonathan B., Nalini Ambady, Nicholas O. Rule, and Kerri L. Johnson. "Will a Category Cue Attract You?" *Journal of Experimental Psychology* 137.4 (2008): 673–90.

Frost, Brian J. *The Essential Guide to Werewolf Literature*. Madison: University of Wisconsin Press, 2003.

Gabriel, Linda. "Brain Wave Basics—What You Need to Know about States of Consciousness." http://thoughtmedicine.com/2011/06/brain-wave-basics-what-you-need-to-know-about-states-of-consciousness/

Gallese, Vittorio. "The Roots of Empathy." *Psychopathology* 36.4 (2003): 171–80.

Gallese, Vittorio. "The Two Sides of Mimesis." *Mimesis and Science*. Ed. Scott R. Garrels. East Lansing: Michigan State University Press, 2011. 87–108.

Gallese, Vittorio, and Alvin Goldman. "Mirror Neurons and the Simulation Theory of Mind-Reading." *Trends in Cognitive Neuroscience* 2.12 (1998): 493–502.

Gazzaniga, Michael S. *Who's in Charge?* New York: HarperCollins, 2011.

Gelder, Ken. *New Vampire Cinema*. London: Palgrave, 2012.

Gelder, Ken. *Reading the Vampire*. London: Routledge, 1994.

Gerbner, George, Larry Gross, Nancy Signorielli, Michael Morgan, and Marilyn Jackson-Beeck. "The Demonstration of Power: Violence Profile No. 10." *Journal of Communications* 29.3 (summer 1979): 177–96.

Gerow, Edwin. "Rasa and Katharsis: A Comparative Study, Aided by Several Films." *The Journal of the American Oriental Society* 122.2 (2002): 264–77.

Gershon, Michael D. *The Second Brain*. New York: Harper, 1998.

Gibbons, Ann. "Bonobos Join Chimps as Closest Human Relatives." *Science*, 13 June 2012. Web.

Gilbert, Daniel. *Stumbling on Happiness*. New York: Knopf, 2006.

Gillit, Cobina. "Indonesian Theatre and Its Double." *The Senses in Performance*. Ed. Sally Banes and Andre Lepecki. New York: Routledge, 2007. 191–99.

Girard, René. *Violence and the Sacred*. Trans. Patrick Gregory. Baltimore: Johns Hopkins University Press, 1977.

Glimcher, Paul W. *Foundations of Neuroeconomic Analysis*. Oxford: Oxford University Press, 2011.

Gnoli, Raniero. *The Aesthetic Experience According to Abhinavagupta*. Varanasi, India: Chowkhamba, 1968.

Goldman, Alvin I. "Simulation Theory and Mental Concepts." *Simulation and Knowledge of Action*. Ed. Jerome Dokic and Joelle Proust. Amsterdam: John Benjamins, 2002. 1–20.

Goldstein, Noah J., and Robert B. Cialdini. "The Spyglass Self: A Model of Vicarious Self-Perception." *Journal of Personality and Social Psychology* 92. 3 (2007): 402–17.

Goleman, Daniel. *The Brain and Emotional Intelligence*. Northampton, MA: More Than Sound, 2011.

Goleman, Daniel. *Destructive Emotions*. New York: Random, 2003.

Goleman, Daniel. *Social Intelligence*. New York: Random, 2006.

Gómez, Juan-Carlos. "The Evolution of Pretence: From Intentional Availability to Intentional Non-Existence." *Mind & Language* 23.5 (2008): 586–606.

Goodwin, Robert E. *The Playworld of Sanskrit Drama*. Delhi: Motilal Banarsidass, 1998.

Gopnik, Alison. *The Philosophical Baby*. New York: Farrar, 2009.

Gopnik, Alison, Andrew N. Meltzoff, and Patricia K. Kuhl. *The Scientist in the Crib*. New York: Morrow, 1999.

Grant, Barry K. "Looking Upward: H.G. Wells, Science Fiction and the Cinema." *Literature Film Quarterly* 14.3 (1986): 154–64.

Graziano, Michael S.A. *Consciousness and the Social Brain*. Oxford: Oxford University Press, 2013.

Great Ape Trust of Iowa. "Bonobos May Have Greater Linguistic Skills Than Previously Thought." *Science Daily*, 2 September 2008. Web. www.sciencedaily.com/releases/2008/08/080828171701.htm

Green, Viviane, ed. *Emotional Development in Psychoanalysis, Attachment Theory and Neuroscience*. New York: Brunner-Routledge, 2003.

Greene, Eric. Planet of the Apes *as American Myth*. Middleton, CT: Wesleyan University Press, 1998.

Greene, Joshua. *Moral Tribes*. New York: Penguin, 2013.

Greenspan, Stanley I., and Stuart G. Shanker. *The First Idea*. Cambridge: Da Capo, 2004.

Greeson, Jeffrey M. "Mindfulness Research Update: 2008." *Complementary Health Practice Review* 14.1 (2009): 10–18.

Grèzes, Julie, Chris Frith, and Richard E. Passingham. "Brain Mechanisms for Inferring Deceit in the Actions of Others." *Journal of Neuroscience* (2004) 24: 5500–05.

Grodal, Torben. *Embodied Visions: Evolution, Emotion, Culture, and Film*. Oxford: Oxford University Press, 2009.

Guthrie, R. Dale. *The Nature of Paleolithic Art*. Chicago, IL: University of Chicago Press, 2005.

Guthrie, Stewart Elliott. *Faces in the Clouds: A New Theory of Religion*. New York: Oxford University Press, 1993.

Haidt, Jonathan. "The Moral Emotions." *Handbook of Affective Sciences*. Ed. Richard J. Davidson, Klaus R. Scherer, and H. Hill Goldsmith. Oxford: Oxford University Press, 2003. 852–70.

Haidt, Jonathan. *The Righteous Mind*. New York: Pantheon, 2012.

Haidt, Jonathan, and Craig Joseph. "Intuitive Ethics: How Innately Prepared Intuitions Generate Culturally Variable Virtues." *Daedalus* 133.4 (2004): 55–66.

Hakemulder, Jèmelian. *The Moral Laboratory*. Amsterdam: John Benjamins, 2000.

Hantke, Steffen. "My Baby Ate the Dingo: The Visual Construction of the Monstrous Infant in Horror Film." *Literature, Interpretation, Theory*. 22.2 (2011): 96–112.

Harris, Sam, Jonas T. Kaplan, Ashley Curiel, Susan Y. Bookheimer, Marco Iacoboni, and Mark S. Cohen. "The Neural Correlates of Religious and Nonreligious Belief." *PLoS ONE* (2009) 4.10: e7272.

Hatfield, Elaine, John T. Cacioppo, and Richard L. Rapson. *Emotional Contagion*. Cambridge: Cambridge University Press, 1994.

Hauser, Marc D. "Minding the Behaviour of Deception." *Machiavellian Intelligence II*. Ed. Andrew Whiten and Richard W. Byrne. Cambridge: Cambridge University Press, 1997. 112–43.

Hay, Mark. "When Mindfulness Goes Wrong." *Good*. 15 April 2015. http://magazine.good.is/articles/when-meditation-is-bad-for-you

Heidegger, Martin. *Being and Time*. 1927. New York: Harper, 1962.

Hejmadi, Ahalya, Richard J. Davidson, and Paul Rozin. "Exploring Hindu Emotion Expressions." *Psychological Science* 11.3 (May 2000): 183–87.

Hellige, Joseph B. *Hemispheric Asymmetry*. Cambridge: Harvard University Press, 1993.

Hersch, Matthew H. "Space Apes Want Our Women! Primate Lust in American Science Fiction." *The Sex Is Out of This World*. Ed. Sherry Ginn and Michael G. Cornelius. Jefferson, NC: McFarland, 2012. 170–86.

Hervé, Pierre-Yves, Laure Zago, Laurent Petit, Bernard Mazoyer, and Nathalie Tzourio-Mazoyer. "Revisiting Human Hemispheric Specialization with Neuroimaging." *Trends in Cognitive Neuroscience*. 17.2 (2013): 69–80.

Hirth, Frank, Lars Kammermeier, Erich Frei, Uwe Walldorf, Markus Noll, and Heinrich Reichert. "An Urbilaterian Origin of the Tripartite Brain." *Development* 130 (2003): 2365–73.

Hogle, Jerrold E. "The Gothic-Romantic Nexus." *Wordsworth Circle* 43.3 (2012): 159–64.

Holte, James Craig. *Dracula in the Dark*. Westport, CT: Greenwood, 1997.

Hopkins, William. "Evolution of Language and Speech from a Neuropsychological Perspective." *Primate Perspectives on Behavior and Cognition*. Ed. David A. Washburn. Washington, DC: American Psychological Association, 2007. 243–53.

Hough, Kirstin J., and Philip G. Erwin. "Children's Attitudes toward Violence on Television." *The Journal of Psychology* 131.4 (1997): 411–15.

Howe, David. *Attachment across the Lifecourse*. London: Palgrave, 2011.

Hrdy, Sarah Blaffer. *Mother Nature*. New York: Ballantine, 1999.

Hrdy, Sarah Blaffer. *Mothers and Others*. Cambridge: Harvard University Press, 2009.

Hu, Jane C. "What Do Talking Apes Really Tell Us?" *Slate*, 20 August 2014. Web.

Huber, Robert, Jules B. Panksepp, Thomas Nathaniel, Antonio Alcaro, and Jaak Panksepp. "Drug-Sensitive Reward in Crayfish." *Neuroscience and Biobehavioral Reviews* 35 (2011): 1847–53.

Huesmann, L. Rowell. "Cross-National Communalities in the Learning of Aggression from Media Violence." *Television and the Aggressive Child*. Ed. L. Rowell Huesmann and Leonard D. Eron. Hillsdale, NJ: Lawrence Erlbaum, 1986. 239–57.

Huesmann, L. Rowell, and Laurie S. Miller. "Long-Term Effects of Media Violence." *Aggressive Behavior: Current Perspectives*. Ed. L. Rowell Huessmann. New York: Plenum Press, 1994. 153–86.

Hughes, James J. "Review of *Limitless*. 105 mins. Relativity Media, USA, 2011 and *Rise of the Planet of the Apes*. 105 min, 20th Century Fox, USA, 2011." *The American Journal of Bioethics* 11.10 (2011): 42–43.

Huh, Jisu, Denise E. DeLorme, Leonard N. Reid, and Soontae An. "Direct-to-Consumer Prescription Drug Advertising: History, Regulation, and Issues." *Minnesota Medicine*, March 2010. Web.

Hurlbut, William B. "Desire, Mimesis, and the Phylogeny of Freedom." *Mimesis and Science*. Ed. Scott R. Garrels. East Lansing: Michigan State University Press, 2011. 175–92.

Iacoboni, Marco. *Mirroring People*. New York: Farrar, Straus and Giroux, 2008.

Inoue, Sana, and Tetsuro Matsuzawa. "Working Memory of Numerals in Chimpanzees." *Current Biology* 17.4 (Dec. 2007): R1004–R1005.

Ionesco, Eugene. *Rhinoceros*. New York: Grove, 1960.

Jablonka, Eva, and Marion J. Lamb. *Evolution in Four Dimensions*. Cambridge: MIT Press, 2005.

Jackson, Kimberly. "*Splice*: The Postmodern Prometheus." *Horror Studies* 3.1 (2012): 125–38.

Jacob, Pierre, and Marc Jeannerod. *Ways of Seeing*. Oxford: Oxford University Press, 2003.

Jenkins, Mark Collins. *Vampire Forensics*. Washington, DC: National Geographic, 2010.

Jensen, Keith. "Punishment and Spite, the Dark Side of Cooperation." *Philosophical Transactions of the Royal Society B* 365 (2010): 2635–50.

Jensen, Keith, Brian Hare, Josep Call, and Michael Tomasello. "What's in It for Me? Self-Regard Precludes Altruism and Spite in Chimpanzees." *Proceedings of the Royal Society B* 273 (2006): 1013–21.

Jha, Alok. "Young Blood Can Reverse Some Effects of Ageing, Study Finds." *The Guardian,* 17 October 2012. Web.

Jowett, Lorna. *Sex and the Slayer*. Middletown, CT: Wesleyan University Press, 2005.

Kahneman, Daniel. *Thinking, Fast and Slow*. New York: Farrar, 2011.

Kalamas, Alicia D., and Mandy L. Gruber. "Electrodermal Responses to Implied Versus Actual Violence on Television." *Journal of General Psychology* 125.1 (1998): 31.

Kale, Pramod. *The Theatric Universe*. Bombay: Popular Prakashan, 1974.

Kandel, Eric R. "Biology and the Future of Psychoanalysis: A New Intellectual Framework for Psychiatry Revisited." *American Journal of Psychiatry* 156.4 (1999): 505–24.

Kandel, Eric R. *In Search of Memory*. New York: Norton, 2006.

Kandel, Eric R. *Psychiatry, Psychoanalysis, and the New Biology of Mind*. New York: American Psychiatric, 2005.

Kane, Tim. *The Changing Vampire of Film and Television*. Jefferson, NC: McFarland, 2006.

Kang, Yoona, June Gruber, and Jeremy R. Gray. "Mindfulness and De-Automatization." *Emotion Review* 5.2 (Apr. 2013): 192–201.

"Kanzi." *Radiolab,* 8 October 2014. Web. 8 Oct. 2014. http://www.radiolab.org/story/91708-kanzi/

Kaplan, Matt. *Medusa's Gaze and Vampire's Bite*. New York: Scribner, 2012.

Kaplan-Solms, Karen, and Mark Solms. *Clinical Studies in Neuro-Psychoanalysis*. London: Karnac, 2000.

Kapurch, Katie. "'I'd Never Given Much Thought to How I Would Die.'" *Genre, Reception, and Adaptation in the "Twilight" Series*. Ed. Anne Morey. Farnham, Surrey, England: Ashgate, 2012. 181–97.

Kawada, Christie L.K., Gabriele Oettingen, Peter M. Gollwitzer, and John A. Bargh. "The Projection of Implicit and Explicit Goals." *Journal of Personality and Social Psychology* 86.4 (2004): 545–59.

Keenan, Julian Paul. *The Face in the Mirror*. New York: Harper, 2003.

Keim, Brandon. "Monkeys See Selves in Mirror, Open a Barrel of Questions." *Wired,* 29 September 2010. Web.

Kemp, Rick. *Embodied Acting*. London: Routledge, 2012.

Kennett, Jeanette, and Cordelia Fine. "Internalism and the Evidence from Psychopaths and 'Acquired Sociopaths.'" *Moral Psychology*. Vol. 3. Ed. Walter Sinnott-Armstrong. Cambridge: MIT Press, 2008. 173–90.

Killackey, Herbert P. "Evolution of the Human Brain: A Neuroanatomical Perspective." *The Cognitive Neurosciences*. Ed. Michael S. Gazzaniga. Cambridge: MIT Press, 1995. 1243–53.

Kim, Joo Ok, and Giselle Liza Anatol. *Bringing Light to Twilight*. Ed. Giselle Liza Anatol. New York: Palgrave, 2011. 191–205.

Kirby, David A. "Are We Not Men? The Horror of Eugenics in *The Island of Dr. Moreau*." *Paradoxa* 17 (2002): 93–108.

Kirk, Ulrich. "The Modularity of Aesthetic Processing and Perception in the Human Brain." *Aesthetic Science*. Ed. Arthur P. Shimamura and Stephen E. Palmer. Oxford: Oxford University Press, 2012. 318–36.

Kirshner, Jonathan. "Subverting the Cold War in the 1960s: *Dr. Strangelove, The Manchurian Candidate*, and *The Planet of the Apes*." *Film and History* 31.2 (2001): 40–44.

Klein, Melanie. *The Selected Melanie Klein*. New York: Free, 1986.

Koechlin, Etienne, Chrystèle Ody, Frédérique Kouneiher. "The Architecture of Cognitive Control in the Human Prefrontal Cortex." *Science* 302 (2003): 1181–85.

Kohler, Evelyne, Christian Keysers, M. Alessandra Umiltà, Leonardo Fogassi, Vittorio Gallese, and Giacomo Rizzolatti. "Hearing Sounds, Understanding Actions: Action Representation in Mirror Neurons." *Science* 297 (2002): 846–48.

Konner, Melvin. "Sacred Violence, Mimetic Rivalry, and War." *Mimesis and Science*. Ed. Scott R. Garrels. East Lansing: Michigan State University Press, 2011. 155–74.

Kosslyn, Stephen Michael. "Einstein's Mental Images." *The Languages of the Brain*. Ed. Albert M. Galaburda, Stephen M. Kosslyn, and Yves Christen. Cambridge: Harvard University Press, 2002. 271–87.

Kosslyn, Stephen Michael. *Ghosts in the Mind's Machine: Creating and Using Images in the Brain*. New York: Norton, 1983.

Kosslyn, Stephen M., and G. Wayne Miller. *Top Brain, Bottom Brain*. New York: Simon, 2013.

Krcmar, Marina. "The Contribution of Family Communication Patterns to Children's Interpretations of Television Violence." *Journal of Broadcasting and Electronic Media* 42.2 (1998): 250–64.

Kristeva, Julia. *Revolution in Poetic Language*. New York: Columbia University Press, 1984.

Kruger, Ann Cale. "Imitation, Communion, and Culture." *Mimesis and Science*. Ed. Scott R. Garrels. East Lansing: Michigan State University Press, 2011. 111–28.

Kurzban, Robert, John Tooby, and Leda Cosmides. "Can Race Be Erased? Coalitional Computation and Social Categorization." *Proceedings of the National Academy of Sciences USA* 98.26 (Dec. 2001): 15387–92.

Lacan, Jacques. *Écrits*. Trans. Bruce Fink. New York: Norton, 2006.

Lacan, Jacques. *The Four Fundamental Concepts of Psycho-Analysis*. Trans. Alan Sheridan. New York: Norton, 1981.

Lakoff, George. *Moral Politics*. Chicago, IL: University of Chicago Press, 1996.

Lamarque, Peter. "Tragedy and Moral Value." *Art and Its Messages*. Ed. Stephen Davies. University Park: Pennsylvania State University Press, 1997. 59–69.

Lammes, Sybil. "So Far, So Close: *Island of Lost Souls* as a Laboratory of Life." *Screen Consciousness*. Ed. Robert Pepperell and Michael Punt. Amsterdam: Rodopi, 2006. 65–84.

Lane, Richard D., and Kateri McRae. "Neural Substrates of Conscious Emotional Experience." *Consciousness, Emotional Self-Regulation, and the Brain*. Ed. Mario Beauregard. Amsterdam: John Benjamins, 2004. 87–122.

Leary, Mark R. "The Sociometer, Self-Esteem, and the Regulation of Interpersonal Behavior." *Handbook of Self-Regulation*. Ed. Roy F. Baumeister and Kathleen D. Vohs. New York: Guilford Press, 2004. 373–91.

LeDoux, Joseph. *The Emotional Brain*. New York: Simon, 1996.

LeDoux, Joseph. *Synaptic Self*. New York: Penguin, 2003.

Leknes, Siri, and Irene Tracey. "A Common Neurobiology for Pain and Pleasure." *Nature* 9 (April 2008): 314–20.

Levine, Elana. "*Buffy* and the 'New Girl Order.'" Levine and Parks 168–89.

Levine, Elana, and Lisa Parks. *Undead TV*. Durham, NC: Duke University Press, 2007.

Lewis, Michael. "Social Cognition and the Self." *Early Social Cognition*. Ed. Philippe Rochat. Mahwah, NJ: Lawrence Erlbaum, 1999. 81–98.

Lewis-Williams, David. *The Mind in the Cave*. London: Thames and Hudson, 2002.

Lewis-Williams, David. "Of People and Pictures." *Becoming Human*. Ed. Colin Renfrew and Iain Morley. Cambridge: Cambridge University Press, 2009.

Lieberman, Matthew D. *Social*. New York: Crown, 2013.

Lieberman, Philip. *Human Language and Our Reptilian Brain*. Cambridge: Harvard University Press, 2000.

Lindahl, Jared R. "Why Right Mindfulness Might Not Be Right for Mindfulness." *Mindfulness* 6 (2015): 57–62.

Lindahl, Jared R., Christopher T. Kaplan, Evan M. Winget, and Willoughby B. Britton. "A Phenomenology of Meditation-Induced Light Experiences." *Frontiers in Psychology* 4 (2013): 973.

Linden, David J. *The Accidental Mind*. Cambridge: Harvard University Press, 2007.

Lindner, Oliver. "'An Entirely Different and New Story': A Case Study of Tim Burton's *Planet of the Apes* (2001)." *Adaptation and Cultural Appropriation*. Ed. Pascal Nicklas and Oliver Lindner. Berlin: De Gruyter, 2012. 117–31.

Luecken, Linda J., Danielle S. Roubinov, and Catherine Purdom. "Clinical Implications of Childhood Stress." *The Handbook of Stress*. Ed. Cheryl D. Conrad. London: Wiley-Blackwell, 2011. 304–23.

Lutterbie, John. *Toward a General Theory of Acting*. New York: Palgrave, 2011.

Lynch, Aaron. *Thought Contagion*. New York: Basic, 1996.

Lyon, Eleanor. "Media Murder and Mayhem: Violence on Network Television." *Marginal Conventions: Popular Culture, Mass Media and Social Deviance*. Ed. Clinton R. Sanders. Bowling Green: Bowling Green State University Popular Press, 1990. 144–54.

Lyons, Derek E., Diana H. Damrosch, Jennifer K. Lin, Deanna M. Macris, and Frank C. Keil. "The Scope and Limits of Overimitation in the Transmission of Artefact Culture." *Philosophical Transactions of the Royal Society B* 366 (2011): 1158–67.

Maillard, Chantal. "The Aesthetic Pleasure of Tragedy in Western and Indian Thought." *The Pursuit of Comparative Aesthetics*. Ed. Mazhar Hussain and Robert Wilkinson. Farnham, Surrey, England: Ashgate, 2006. 29–38.

Maisano, Scott. "Rise of the Poet of the Apes." *Shakespeare Studies* 41 (2013): 64–76.

Makin, Tamar R., Nicholas P. Holmes, and H. Henrik Ehrsson. "On the Other Hand." *Behavioural Brain Research* 191 (2008): 1–10.

Mars, Rogier B., Franz-Xaver Neubert, Mary Ann P. Noonan, Jerome Sallet, IvanToni, and Matthew F. S. Rushworth. "On the Relationship between the 'Default Mode Network' and the 'Social Brain.'" *Frontiers in Human Neuroscience* 6.189 (June 2012): 1–9.

Marx, Karl. *Capital*. https://www.marxists.org/archive/marx/works/1867-c1/ch10.htm#4

Mascaro, Jennifer S., Alana Darcher, Lobsang T. Negi, and Charles L. Raison. "The Neural Mediators of Kindness-Based Meditation." *Frontiers in Psychology*, 12 February 2015. http://dx.doi.org/10.3389/fpsyg.2015.00109

McAvan, Emily. *The Postmodern Sacred*. Jefferson, NC: McFarland, 2012.

McClelland, Bruce A. *Slayers and Their Vampires*. Ann Arbor: University of Michigan Press, 2006.

McConachie, Bruce. *Engaging Audiences: A Cognitive Approach to Spectating in the Theatre*. New York: Palgrave, 2008.

McGilchrist, Iain. *The Master and His Emissary*. New Haven: Yale University Press, 2009.

McHugh, Susan Bridget. "Horses in Blackface: Visualizing Race as Species Difference in '*Planet of the Apes*.'" *South Atlantic Review* 65.2 (2000): 40–72.

McLeod, Susan. "Pygmalion or Golem? Teacher Affect and Efficacy." *College Composition and Communication* 46.3 (1995): 369–86.

McNamara, Patrick. *Spirit Possession and Exorcism*. Vol. 1 and 2. Santa Barbara: Praeger, 2011.

McQueen, Sean. "Biocapitalism and Schizophrenia." *Science Fiction Studies* 41.1 (2014): 120–35.

Meltzoff, Andrew N. "Elements of a Developmental Theory of Imitation." *The Imitative Mind: Development, Evolution, and Brain Bases*. Ed. Andrew N. Meltzoff and Wolfgang Prinz. Cambridge: Cambridge University Press, 2002. 1941.

Menary, Richard, ed. *The Extended Mind*. Cambridge: MIT Press, 2010.

Mikulincer, Mario, and Phillip R. Shaver. *Attachment in Adulthood*. New York: Guilford, 2010.

Miller, Geoffrey F. "Protean Primates: The Evolution of Adaptive Unpredictability in Competition and Courtship." *Machiavellian Intelligence II*. Ed. Andrew Whiten and Richard W. Byrne. Cambridge: Cambridge University Press, 1997. 312–40.

Mithen, Stephen. "The Evolution of Imagination: An Archaeological Perspective." *Sub-Stance* 94/95 (2001): 48–50.

Mithen, Stephen. *The Prehistory of the Mind*. London: Thames and Hudson, 1996.

Moise, Jessica F., and L. Rowell Huesmann. "Television Violence Viewing and Aggression in Females." *Annals of the New York Academy of Sciences* 794 (1996): 382.

Monastra, V.J., D.M. Monastra, and S. George. "The Effects of Stimulant Therapy, EEG Biofeedback, and Parenting Style on the Primary Symptoms of Attention-Deficit/Hyperactivity Disorder." *Applied Psychophysiology & Biofeedback* 27 (2002): 231–49.

Montelle, Yann-Pierre. "Mimicry, Deception, Mimesis." *Rock Art Research* 23.1 (2006): 23–24.

Montelle, Yann-Pierre. *Palaeoperformance*. London: Seagull, 2009.

Montelle, Yann-Pierre. "Paleoperformance: Investigating the Human Use of Caves in the Upper Paleolithic." *New Perspectives on Prehistoric Art*. Ed. Günter Berghaus. Westport, CT: Praeger, 2004. 131–52.

Moore, Chris. "Gaze Following and the Control of Attention." *Early Social Cognition*. Ed. Philippe Rochat. Mahwah, NJ: Lawrence Erlbaum, 1999. 241–56.

Moses, Kara. "US Pledges to Reduce Use of Chimpanzees in Scientific Research." *The Guardian*, 26 June 2013. Web.

Mukherjea, Ananya. "Team Bella." *Theorizing* Twilight. Ed. Maggie Parke and Natalie Wilson. Jefferson, NC: McFarland, 2011. 70–83.

Mulvey, Laura. "Visual Pleasure and Narrative Cinema." *Narrative, Apparatus, Ideology: A Film Theory Reader*. Ed. Philip Rosen. New York: Columbia University Press, 1986. 198–209.

Murphy, Colette. "Someday My *Vampire* Will Come?" *Theorizing* Twilight. Ed. Maggie Parke and Natalie Wilson. Jefferson, NC: McFarland, 2011. 56–69.

Murray, John P. "The Violent Face of Television: Research and Discussion." *The Faces of Televisual Media*. 2nd ed. Ed. Edward L. Palmer and Brian M. Young. Mahwah, NJ: Lawrence Erlbaum, 2003. 143–60.

Murray, Sandra L., John G. Holmes, Gina Bellavia, Dale W. Griffin, and Dan Dolderman. "Kindred Spirits? The Benefits of Egocentrism in Close Relationships." *Journal of Personality and Social Psychology* 82.4 (2002): 563–81.

Nadal, Marcos, Miquel Capo, Enric Munar, Gisele Marty, and Camilo Jose Cela-Conde. "Constraining Hypotheses on the Evolution of Art and Aesthetic Appreciation." *Neuroaesthetics*. Ed. Martin Skov and Oshin Vartanian. Amityville, NY: Baywood, 2009. 103–29.

Nakamichi, M. "Mother-Offspring Relationship in Macaques." *Primate Origins of Human Cognition and Behavior*. Ed. Tetsuro Matsuzawa. Tokyo: Springer, 2001. 418–40.

Nauert, R. "Neurofeedback Trains Brain Waves, Restores Brain Function." *Psych Central* (2012). http://psychcentral.com/news/2012/10/26/neurofeedback-trains-brain-waves-restores-brain-function/46720.html

Nelson, Victoria. *The Secret Life of Puppets*. Cambridge: Harvard University Press, 2001.

Newberg, Andrew, Eugene d'Aquili, and Vince Rause. *Why God Won't Go Away: Brain Science and the Biology of Belief*. New York: Ballantine, 2002.

Nicol, Rhonda. "'When You Kiss Me I Want to Die.'" *Bringing Light to Twilight*. Ed. Giselle Liza Anatol. New York: Palgrave, 2011. 113–23.

Niehoff, Debra. *The Biology of Violence*. New York: Free Press, 1999.

Nielsen, Mark, and Keyan Tomaselli. "Overimitation in Kalahari Bushman Children and the Origins of Human Cultural Cognition." *Psychological Science* 21.5 (2010): 729–36.

Nili, Uri, Hagar Goldberg, Abraham Weizman, and Yadin Duda. "Fear Thou Not: Activity of Frontal and Temporal Circuits in Moments of Real-Life Courage." *Neuron* 66.6 (June 2010): 949–62.

Nisbett, Richard E. *The Geography of Thought*. New York: Free Press, 2003.

Noë, Alva. *Out of Our Heads*. New York: Hill and Wang, 2009.

Nussbaum, Martha C. *The Fragility of Goodness: Luck and Ethics in Greek Tragedy and Philosophy*. Cambridge: Cambridge University Press, 1986.

Oatley, Keith. *Such Stuff as Dreams*. Chichester, UK: Wiley and Sons, 2011.

Ochsner, Kevin N., Silvia A. Bunge, James J. Gross, and John D.E. Gabrieli. "Rethinking Feelings: An fMRI Study of the Cognitive Regulation of Emotion." *Journal of Cognitive Neuroscience* 14.8 (2002): 1215–29.

Ochsner, Kevin N., and James J. Gross. "Thinking Makes It So." *Handbook of Self-Regulation*. Ed. Roy F. Baumeister and Kathleen D. Vohs. New York: Guilford Press, 2004. 229–55.

Oehman, Arne, and Susan Mineka. "Fears, Phobias, and Preparedness." *Psychological Review* 108.3 (2001): 483–522.

Ogawa, Masaru. "Integration in Franco-Cantabrian Parietal Art." *Aesthetics and Rock Art*. Ed. Thomas Heyd and John Clegg, Hampshire, England: Ashgate, 2005. 117–30.

O'Hehir, Andrew. "Gorilla Warfare." *Sight and Sound* 11.9 (2001): 12–15.

Ohio State University. "Ancient Raptors Likely Feasted on Early Man, Study Suggests." *Science Daily*, 30 August 2006. Web. www.sciencedaily.com/releases/2006/08/060830005634.htm

Ouellette, Jennifer. *The Physics of the Buffyverse*. New York: Penguin, 2006.

Oughourlian, Jean-Michel. *The Puppet of Desire*. Stanford, CA: Stanford University Press, 1991.

Packer, Sharon. *Neuroscience in Science Fiction Films*. Jefferson, NC: McFarland, 2015.

Paladino, Maria-Paola, Mara Mazzurega, Francesco Pavani, and Thomas W. Schubert. "Synchronous Multisensory Stimulation Blurs Self-Other Boundaries." *Psychological Science* 21.9 (2010): 1202–7.

Panksepp, Jaak. *Affective Neuroscience: The Foundations of Human and Animal Emotions*. Oxford: Oxford University Press, 1998.

Panksepp, Jaak. "Emotions as Viewed by Psychoanalysis and Neuroscience." *NeuroPsychoanalysis* 1 (1999): 15–38.

Panksepp, Jaak, and Lucy Biven. *The Archaeology of Mind*. New York: Norton, 2012.

Panksepp, Jules B., and Garet P. Lahvis. "Rodent Empathy and Affective Neuroscience." *Neuroscience and Biobehavioral Reviews* 35 (2011): 1864–75.

Papousek, Ilona, H. Harald Freudenthaler, and Günter Schulter. "The Interplay of Perceiving and Regulating Emotions in Becoming Infected with Positive and Negative Moods." *Personality and Individual Differences* 45 (2008): 463–67.

Pedersen, Cort A. "Biological Aspects of Social Bonding and the Roots of Human Violence." *Annals of the New York Academy of Sciences* 1036 (2004): 106–27.

Petkova, Valeria I., and H. Henrik Ehrsson. "If I Were You: Perceptual Illusion of Body Swapping." *PLoS ONE* 3.12 (Dec. 2008): 1–9.

Phillips, Kendall R. "You Said Forever." *Dracula's Daughters*. Ed. Douglas Brode and Leah Deynera. Lanham, MD: Scarecrow, 2014. 253–65.

Philo, Greg. "The Media and Public Belief." *Media and Mental Distress*. Ed. Greg Philo. London: Longman, 1996. 82–104.

Pinker, Steven. *The Better Angels of Our Nature*. New York: Viking, 2011.

Pizzato, Mark. *Edges of Loss: From Modern Drama to Postmodern Theory*. Ann Arbor: University of Michigan Press, 1998.

Pizzato, Mark. *Ghosts of Theatre and Cinema in the Brain*. New York: Palgrave: 2006.

Pizzato, Mark. *Inner Theatres of Good and Evil: The Mind's Staging of Gods, Angels and Devils*. Jefferson, NC: McFarland, 2011.

Pizzato, Mark. "Soyinka's *Bacchae*, African Gods, and Postmodern Mirrors." *Journal of Religion and Theatre* 2.1 (Fall 2003). http://apollo.fa.mtu.edu/~dlbruch/rtjournal_.

Pizzato, Mark. *Theatres of Human Sacrifice: From Ancient Ritual to Screen Violence*. Albany: State University of New York Press, 2005.

Plantinga, Car. *Moving Viewers: American Film and the Spectator's Experience*. Berkeley: University of California Press, 2009.

Plato. *Phaedrus*. Trans. Benjamin Jowett. The Internet Classics Archive. Web.

Plato. *The Republic*. Trans. Benjamin Jowett. University of Adelaide: eBooks, 2014. Web.

Politser, Peter. *Neuroeconomics*. Oxford: Oxford University Press, 2008.

Porges, Stephen W. "The Polyvagal Theory." *International Journal of Psychophysiology*: 42 (2001): 123–46.

Porter, Pete. "Rise of the Consciousness of the Apes." *Society & Animals*. 21.5 (2013): 497–500.

Poulin-Dubois, Diane. "Infants' Distinction between Animate and Inanimate Objects." *Early Social Cognition*. Ed. Philippe Rochat. Mahwah, NJ: Lawrence Erlbaum, 1999. 257–80.

Povinelli, Daniel J. *Folk Physics for Apes*. Oxford: Oxford University Press, 2000.

Preston, Stephanie D., and Frans B.M. de Waal. "Empathy." *Behavioral and Brain Sciences* 25 (2002): 1–20.

Prinz, Jesse J. *The Emotional Construction of Morals*. Oxford: Oxford University Press, 2007.

Prior, Helmut, Ariane Schwarz, Onur Güntürkün, and Frans De Waal. "Mirror-Induced Behavior in the Magpie (Pica pica): Evidence of Self-Recognition." *PLoS Biology* 6.8 (2008): e202.

Radulescu, Anca R., and Lilianne R. Mujica-Parodi. "Human Gender Differences in the Perception of Conspecific Alarm Chemosensory Cues." *PLoS ONE* 8.7 (2013): e68485.

Raffaele, Paul. "Speaking Bonobo." *Smithsonian Magazine*, November 2006. Web.

Ramachandran, V.S. *The Tell-Tale Brain*. New York: Norton, 2011.

Ramachandran, V.S., and Sandra Blakeslee. *Phantoms in the Brain: Probing the Mysteries of the Human Mind*. New York: William Morrow, 1998.

Ramachandran, Vilayanur S., and Elizabeth Seckel. "Neurology of Visual Aesthetics." *Aesthetic Science*. Ed. Arthur P. Shimamura and Stephen E. Palmer. Oxford: Oxford University Press, 2014. 375–89.

Rankin, Sandy. "Disalienation and the Irrepressible Revolutionary Wish: Apes, Heston, Ludics, Home." *The Journal of Popular Culture* 40.6 (2007): 1019–31.

Raz, Gal, Boaz Hagin, and Talma Hendler. "E-Motion Pictures of the Brain." *Psychocinematics: Exploring Cognition at the Movies*. Ed. Arthur P. Shimamura. Oxford: Oxford University Press, 2013. 285–313.

Reeder, Glenn D., and David Trafimow. "Attributing Motives to Other People." *Other Minds*. Ed. Bertram F. Malle and Sara D. Hodges. New York: Guilford, 2005. 106–23.

Reiss, Diana. *The Dolphin in the Mirror*. Boston, MA: Harcourt, 2011.

Renfrew, Colin. "Situating the Creative Explosion." *Becoming Human*. Ed. Colin Renfrew and Iain Morley. Cambridge: Cambridge University Press, 2009. 74–92.

Revonsuo, Antti. "The Reinterpretation of Dreams: An Evolutionary Hypothesis of the Function of Dreaming." *Sleeping and Dreaming*. Ed. Edward F. Pace-Schott, Mark Solms, Mark Blagrove, and Stevan Harnad. Cambridge: Cambridge University Press, 2003. 85–111.

Richards, Janet Radcliffe. *Human Nature after Darwin*. London: Routledge, 2000.

Richmond, Farley P. "Characteristics of Sanskrit Theatre and Drama." *Indian Theatre*. Ed. Farley P. Richmond, Darius l. Swann, and Phillip B. Zarrilli. Honolulu: University of Hawaii Press, 1990. 33–86.

Riess, Jana. *What Would Buffy Do?* San Francisco, CA: Wiley, 2004.

Riters, Lauren V. "Pleasure Seeking and Birdsong." *Neuroscience and Biobehavioral Reviews* 35 (2011): 1837–45.

Rizzolatti, Giacomo, and Corrado Sinigaglia. *Mirrors in the Brain—How Our Minds Share Our Actions*. Oxford: Oxford University Press, 2008.

Rocha, Tomas. "The Dark Knight of the Soul." *The Atlantic*. 25 June 2014. http://www.theatlantic.com/health/archive/2014/06/the-dark-knight-of-the-souls/372766/

Rochat, Philippe. "Ego Function of Early Imitation." *The Imitative Mind: Development, Evolution, and Brain Bases*. Ed. Andrew N. Meltzoff and Wolfgang Prinz. Cambridge: Cambridge University Press, 2002. 85–97.

Rochat, Philippe, and Tricia Striano. "Social-Cognitive Development in the First Year." *Early Social Cognition*. Ed. Philippe Rochat. Mahwah, NJ: Lawrence Erlbaum, 1999. 3–34.

Rochat, Philippe, and Tricia Striano. "Who's in the Mirror? Self–Other Discrimination in Specular Images by Four- and Nine-Month-Old Infants." *Child Development* 73.1 (2002): 35–46.

Rogers, Lesley J., and Richard J. Andrew, eds. *Comparative Vertebrate Lateralization*. Cambridge: Cambridge University Press, 2002.

Rokotnitz, Naomi. *Trusting Performance*. New York: Palgrave, 2011.

Rony, Fatimah Tobing. *The Third Eye*. Durham, NC: Duke University Press, 1996.

Rosenthal, R., and E. Y. Babad. "Pygmalion in the Gymnasium." *Educational Leadership* 43.1 (1985): 36–39.

Roskies, Adina L. "Internalism and the Evidence from Pathology." *Moral Psychology*. Vol. 3. Ed. Walter Sinnott-Armstrong. Cambridge: MIT Press, 2008. 191–206.

Roth, Lillian, Tina Kaffenberger, Uwe Herwig, and Annette B. Brühl. "Brain Activation Associated with Pride and Shame." *Neuropsychobiology* 69.2 (2014): 95–106.

Ruby, Perrine, and Jean Decety. "Effect of Subjective Perspective Taking During Simulation of Action." *Nature Neuroscience* 4.5 (2001): 546–50.

Ruby, Perrine, and Jean Decety. "How Would You Feel versus How Do You Think She Would Feel?" *Journal of Cognitive Neuroscience* 16.6 (2004): 988–99.

Rutte, Claudia, and Michael Taborsky. "Generalized Reciprocity in Rats." *PLoS Biology* 5.7 (2007): 1421–25.

Sacks, Oliver W. "Sigmund Freud: The Other Road." *Freud and the Neurosciences*. Ed. Giselher Guttman and Inge Scholz-Strasser. Vienna: Austrian Academy of Sciences, 1998. 11–22.

Sagan, Carl. *The Dragons of Eden: Speculations on the Evolution of Human Intelligence*. New York: Ballantine, 1977.

Sapolsky, Robert M. *Why Zebras Don't Get Ulcers*. New York: Freeman, 1994.

Saxe, Rebecca. "The Right Temporo-Parietal Junction: A Specific Brain Region for Thinking about Thoughts." *Handbook of Theory of Mind*. Ed. Alan Leslie and Tamsin German. London: Taylor, 2010.

Schechner, Richard. *Performance Studies: An Introduction*. 2002. New York: Routledge, 2013.

Schechner, Richard. *Performance Theory*. 1988. New York: Routledge, 2003.

Schechner, Richard. "Rasaesthetics." *The Senses in Performance*. Ed. Sally Banes and Andre Lepecki. New York: Routledge, 2007. 10–28.

Schechner, Richard. "Thoughts on Rasaesthetics." 2 February 1998. Web. http://www.nyu.edu/classes/bkg/rasa

Scheff, Thomas J. *Catharsis in Healing, Ritual, and Drama*. Berkeley: University of California Press, 1979.

Schmidt, Marco F. H., and Jessica A. Sommerville. "Fairness Expectation and Altruistic Sharing in 15-Month-Old Infants." *PlosONE* 6.10 (2011): 1–7.

Schulze, Lars, Alexander Lischke, Jonas Greif, Sabine C. Herpertz, Markus Heinrichs, and Gregor Domes. "Oxytocin Increases Recognition of Masked Emotional Faces." *Psychoneuroendocrinology* 36.9 (2011): 1378–82.

Schwartz, Jeffrey M., Henry P. Stapp, and Mario Beauregard. "The Volitional Influence of the Mind on the Brain, with Special Reference to Emotional Self-Regulation." *Consciousness, Emotional Self-Regulation, and the Brain*. Ed. Mario Beauregard. Amsterdam: John Benjamins, 2004. 195–238.

Sen, Ramendra Kumar. "A Comparative Study of Greek and Indian Poetics and Aesthetics." *The Pursuit of Comparative Aesthetics: An Interface between East and West*. Ed. Mazhar Hussain and Robert Wilkinson. Farnham, Surrey, England: Ashgate, 2006. 85–112.

Sforza, Anna, Ilaria Bufalari, Patrick Haggard, and Salvatore M. Aglioti. "My Face in Yours." *Social Neuroscience* (2009): 1–15.

Shachar, Hila. "A Post-Feminist Romance." *Theorizing* Twilight. Ed. Maggie Parke and Natalie Wilson. Jefferson, NC: McFarland, 2011. 147–61.

Shamay-Tsoory, Simone G., Meytal Fischer, Jonathan Dvash, Hagai Harari, Nufar Perach-Bloom, and Yechiel Levkovitz. "Intranasal Administration of Oxytocin Increases Envy and Schadenfreude (Gloating)." *Biological Psychiatry* 66.9 (2009): 864–70.

Shiller, Robert J. *Irrational Exhuberance*. Princeton: Princeton University Press, 2000.

Shimamura, Arthur P., ed. *Psychocinematics: Exploring Cognition at the Movies*. Oxford: Oxford University Press, 2013.

Shlain, Leonard. *Sex, Time and Power: How Women's Sexuality Shaped Evolution*. New York: Penguin, 2003.

Shrum, L. J. "Effects of Television Portrayals of Crime and Violence on Viewers' Perceptions of Reality: A Psychological Process Perspective." *Legal Studies Forum* 22.1–3 (1998): 257–68.

Shubin, Neil. *Your Inner Fish*. New York: Vintage, 2008.

Siegel, Daniel J. *The Developing Mind*. New York: Guilford, 1999.

Siegel, Daniel J. *The Mindful Brain*. New York: Norton, 2007.

Singer, Ben. *Melodrama and Modernity: Early Sensational Cinema and Its Contexts*. New York: Columbia University Press, 2001.

Slingerland, Edward, and Mark Collard, eds. *Creating Consilience: Integrating the Sciences and the Humanities*. Oxford: Oxford University Press, 2012.

Smith, Greg M. *Film Structure and the Emotion System*. Cambridge: Cambridge University Press, 2003.

Smith, Jeff. "Movie Music as Moving Music." *Passionate Views*. Ed. Carl Plantinga and Greg M. Smith. Baltimore: Johns Hopkins University Press, 1999. 146–67.

Smith, John Maynard. "The Concept of Information in Biology." *Information and the Nature of Reality*. Ed. Paul Davies and Niels Henrik Gregersen. Cambridge: Cambridge University Press, 2010. 123–45.

Smith, Murray. "Gangsters, Cannibals, Aesthetes, or Apparently Perverse Allegiances." *Passionate Views*. Ed. Carl Plantinga and Greg M. Smith. Baltimore: Johns Hopkins University Press, 1999. 217–38.

Smith, Murray. "The Logic and Legacy of Brechtianism." *Post-Theory: Reconstructing Film Studies*. Ed. David Bordwell and Noël Carroll. Madison: University of Wisconsin Press, 1996. 130–48.

Smith, Stephen D., Michael J. Dixon, William J. Tays, and M. Barbara Bulman-Fleming. "Anomaly Detection in the Right Hemisphere." *Brain and Cognition* 55 (2004): 458–62.

Sodian, B. "Early Deception and the Conceptual Continuity Claim." *Children's Early Understanding of Mind*. Ed. C. Lewis and P. Mitchell. Hillsdale, NJ: Lawrence Erlbaum, 1994. 385–401.

Solms, Mark. "The Conscious Id." *Neuropsychoanalysis* 15.1 (2013): 5–19.

Solms, Mark. "Interview: Wo Sitzt das Unbewusste?" 3satMediathek. Web. http://www.3sat.de/mediathek/index.php?mode=play&obj=45912

Solms, Mark, and Jaak Panksepp. "The 'Id' Knows More Than the 'Ego' Admits." *Brain Science* 2 (2012): 147–75.

Solms, Mark, Eleni Pantelis, and Jaak Panksepp. "Neuropsychoanalytic Notes on Addiction." *The Feeling Brain*. Ed. Mark Solms. London: Karnac, 2015. 109–22.

Solms, Mark, and Oliver Turnbull. *The Brain and the Inner World*. New York: Other Press, 2002.

Somel, Mehmet, et al. "Transcriptional Neoteny in the Human Brain." *Proceedings of the National Academy of Sciences USA* 106.14 (2009): 5743–48.

Sorenson, John. "Two Films: A Review of *Rise of the Planet of the Apes* and *Project Nim*." *Journal of Applied Animal Welfare Science* 15.3 (2012): 289–93.

Soyinka, Wole. *The Bacchae of Euripides: A Communion Rite*. In *Collected Plays 1*. Oxford: Oxford University Press, 1973.

Staub, Ervin. *The Psychology of Good and Evil*. Cambridge: Cambridge University Press, 2003.

Sternberg, Robert J., and James C. Kaufman, eds. *The Evolution of Intelligence*. Mahwah, NJ: Lawrence Erlbaum, 2002.

Stoker, Bram. *Dracula*. Oxford: Oxford University Press, 1988.

Sullins, Ellen S. "Emotional Contagion Revisited." *Personality and Social Psychology Bulletin* 17.2 (1991): 166–74.

"Supervolcano Eruption—in Sumatra—Deforested India 73,000 Years Ago." *ScienceDaily*, 23 November, 2009. Web.

Tabor, James. *Paul and Jesus*. New York: Simon, 2012.

Tajadura-Jiménez, Ana, Matthew R. Longob, Rosie Colemana, and Manos Tsakiris. "The Person in the Mirror." *Conscious Cognition* 21.4 (Dec. 2012): 1725–38.

Tan, Ed S. *Emotion and the Structure of Narrative Film*. Mahwah, NJ: Lawrence Erlbaum, 1996.

Taylor, M., A.C. Hulette, and T.J. Dishion. "Longitudinal Outcomes of Young High-Risk Adolescents with Imaginary Companions." *Developmental Psychology* 46.6 (2010): 1632–36.

Temerlin, Maurice K. *Lucy: Growing Up Human*. Palo Alto, CA: Science and Behavior Books, 1975.

Tenga, Angela. "Read Only as Directed." *Theorizing* Twilight. Ed. Maggie Parke and Natalie Wilson. Jefferson, NC: McFarland, 2011. 102–16.

Tervaniemi, Mari. "Musical Sounds in the Human Brain." *Neuroaesthetics*. Ed. Martin Skov and Oshin Vartanian. Amityville, NY: Baywood, 2009. 221–31.

Thatcher, R. W., R. A. Walker, and S. Giudice. "Human Cerebral Hemispheres Develop at Different Rates and Ages." *Science* 236 (1987): 1110–13.

Toga, Arthur W., and Paul M. Thompson. "Mapping Brain Asymmetry." *Nature Reviews Neuroscience* 4 (Jan. 2003): 37–48.

Tomasello, Michael. "Social Cognition before the Revolution." *Early Social Cognition*. Ed. Philippe Rochat. Mahwah, NJ: Lawrence Erlbaum, 1999. 301–14.

Tomasello, Michael. *Why We Cooperate*. Cambridge: Massachusetts Institute of Technology Press, 2009.

Tomasello, Michael, Brian Hare, Hagen Lehmann, and Josep Call. "Reliance on Head Versus Eyes in the Gaze Following of Great Apes and Human Infants." *Journal of Human Evolution* 52 (2007): 314–20.

Torey, Zoltan. *The Crucible of Consciousness*. Cambridge: MIT Press, 1999.

Torkelson, Anne. "Violence, Agency, and the Women of *Twilight*." *Theorizing* Twilight. Ed. Maggie Parke and Natalie Wilson. Jefferson, NC: McFarland, 2011. 209–23.

Trivers, Robert. *Social Evolution*. Menlo Park, CA: Benjamin/Cummings, 1985.

Tsakiris, Manos. "Looking for Myself: Current Multisensory Input Alters Self-Face Recognition." *PLoS ONE* 3.12 (2008): 1–6.

Tsakiris, Manos, and Patrick Haggard. "The Rubber Hand Illusion Revisited." *Journal of Experimental Psychology: Human Perception and Performance* 31.1 (2005): 80–91.

Tsakiris, Manos, Maike D. Hesse, Christian Boy, Patrick Haggard, and Gereon R. Fink. "Neural Structures of Body Ownership." *Cerebral Cortex* 17 (Oct. 2007): 2235–44.

Van Boven, Leaf, and George Loewenstein. "Empathy Gaps in Emotional Perspective Taking." *Other Minds*. Ed. Bertram F. Malle and Sara D. Hodges. New York: Guilford, 2005. 284–97.

van der Kolk, Bessel A. *The Body Keeps the Score*. New York: Penguin, 2014.

Vartanian, Oshin. "Conscious Experience of Pleasure in Art." *Neuroaesthetics*. Ed. Martin Skov and Oshin Vartanian. Amityville, NY: Baywood, 2009. 261–74.

Vaughan, Susan C. *The Talking Cure: The Science Behind Psychotherapy*. New York: Henry Holt, 1998.

Villeda, Saul A., and Tony Wyss-Coray. "The Circulatory Systemic Environment as a Modulator of Neurogenesis and Brain Aging." *Autoimmunity Reviews* 12.6 (April 2013): 674–77.

Vint, Sherryl. "Simians, Subjectivity and Sociality: *2001: A Space Odyssey* and Two Versions of *Planet of the Apes*." *Science Fiction Film and Television* 2.2 (2009): 225–50.

Virno, Paolo. *Multitude*. Los Angeles: Semiotext(e), 2008.

Virtanen, Keijo. "The Concept of Purification in the Greek and Indian Theories of Drama." *The Pursuit of Comparative Aesthetics*. Ed. Mazhar Hussain and Robert Wilkinson. Farnham, Surrey, England: Ashgate, 2006. 55–84.

Volkan, Vamik D. *The Need to Have Enemies and Allies*. North Vale, NJ: Jacob Aronson, 1988.

Vrticka, Pascal, Jessica M. Black, and Allan L. Reiss. "The Neural Basis of Humour Processing." *Nature Reviews Neuroscience* 14 (2013): 860–68.

Vygotsky, Lev S. *Thought and Language*. Cambridge: MIT Press, 1962.

Wagner, Ullrich, Karim N'Diaye, Thomas Ethofer, and Patrik Vuilleumier. "Guilt-Specific Processing in the Prefrontal Cortex." *Cerebral Cortex* 21 (Nov. 2011): 2461–70.

Warneken, Felix, and Michael Tomasello. "Altruistic Helping in Human Infants and Young Chimpanzees." *Science* 311 (2006): 1301–3.

Wasik, Bill, and Monica Murphy. *Rabid*. New York: Penguin, 2012.

Watson, Jeanne C., and Leslie S. Greenberg. "Empathic Resonance." *The Social Neuroscience of Empathy*. Ed. Jean Decety and William Ickes. Cambridge: MIT Press, 2009. 125–37.

Wegner, Daniel M. *The Illusion of Conscious Will*. Cambridge: MIT Press, 2002.

Wexler, Bruce E. *Brain and Culture*. Cambridge: MIT Press, 2006.

Wheeler, Robert E., Richard J. Davidson, and Andrew J. Tomarken. "Frontal Brain Asymmetry and Emotional Reactivity." *Psychopathology* 30 (1993): 82–89.

Whiten, Andrew. "The Machiavellian Mindreader." *Machiavellian Intelligence II*. Ed. Andrew Whiten and Richard W. Byrne. Cambridge: Cambridge University Press, 1997. 144–73.

Whiten, Andrew, Nicola McGuigan, Sarah Marshall-Pescini, and Lydia M. Hopper. "Emulation, Imitation, Over-Imitation and the Scope of Culture for Child and Chimpanzee." *Philosophical Transactions of the Royal Society B* 364 (2009): 2417–28.

Whitley, David S. *Cave Paintings and the Human Spirit*. New York: Prometheus, 2009.

Whitton, Merrine. "One is Not Born a Vampire But Becomes One." *Bringing Light to Twilight*. Ed. Giselle Liza Anatol. New York: Palgrave, 2011. 125–37.

Wicker, Bruno, Christian Keysers, Jane Plailly, Jean-Pierre Royet, Vittorio Gallese, and Giacomo Rizzolatti. "Both of Us Disgusted in *My* Insula: The Common Neural Basis of Seeing and Feeling Disgust." *Neuron* 40 (2003): 655–64.

Williamson, Milly. *The Lure of the Vampire*. London: Wallflower, 2005.

Wilson, Edward O. *Consilience: The Unity of Knowledge*. New York: Vintage, 1998.

Wilson, Edward O. *The Social Conquest of Earth*. New York: Norton, 2012.

Wilson, Natalie. *Seduced by* Twilight. Jefferson, NC: McFarland, 2011.

Wilson, Timothy D. *Strangers to Ourselves*. Cambridge: Harvard University Press, 2002.

Winkielman, Piotr, Kent C. Berridge, and Julia L. Wilbarger. "Emotion, Behavior, and Conscious Experience." *Emotion and Consciousness*. Ed. Lisa Feldman Barrett, Paul M. Niedenthal, and Piotr Winkielman. New York: Guilford, 2005. 335–62.

Winnicott, D. W. *Playing and Reality*. London: Tavistock, 1971.

Winslade, Jason. "Oh . . . My . . . Goddess." *Joss Whedon and Religion*. Ed. Anthony R. Mills, John W. Morehead, and J. Ryan Parker. Jefferson, NC: McFarland, 2013. 51–66.

Wise, Steven M. *Drawing the Line: Science and the Case for Animal Rights*. New York: Basic, 2003.

Wrangham, Richard, and Dale Peterson. *Demonic Males: Apes and the Origins of Human Violence*. Boston: Houghton Mifflin, 1996.

Zack, Naomi. "*The Island of Dr. Moreau*." *SciFi in the Mind's Eye*. Ed. Margaret Grebowicz. Chicago: Open Court, 2007. 25–37.

Zacks, Jeffrey M. *Flicker: Your Brain on Movies*. Oxford: Oxford University Press, 2015.

Zak, Paul J. "The Neurobiology of Trust." *Scientific American* (June 2008): 88–95.

Zhu, Y., L. Zhang, J. Fan, and S. Han. "Neural Basis of Cultural Influence on Self-Representation." *Neuroimage* 34.3 (Feb. 2007): 1310–6.

Zillmann, Dolf. "Anatomy of Suspense." *The Entertainment Functions of Television*. Ed. Percy H. Tannenbaum. Hillsdale, NJ: Lawrence Erlbaum, 1980. 133–63.

Zillmann, Dolf. "Mechanisms of Emotional Involvement with Drama." *Poetics* 23 (1994): 33–51.

Zimbardo, Philip. *The Lucifer Effect*. New York: Random, 2008.

Zizek, Slavoj. *Enjoy Your Symptom! Jacques Lacan in Hollywood and Out*. New York: Routledge, 1992.

Zizek, Slavoj. *For They Know Not What They Do: Enjoyment as a Political Factor*. London: Verso, 1991.

Zunshine, Lisa. *Why We Read Fiction*. Columbus: Ohio University Press, 2006.

Index

About the Author

Mark Pizzato, MFA, PhD, is Professor of Theatre and Film at UNC–Charlotte, where he teaches theater history, dramaturgy, playwriting, screenwriting, and various topics courses in film. His writings include *Inner Theatres of Good and Evil: The Mind's Staging of Gods, Angels and Devils* (2011), *Ghosts of Theatre and Cinema in the Brain* (2006), *Theatres of Human Sacrifice: From Ancient Ritual to Screen Violence* (2005), and *Edges of Loss: From Modern Drama to Postmodern Theory* (1998). He also co-edited, with Lisa K. Perdigao, *Death in American Texts and Performances: Corpses, Ghosts, and the Reanimated Dead* (2010). Short films, produced from his screenplays, have won New York Film Festival and Minnesota Community Television awards. He blogs at mpizzato.wordpress.com, writes theater and film reviews for *Charlotte Viewpoint*, an online magazine, and participates in the podcast "Conversations about Buddha, Jesus, and Mohammed."